WHAT EVERY CHEMICAL TECHNOLOGIST WANTS TO KNOW ABOUT...

Volume III

PLASTICIZERS, STABILIZERS AND THICKENERS

Compiled by

Michael and Irene Ash

Chemical Publishing Co., Inc.
New York, N.Y.

Plasticizers, Stabilizers, and Thickeners Volume 3

ISBN: 978-0-8206-0051-2

Chemical Publishing Company:
www.chemical-publishing.com
www.chemicalpublishing.net

First Edition:

© **Chemical Publishing Company, Inc.** - New York 1989

Second Impression:

Chemical Publishing Company, Inc. - 2011

Printed in the United States of America

PREFACE

This reference book is the third volume in the set of books entitled WHAT EVERY CHEMICAL TECHNOLOGIST WANTS TO KNOW . . . SERIES. This compendium serves a unique function for those involved in the chemical industry—it provides the necessary information for making the decision as to which trademark chemical product is most suitable for a particular application.

The chemicals included in this third book of the series have their major function as plasticizers, stabilizers, and thickeners, however, complete cross-referencing is provided for the multiple functions of all the chemicals.

The first section which is the major portion of each volume contains the most common generic name of the chemicals as the main entry. All these generic entries are in alphabetical order. Synonyms for these chemicals are then listed. The CTFA name appears alongside the appropriate generic name. The structural and/or molecular formula of the chemical is listed whenever possible. The generic chemical is sold under various tradenames and these are listed here in alphabetical order for ease of reference along with their manufacturer in parentheses. The *Category* subheading lists all the possible functions that the chemical can serve. Because of differences in form, activity, etc., individual tradenames of the generic chemical are used in particular applications more frequently. These are delineated in the *Applications* section. The differences in properties, toxicity/handling, storage/handling, and standard packaging are specified in the subsequent sections wherever distinguishing characteristics are known.

The second section of the volume TRADENAME PRODUCTS AND GENERIC EQUIVALENTS helps the user who only knows a chemical by one tradename to locate its main entry in section 1. The user can look up this tradename in this section of the book and be referred to the appropriate, main-entry, generic chemical name.

The third section GENERIC CHEMICAL SYNONYMS AND CROSS REFER-ENCES provides a way of locating the main entries by knowing only one of the synonyms. If the generic chemical is not in the volume, it will refer you to the volume in which it is contained.

The fourth section TRADENAME PRODUCT MANUFACTURERS lists the full addresses of the companies that manufacture or distribute the tradename products found in the first section.

The following is a list of the six volumes that comprise this series:

Volume I	Emulsifiers and Wetting Agents
Volume II	Dispersants, Solvents and Solubilizers
Volume III	Plasticizers, Stabilizers and Thickeners
Volume IV	Conditioners, Emollients and Lubricants
Volume V	Resins
Volume VI	Polymers and Plastics

This series has been made possible through long hours of research and compilation and the dedication and tireless efforts of Roberta Dakan who helped make this distinctive series possible. Our appreciation is extended to all the chemical manufacturers and distributors who supplied the technical information.

M. and I. Ash

NOTE

The information contained in this series is accurate to the best of our knowledge; however, no liability will be assumed by the publisher for the correctness or comprehensiveness of such information. The determination of the suitability of any of the products for prospective use is the responsibility of the user. It is herewith recommended that those who plan to use any of the products referenced seek the manufacturer's instructions for the handling of that particular chemical.

OTHER BOOKS BY MICHAEL AND IRENE ASH

A Formulary of Paints and Other Coatings, Volumes I and II
A Formulary of Detergents and Other Cleaning Agents
A Formulary of Adhesives and Sealants
A Formulary of Cosmetic Preparations
The Thesaurus of Chemical Products, Volumes I and II
Encyclopedia of Industrial Chemical Additives, Volumes I–IV
Encyclopedia of Surfactants, Volumes I–IV
Encyclopedia of Plastics, Polymers and Resins, Volumes I–IV
What Every Chemical Technologist Wants to Know About. . .
    Volume I—Emulsifiers and Wetting Agents
    Volume II—Dispersants, Solvents and Solubilizers

ABBREVIATIONS

@ .... at
anhyd. .... anhydrous
APHA .... American Public Health Association
approx. .... approximately
aq. .... aqueous
ASTM .... American Society for Testing and Materials
avg. .... average
B.P. .... boiling point
Btu .... British thermal unit
C .... degrees Centigrade
CAS .... Chemical Abstracts Service
cc .... cubic centimeter(s)
CC .... closed cup
cm .... centimeter(s)
cm³ .... cubic centimeter(s)
COC .... Cleveland Open Cup
compd. .... compound, compounded
conc. .... concentrated, concentration
cP, cps .... centipoise
cs, cSt .... centistokes
CTFA .... Cosmetic, Toiletry and Fragrance Association
DEA .... diethanolamine
disp .... dispersible, dispersion
dist .... distilled
DOT .... Department of Transportation
DW .... distilled water
EO .... ethylene oxide
equiv. .... equivalent
F .... degrees Fahrenheit
F.P. .... freezing point
FDA .... Food and Drug Administration
ft³ .... cubic foot, cubic feet
g .... gram(s)
gal .... gallon(s)
HLB .... hydrophile-lipophile balance
insol. .... insoluble
IPA .... isopropyl alcohol
kg .... kilogram(s)
l, L .... liter(s)
lb .... pound(s)
M.P. .... melting point
M.W. .... molecular weight
max .... maximum
MEA .... monoethanolamine
MEK .... methyl ethyl ketone
mfg. .... manufacture
MIBK .... methyl isobutyl ketone
min .... minute(s)
min. .... mineral, minimum
MIPA .... monoisopropanolamine

misc. .... miscible
ml .... milliliter(s)
mm .... millimeter(s)
NF .... National Formulary
no. .... number
o/w .... oil-in-water
OC .... open crucible
PEG .... polyethylene glycol
pH .... hydrogen-ion concentration
pkgs .... packages
PMCC .... Pensky Marten closed cup
POE .... polyoxyethylene, polyoxyethylated
POP .... polyoxypropylene
PPG .... polypropylene glycol
pt. .... point
R&B .... Ring & Ball
RD .... Recognized Disclosure
ref. .... refractive
rpm .... revolutions per minute
R.T. .... room temperature
s .... second(s)
sol. .... soluble, solubility
sol'n. .... solution
sp.gr. .... specific gravity
SS .... stainless steel
std. .... standard
SUS .... Saybolt Universal seconds
TCC .... Taggart closed cup
TEA .... triethanolamine
tech. .... technical
temp. .... temperature
theoret. .... theoretical
TLV .... threshold limit value
TOC .... Taggart open cup
UL .... Underwriter's Laboratory
USP .... United States Pharmacopoeia
uv, UV .... ultraviolet
veg .... vegetable
visc. .... viscosity, viscous
w/o .... water-in-oil
wt .... weight
≈ .... approximately equal to
< .... less than
> .... greater than
≤ .... less than or equal to
≥ .... greater than or equal to

TABLE OF CONTENTS

Acetylated lanolin alcohol (CTFA)

SYNONYMS:
Lanolin, alcohols, acetates
CAS No.:
61788-49-6
TRADENAME EQUIVALENTS:
Acetol 1706 [Emery]
Acetulan [Amerchol; Amerchol Europe]
Crodalan LA [Croda Inc.]
Fancol ALA [Fanning]
Hetlan AC [Heterene]
Ritawax ALA [Rita]
CATEGORY:
Plasticizer, penetrant, binder, emollient, lubricant, cosolvent, solvent, stabilizer, solubilizer, suspending agent, water repellent
APPLICATIONS:
Cosmetic industry preparations: (Acetulan; Fancol ALA); baby products (Acetol 1706); body rubs (Acetulan); creams and lotions (Hetlan AC); hair sprays and preparations (Acetulan); makeup (Acetulan); nail polish removers (Acetulan); pressed powders (Acetulan); toiletries (Acetulan; Fancol ALA)
Industrial applications: aerosols (Acetulan); silicone emulsions (Acetulan)
Pharmaceutical applications: (Acetulan); acne products (Acetulan); antiseptics (Acetulan); ointments (Acetulan); sunscreens (Acetol 1706; Acetulan); suppositories (Acetulan)

PROPERTIES:
Form:
Liquid (Hetlan AC)
Thin, oily liquid (Acetulan)
Oily liquid (Fancol ALA)

1

Acetylated lanolin alcohol *(cont'd.)*

Color:
  Pale yellow (Acetulan; Hetlan AC)
Odor:
  Practically odorless (Acetulan)
  Faint, characteristic (Hetlan AC)
Solubility:
  Sol. in aerosol propellants (Acetulan)
  Sol. in alcohols (Ritawax ALA)
  Sol. in butyl stearate (Acetulan)
  Sol. in castor oil (Acetol 1706; Acetulan)
  Sol. in 95% ethanol (Acetulan)
  Sol. in ethyl acetate (Acetulan)
  Sol. in most fatty chemicals (Crodalan LA)
  Sol. in isopropanol (Acetulan)
  Sol. in isopropyl myristate (Acetulan)
  Sol. in isopropyl palmitate (Acetol 1706; Acetulan)
  Sol. in min. oil (Acetol 1706; Acetulan; Ritawax ALA)
  Sol. in olive oil (Acetol 1706)
  Sol. in common organic solvents (Acetulan)
  Sol. in silicones (Acetulan)
  Sol. in sulfonated castor oil (Acetulan)
  Sol. in veg. oil (Acetulan)
  Insol. in water (Acetol 1706; Crodalan LA); insol. in water but easily dispersed in
    presence of emulsifiers and surfactants (Acetulan)
Sp.gr.:
  0.850–0.880 (Acetulan)
Visc.:
  10 cps (Acetulan)
Acid No.:
  1.0 max. (Acetulan; Hetlan AC)
Iodine No.:
  8–12 (Acetulan)
Saponification No.:
  180–200 (Acetulan; Hetlan AC)
Hydroxyl No.:
  8.0 max. (Acetulan; Hetlan AC)
Stability:
  Excellent resistance to hydrolysis within broad pH range (Acetulan)
pH:
  Neutral (Acetulan)

Acetyl tributyl citrate (CTFA)

SYNONYMS:
2-(Acetyloxy)-1,2,3-propanetricarboxylic acid, tributyl ester
1,2,3-Propanetricarboxylic acid, 2-(acetyloxy)-, tributyl ester
EMPIRICAL FORMULA:
$C_{20}H_{34}O_8$
STRUCTURE:

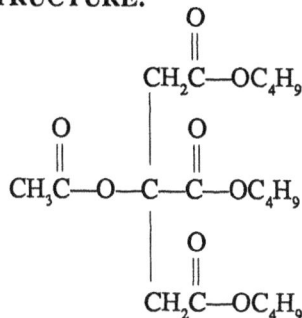

CAS No.:
77-90-7
TRADENAME EQUIVALENTS:
ATBC [Croda Chem. Ltd.]
Citroflex A-4 [Morflex]
Generically sold by: [Unitex]
CATEGORY:
Plasticizer
APPLICATIONS:
Food applications: food packaging (ATBC)
Industrial applications: cellulosics (generic; Citroflex A-4); plastics (generic; Citroflex A-4)
PROPERTIES:
Form:
Liquid (generic; ATBC)
Color:
Colorless (generic)
Odor:
Odorless (generic)
Solubility:
Insol. in water (generic)
Sp.gr.:
1.046 (generic)
Density:
8.74 lb/gal (generic)
Visc.:
42.7 cps (generic)

Acetyl tributyl citrate (cont'd.)

B.P.:
172–174 C (1 mm) (generic)
M.P.:
–80 C (generic)
Pour Pt.:
–60 C (generic)
Flash Pt.:
204 C (generic)
Ref. Index:
1.4408 (generic)
STORAGE/HANDLING:
Combustible (generic)

Acetyl triethyl citrate (CTFA)

SYNONYMS:
2-(Acetyloxy)-1,2,3-propanetricarboxylic acid, triethyl ester
1,2,3-Propanetricarboxylic acid, 2-(acetyloxy)-, triethyl ester
EMPIRICAL FORMULA:
$C_{14}H_{22}O_8$
STRUCTURE:

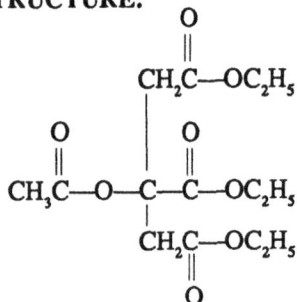

CAS No.:
77-89-4
TRADENAME EQUIVALENTS:
Citroflex A-2 [Morflex]
Generically sold by: [Unitex]
CATEGORY:
Solvent plasticizer
APPLICATIONS:
Industrial applications: cellulosics (generic; Citroflex A-2); plastics (generic; Citroflex A-2)

4

PROPERTIES:
Form:
  Liquid (generic)
Color:
  Colorless (generic)
Odor:
  Odorless (generic)
Solubility:
  Slightly sol. in water (generic)
Sp.gr.:
  1.135 (23 C) (generic)
Density:
  9.47 lb/gal (generic)
Visc.:
  53.7 cps (generic)
B.P.:
  131–132 C (1 mm) (generic)
M.P.:
  −50 C (generic)
Pour Pt.:
  −47 C (generic)
Flash Pt.:
  187 C (generic)
Ref. Index:
  1.4386 (23 C) (generic)
STORAGE/HANDLING:
  Combustible (generic)

Algin (CTFA)

SYNONYMS:
  Alginic acid, sodium salt
  Sodium alginate
CAS No.:
  9005-38-3
TRADENAME EQUIVALENTS:
  Dariloid QH [Kelco/Div. of Merck & Co.]
  Kelco-Gel HV, LV [Kelco/Div. of Merck & Co.] (specially clarified, low-calcium)
  Kelco-Pac [Kelco/Div. of Merck & Co.] (specially clarified, low-calcium)
  Kelcosol [Kelco/Div. of Merck & Co.]

Algin *(cont'd.)*

Kelgin F, HV, LV, MV, RL, XL [Kelco/Div. of Merck & Co.] (refined)
Kelgin QH, QL, QM [Kelco/Div. of Merck & Co.] (treated for improved dispersion)
Keltex, P, S [Kelco/Div. of Merck & Co.] (industrial)
Keltone [Kelco/Div. of Merck & Co.]
Kelvis [Kelco/Div. of Merck & Co.] (refined)

CATEGORY:
Stabilizer, thickener, gelling agent, gum, emulsifier

APPLICATIONS:
Cosmetic industry preparations: (Kelcosol)

Food applications: (Dariloid QH; Kelco-Gel HV, LV; Kelco-Pac; Kelcosol; Kelgin F, HV, LV, MV, QH, QL, QM, RL, XL; Keltone; Kelvis); dairy (Dariloid QH); food emulsifying (Kelco-Gel HV, LV; Kelco-Pac; Kelcosol; Kelgin F, HV, LV, MV, QH, QL, QM, RL, XL; Keltone; Kelvis); food packaging (Kelgin)

Industrial applications: adhesives (Kelco-Gel HV, LV; Kelco-Pac; Kelcosol; Kelgin F, HV, LV, MV, QH, QL, QM, RL, XL; Keltex, P, S; Keltone; Kelvis); ceramics (Kelco-Gel HV, LV; Kelco-Pac; Kelcosol; Kelgin F, HV, LV, MV, QH, QL, QM, RL, XL; Keltex, P, S; Keltone; Kelvis); explosives (Kelco-Gel HV, LV; Kelco-Pac; Kelcosol; Kelgin F, HV, LV, MV, QH, QL, QM, RL, XL; Keltex, P, S; Keltone; Kelvis); industrial processing (Kelco-Gel HV, LV; Kelco-Pac; Kelcosol; Kelgin F, HV, LV, MV, QH, QL, QM, RL, XL; Keltex, P, S; Keltone; Kelvis); latexes (Kelco-Gel HV, LV; Kelco-Pac; Kelcosol; Kelgin F, HV, LV, MV, QH, QL, QM, RL, XL; Keltex, P, S; Keltone; Kelvis); paper mfg. (Kelco-Gel HV, LV; Kelco-Pac; Kelcosol; Kelgin F, HV, LV, MV, QH, QL, QM, RL, XL; Keltex, P, S; Keltone; Kelvis); polishes (Kelco-Gel HV, LV; Kelco-Pac; Kelcosol; Kelgin F, HV, LV, MV, QH, QL, QM, RL, XL; Keltex, P, S; Keltone; Kelvis); textile printing/dyeing (Kelco-Gel HV, LV; Kelco-Pac; Kelcosol; Kelgin F, HV, LV, MV, QH, QL, QM, RL, XL; Keltex, P, S; Keltone; Kelvis); water treatment (Kelco-Gel HV, LV; Kelco-Pac; Kelcosol; Kelgin F, HV, LV, MV, QH, QL, QM, RL, XL; Keltex, P, S; Keltone; Kelvis)

Industrial cleaners: (Kelco-Gel HV, LV; Kelco-Pac; Kelcosol; Kelgin F, HV, LV, MV, QH, QL, QM, RL, XL; Keltex, P, S; Keltone; Kelvis)

PROPERTIES:
Form:
Fibrous particles (Kelco-Gel HV, LV; Kelcosol; Keltone)
Granular (Dariloid QH; Kelco-Pac; Kelgin F, HV, LV, MV, QH, QL, QM, RL, XL; Keltex, P, S; Kelvis)

Color:
Light ivory (Dariloid QH)
Ivory (Kelco-Pac; Kelgin F, HV, LV, MV, QH, QL, QM, RL, XL; Kelvis)
Cream (Kelco-Gel HV, LV; Kelcosol; Keltone)
Tan (Keltex, P, S)

Solubility:
 Sol. in milk (Dariloid QH)
 Sol. in cold or hot water (Kelcosol; Kelgin F, HV, LV, MV, QH, QL, QM, RL, XL; Keltex; Keltone; Kelvis)
Sp.gr.:
 1.59 (Kelgin F, HV, LV, MV, RL, XL; Kelvis)
 1.64 (Kelco-Gel HV, LV; Kelco-Pac; Kelcosol; Keltone)
Bulk Density:
 43.38 lb/ft³ (Kelco-Gel HV, LV; Kelco-Pac; Kelcosol; Keltone)
 54.62 lb/ft³ (Kelgin F, HV, LV, MV, RL, XL; Kelvis)
Visc.:
 10 cps (Brookfield, 1%, LVF, 60 rpm) (Kelgin RL)
 30 cps (Brookfield, 1%, LVF, 60 rpm) (Kelgin QL, XL)
 50 cps (Brookfield, 1%, LVF, 60 rpm) (Kelco-Gel LV)
 55 cps (Brookfield, 1%, LVF, 60 rpm) (Kelco-Pac)
 60 cps (Brookfield, 1%, LVF, 60 rpm) (Kelgin LV)
 180 cps (Brookfield, 1%, LVF, 60 rpm) (Kelgin QM)
 300 cps (Brookfield, 1%, LVF, 60 rpm) (Kelgin F)
 400 cps (Brookfield, 1%, LVF, 60 rpm) (Kelco-Gel HV; Kelgin MV, QH; Keltone)
 760 cps (Brookfield, 1%, LVF, 60 rpm) (Kelvis)
 765 cps (Brookfield, 1%, LVF, 60 rpm) (Keltex P)
 800 cps (Brookfield, 1%, LVF, 60 rpm) (Kelgin HV; Keltex)
 1300 cps (Brookfield, 1%, LVF, 60 rpm) (Kelcosol; Keltex S)
Ref. Index:
 1.3342 (Kelco-Gel HV, LV)
 1.3343 (Kelgin F, HV, LV, MV, RL, XL; Kelvis)
pH:
 Neutral (Kelgin QH, QL, QM; Keltex, P, S)
 7.2 (Kelco-Gel HV, LV; Kelco-Pac; Kelcosol; Keltone)
 7.5 (Kelgin F, HV, LV, MV, RL, XL; Kelvis)
Surface Tension:
 62 dynes/cm (Kelgin F, HV, LV, MV, RL, XL; Kelvis)
 70 dynes/cm (Kelco-Gel HV, LV; Kelco-Pac; Kelcosol; Keltone)

Alginic acid (CTFA)

SYNONYMS:
 Norgine
CAS No.:
 9005-32-7

Alginic acid *(cont'd.)*

TRADENAME EQUIVALENTS:
Kelacid [Kelco Div. of Merck]
CATEGORY:
Thickener
APPLICATIONS:
Food applications
Pharmaceutical applications; tablet mfg.
PROPERTIES:
Form:
Fibrous
Solubility:
Sol. in alkaline solution
Insol. in organic solvents
Insol. in water, swells in water

Aluminum distearate (CTFA)

SYNONYMS:
Aluminum, hydroxybis(octadecanoato-O)-
Hydroxybis(octadecanoato-O) aluminum
TRADENAME EQUIVALENTS:
Witco Aluminum Stearate 22, 30 [Witco/Organics]
Witco Aluminum Stearate EA [Witco/Organics] (food grade)
CATEGORY:
Gelation agent, thickening agent, lubricant, suspending agent, water repellent, binder,
emulsifier, anticaking agent
APPLICATIONS:
Food applications (Witco Aluminum Stearate EA)
Industrial applications: cements (Witco Aluminum Stearate 22); dyes and pigments
(Witco Aluminum Stearate 22); greases and waxes (Witco Aluminum Stearate 22,
30); plastics (Witco Aluminum Stearate 22); paints (Witco Aluminum Stearate 22)
PROPERTIES:
Form:
Powder (Witco Aluminum Stearate 22, 30, EA)
Color:
White (Witco Aluminum Stearate 22, 30, EA)
Solubility:
Insol. in lower alcohols (Witco Aluminum Stearate 22, 30, EA)
Sol. with gelation on cooling in benzene (Witco Aluminum Stearate 22, 30, EA)
Sol. with gelation on cooling in carbon tetrachloride (Witco Aluminum Stearate 22, 30,
EA)

8

Aluminum distearate (cont'd.)

Insol. in ethers (Witco Aluminum Stearate 22, 30, EA)
Insol. in ketones (Witco Aluminum Stearate 22, 30, EA)
Sol. with gelation on cooling in min. oils (Witco Aluminum Stearate 22, 30, EA)
Sol. with gelation on cooling in oleic acid (Witco Aluminum Stearate 22, 30, EA)
Sol. with gelation on cooling in toluene (Witco Aluminum Stearate 22, 30, EA)
Sol. with gelation on cooling in hot turpentine (Witco Aluminum Stearate 22, 30, EA)
Sol. with gelation on cooling in veg. oils (Witco Aluminum Stearate 22, 30, EA)
Insol. in water (Witco Aluminum Stearate 22, 30, EA)
Sol. with gelation on cooling in waxes (Witco Aluminum Stearate 22, 30, EA)
Sol. with gelation on cooling in xylene (Witco Aluminum Stearate 22, 30, EA)
Sp.gr.:
1.01 (Witco Aluminum Stearate 22, 30, EA)
Fineness:
98% thru 200 mesh (Witco Aluminum Stearate 22, 30, EA)
Softening Pt.:
160 C (Witco Aluminum Stearate 22, EA)
162 C (Witco Aluminum Stearate 30)

Aluminum di/tristearate (CTFA)

SYNONYMS:
50:50 mixture of aluminum distearate and aluminum tristearate
TRADENAME EQUIVALENTS:
Witco Aluminum Stearate 18 [Witco/Organics]
CATEGORY:
Thickening agent, lubricant, suspending agent
APPLICATIONS:
Industrial applications: dyes and pigments; metal processing; paints
PROPERTIES:
Form:
Powder
Color:
White
Solubility:
Insol. in lower alcohols
Sol. with gelation on cooling in benzene
Sol. with gelation on cooling in carbon tetrachloride
Insol. in ethers
Insol. in ketones
Sol. with gelation on cooling in min. oils

Aluminum di/tristearate *(cont'd.)*

    Sol. with gelation on cooling in oleic acid
    Sol. with gelation on cooling in toluene
    Sol. with gelation on cooling in hot turpentine
    Sol. with gelation on cooling in veg. oils
    Insol. in water
    Sol. with gelation on cooling in waxes
    Sol. with gelation on cooling in xylene
Sp.gr.:
    1.01
Fineness:
    97% thru 200 mesh
Softening Pt.:
    145 C

Aluminum hydroxide *(CTFA)*

SYNONYMS:
    Alumina, hydrated
    Alumina trihydrate
    Aluminum hydrate
    C-Weiss 1 (Germany)
    Hydrated alumina
    Hydrated aluminum oxide
EMPIRICAL FORMULA:
    AlH_3O_3
STRUCTURE:
    $Al(OH)_3 \cdot xH_2O$
CAS No.:
    21645-51-2
TRADENAME EQUIVALENTS:
    Akrochem Hydrated Alumina [Akron]
    Alumina Hydrate 753, 983 [Harwick]
    C-30, -30BF, -31, -230, -330, -331, -333, -430, -433 [Alcoa]
    H-30, H-36 [Kaiser]
    Hydral 705, 710 [Alcoa]
    Lubral 710 [Alcoa]
    Micral 916, 932 [Solem]
    SB 100, 30-0, 400, 500, 600, 700 Series [Solem]
    Techfill A-100 Series, A-200 Series [Great Lakes Minerals]
    Techfill AS-101, AS-1005 [Great Lakes Minerals] (precipitated grades)
    Generic

CATEGORY:
Flame retardant, smoke suppressor, filler, reinforcing agent

APPLICATIONS:
Cosmetic preparations: (generic)

Industrial applications: adhesives (Micral 916, 932); carpet backing (Akrochem Hydrated Alumina); ceramics (generic); chemical production (generic); coatings (Micral 916, 932); glass (generic); latex foam (Akrochem Hydrated Alumina); paper mfg. (generic; Micral 916, 932); plastics (Akrochem Hydrated Alumina; Alumina Hydrate 753, 983; C-30, -30BF, -31, -230, -330, -331, -333, -430, -433; H-30, H-36; Hydral 705, 710; Lubral 710; Micral 916, 932; SB 100, 30-0, 400, 500, 600, 700 Series; Techfill A-100 Series, A-200 Series, AS-101, AS-1005); rubber (generic; Akrochem Hydrated Alumina; Alumina Hydrate 753, 983; Micral 916, 932); wire and cable insulation (Micral 916, 932)

PROPERTIES:

Form:
Powder, balls or granules (generic)
Free-flowing powder (Alumina Hydrate 753, 983)
Powder (1.5 μ particle size) (Micral 932)
Powder (avail. in 3.0 μ, 4.0 μ, 8.0 μ particle size grades) (Akrochem Hydrated Alumina)
Ultrafine powder (0.8 μ particle size) (Micral 916)

Color:
White (generic; Alumina Hydrate 753, 983)
White (also avail. in extra white grades) (Akrochem Hydrated Alumina)

Composition:
64.9% Al_2O_3 (Micral 916, 932)
65% alumina (Akrochem Hydrated Alumina)

Solubility:
Sol. in caustic soda (generic)
Sol. in min. acids (generic)
Insol. in water (generic)

Sp.gr.:
2.42 (Akrochem Hydrated Alumina; Alumina Hydrate 753, 983; Micral 916, 932)

Bulk Density:
0.15 g/cm³ (loose) (Micral 916)
0.3 g/cm³ (loose) (Micral 932)
0.35 g/cm³ (packed) (Micral 916)
0.5 g/cm³ (packed) (Micral 932)
40 lb/ft³ (Alumina Hydrate 983)
55 lb/ft³ (Alumina Hydrate 753)

Fineness:
75% thru 325 mesh (Alumina Hydrate 753)

Aluminum hydroxide *(cont'd.)*

98% thru 325 mesh (Alumina Hydrate 983)
100% thru 325 mesh (Micral 916, 932)
Surface Area:
7 m²/g (Micral 932)
13 m²/g (Micral 916)
Oil Absorption:
38 m./100 g (Micral 932)
46 ml/100 g (Micral 916)
Brightness:
92–93 (GE) (Akrochem Hydrated Alumina, regular grades)
95–97 (GE) (Akrochem Hydrated Alumina, extra white grades)
97+ (Photovolt) (Micral 916, 932)
STD. PKGS.:
50-lb bags on stretch-wrapped 2500-lb pallets (Micral 916, 932)

Aluminum monostearate

SYNONYMS:
Aluminum, dihyroxy (octadecanoate-O)-
Aluminum stearate (CTFA)
Dihydroxy (octanoateo-O) aluminum
Octadecanoic acid, aluminum salt
EMPIRICAL FORMULA:
$C_{18}H_{37}AlO_4$
STRUCTURE:
$CH_3(CH_2)_{16}COOAl(OH)_2$
CAS No.:
7047-84-9
TRADENAME EQUIVALENTS:
Witco Aluminum Monostearate USP [Witco/Organics]
Generically sold by:
[C.P. Hall; Harwick]
CATEGORY:
Gelling agent, thickening agent, lubricant
APPLICATIONS:
Cosmetic industry preparations: toiletries (Witco Aluminum Monostearate USP)
Industrial applications: rubber (generic)
Pharmaceutical applications: (Witco Aluminum Monostearate USP)
PROPERTIES:
Form:
Powder (generic; Witco Aluminum Monostearate USP)

12

Aluminum monostearate (cont'd.)

Color:
White (generic; Witco Aluminum Monostearate USP)
Odor:
Odorless (generic)
Solubility:
Insol. in lower alcohols (Witco Aluminum Monostearate USP)
Sol. with gelation on cooling in benzene (Witco Aluminum Monostearate USP)
Sol. with gelation on cooling in carbon tetrachloride (Witco Aluminum Monostearate USP)
Insol. in ethers (Witco Aluminum Monostearate USP)
Insol. in ketones (Witco Aluminum Monostearate USP)
Sol. with gelation on cooling in min. oils (Witco Aluminum Monostearate USP)
Sol. with gelation on cooling in oleic acid (Witco Aluminum Monostearate USP)
Sol. with gelation on cooling in toluene (Witco Aluminum Monostearate USP)
Sol. with gelation on cooling in hot turpentine (Witco Aluminum Monostearate USP)
Sol. with gelation on cooling in veg. oils (Witco Aluminum Monostearate USP)
Insol. in water (Witco Aluminum Monostearate USP)
Sol. with gelation on cooling in waxes (Witco Aluminum Monostearate USP)
Sol. with gelation on cooling in xylene (Witco Aluminum Monostearate USP)
Sp.gr.:
1.14 (Witco Aluminum Monostearate USP)
Fineness:
99% thru 200 mesh (Witco Aluminum Monostearate USP)
M.P.:
152–155 C (generic)
Softening Pt.:
Decomposes at 240 C (Witco Aluminum Monostearate USP)
STD. PKGS.:
50-lb cartons (generic)

Attapulgite (CTFA)

SYNONYMS:
Variety of Fuller's earth characterized by chain structure
CAS No.:
1337-76-4
TRADENAME EQUIVALENTS:
Attacote [Engelhard Min. & Chem.]
Attaflow [Engelhard Min. & Chem.]
Attagel 40, 50 [Engelhard Min. & Chem.] (colloidal)

13

Attapulgite (cont'd.)

TRADENAME EQUIVALENTS *(cont'd.):*
  Attagel 150, 350 [Engelhard Min. & Chem.]
  Diluex [Floridin]
  Florco X [Floridin]
  Florex [Floridin]
  Florex Granular Grades [Floridin]
  Minugel 400, LF [Floridin] (colloidal)
  Pharmasorb Reg. [Engelhard Min. & Chem.]
  Pharmasorb Colloidal [Engelhard Min. & Chem.] (colloidal)
  Refinex [Floridin]
  Sol-Speedi-Dri, Auto-Dri [Engelhard Min. & Chem.]
CATEGORY:
  Thickener, stabilizer, suspending agent, flatting agent, bodying agent, sag control
    agent, gelling agent, thixotropic agent, anticaking agent, antiagglomerating agent,
    absorbent, adsorbent, carrier
APPLICATIONS:
  Farm products: animal feed (Attagel 350; Minugel LF); fertilizers (Attacote; Attagel
    350); insecticides/pesticides (Diluex; Florex)
  Industrial applications: adhesives (Attagel 40, 50); catalyst stripping (Florex Granular
    Grades); oil treatment (Diluex; Florco X; Florex Granular Grades; Refinex); paint
    mfg. (Attagel 40, 50; Minugel 400); petroleum industry (Attagel 150, 350; Florex
    Granular Grades); water treatment (Diluex; Florco X)
  Industrial cleaners: floor absorbent for metalworking, automotive industries, animal
    barns, butcher shops (Sol-Speedi-Dri, Auto-Dri)
  Pet absorbent: (Florco X)
  Pharmaceutical applications: (Pharmasorb Reg. & Colloidal)

PROPERTIES:
Form:
  Liquid (Attaflow)
  Very fine powder (Minugel 400)
  Fine powder (Diluex; Refinex)
  Fine powder; avg. particle size 0.12 μ (Attagel 150, 350)
  Fine powder; avg. particle size 0.14 μ (Attagel 40, 50; Pharmasorb Colloidal)
  Fine powder; avg. particle size 2.9 μ (Pharmasorb Reg.)
  Fine powder; avg. particle size 5.3 μ (Attacote)
  Coarse ground (Minugel LF)
  Granular (Florex; Florex Granular Grades)
Color:
  Light cream (Attagel 40, 50, 150, 350; Pharmasorb Colloidal)
  Cream (Attacote; Pharmasorb Regular)
  Gray (Attaflow)

Sp.gr.:
2.36 (Attagel 40, 50, 150, 350; Pharmasorb Colloidal)
2.47 (Attacote; Pharmasorb Reg.)
Density:
10 lb/gal (Attaflow)
Bulk Density:
15–18 lb/ft^3 (tamped) (Attacote; Pharmasorb Reg.)
18–21 lb/ft^3 (tamped) (Pharmasorb Colloidal)
19–22 lb/ft^3 (tamped) (Attagel 40, 50)
29–33 lb/ft^3 (Sol-Speedi-Dri)
33–36 lb/ft^3 (Sol-Auto-Dri)
38–45 lb/ft^3 (tamped) (Attagel 150, 350)
Bulking Value:
0.0486 gal/lb (Attacote; Pharmasorb Reg.)
0.0507 gal/lb (Attagel 40, 50, 150, 350; Pharmasorb Colloidal)
0.10 gal/lb (Attaflow)
Fineness:
0.01% 325-mesh residue (Attagel 50)
0.10% 325-mesh residue (Pharmasorb Reg.)
0.3% 325-mesh residue (Attagel 40; Pharmasorb Colloidal)
0.4% 325-mesh residue (Attaflow)
1.0% 325-mesh residue (Attacote)
8.0% 325-mesh residue (Attagel 150, 350)
8/30 mesh (Florco X; Sol-Auto-Dri)
12/45 mesh (Sol-Speedi-Dri)
Surface Area:
210 m^2/g (Attagel 40, 50, 150, 350)
pH:
7.5–9.5 (Attacote; Attagel 40, 50, 150, 350; Pharmasorb Reg. & Colloidal)
8.5 (Attaflow)

Barium-cadmium compound

TRADENAME EQUIVALENTS:

Interstab BC-100S [Interstab]

Mark 99, 180, 1314, 1330C, 1413, TT, WS [Argus]

Nuostabe V-133, V-134, V-1099, V-1399, V-1728, V-1760, V-1764, V-1767, V-1786, V-1785 [Tenneco]

Vanstay 4017, 4030, 4039, 7024, 7025, 7032, HT, HTA, RR [R.T. Vanderbilt]

CATEGORY:

Chemical, heat, and light stabilizer

APPLICATIONS:

Industrial applications: coating processes (Mark 180, 1314, 1413, TT, WS); filled systems (Interstab BC-100S); film and sheet (Interstab BC-100S); hose and profile (Interstab BC-100S); pipe and pipe fittings (Mark 1330C, 1413); plastics/plastisols (Interstab BC-100S; Mark 99, 180, 1314, 1330C, 1413, TT, WS; Nuostabe V-133, V-134, V-1099, V-1399, V-1728, V-1760, V-1764, V-1767, V-1785, V-1786; Vanstay 4017, 4030, 4039, 7024, 7025, 7032, HT, HTA, RR); wire and cable (Vanstay 4030, 4039)

PROPERTIES:

Form:

Liquid (Mark 180; Vanstay 7024, 7025, 7032, RR)

Solution (Nuostabe V-134, V-1099, V-1399, V-1728, V-1760, V-1764, V-1785, V-1786)

Solid (Mark 1314, TT, WS)

Powder (Mark 99; Nuostabe V-133, V-1767; Vanstay 4017, 4030, 4039, HT, HTA)

Color:

Clear (Nuostabe V-134, V-1099, V-1399, V-1728, V-1760, V-1764, V-1785, V-1786)

White (Mark 99; Vanstay 4017, 4030, 4039, HT)

Cream white (Vanstay HTA)

Amber (Mark 180)

Odor:

Mild phenolic (Mark 180)

Sp.gr.:

0.995 (Mark 180)

1.43 (Mark 99)

Barium-cadmium-zinc compound

SYNONYMS:

Ba-Cd-Zn stabilizer

TRADENAME EQUIVALENTS:

Ferro 1288 [Ferro]

Interstab 761-28, BC-109, BC-110 [Interstab] (high zinc)

Interstab 761-28A, BC-103, BC-103L, R-4109, R-4114, R-4137 [Interstab] (low zinc)

Interstab BC-103A, BC-103C, R-4101 [Interstab] (medium zinc)

Interstab BC-4362 [Interstab]

Mark 503, 507, 755, 1014, 2109, 2114, 2115 Series, BB [Argus]

Nuostabe V-1207, V-1541 [Tenneco]

Synpron 1343, 1434 [Synthetic Prod.]

Therm-Chek 6-V-6A, 1292, 5918 [Ferro]

Vanstay 162-B, 246, 3027, 6032, 6040, 6053, 6055, 6074, 6078, 6133, 6172, 6191, 6201, HA, HTF, RRE, RRZ [R.T. Vanderbilt]

Vanstay RZ-25 [R.T. Vanderbilt] (with chelating agents)

CATEGORY:

Heat and light stabilizer

APPLICATIONS:

Automotive products: (Vanstay 6053, 6055, 6074, 6078)

Industrial applications: coating processes (Mark 755); filled systems (Ferro 1288; Interstab 761-28, BC-103A, BC-103C, R-4101; Vanstay 246, HTF); film and sheet (Interstab 761-28A, BC-103, BC-103A, BC-103C, BC-103L, R-4101, R-4114, R-4137; Vanstay 246, 6053, 6191, RRE, RRZ); footwear (Interstab BC-103A, BC-103L; Vanstay 246); hose and profile (Interstab 761-28, BC-103A, BC-103C, BC-103L, R-4101, R-4109); pipe and pipe fittings (Mark 2115 Series); plastics/plastisol (Ferro 1288; Interstab 761-28, 761-28A, BC-103, BC-103A, BC-103C, BC-103L, BC-109, BC-110, BC-4362, R-4101, R-4109, R-4114, R-4137; Mark 755, 2109, 2114, BB; Nuostabe V-1207, V-1541; Synpron 1343, 1434; Therm-Chek 6-V-6A, 1292, 5918; Vanstay 162-B, 246, 3027, 6032, 6040, 6053, 6133, 6172, 6201, HA, HTF, RRE, RRZ, RZ-25); tile (Synpron 1434; Vanstay HA, HTF)

PROPERTIES:

Form:

Liquid (Ferro 1288; Interstab 761-28, 761-28A, BC-103, BC-103A, BC-103C, BC-103L, BC-109, BC-110, BC-4362, R-4101, R-4109, R-4114, R-4137; Mark 755, 2109, 2114, BB; Therm-Chek 6-V-6A, 1292, 5918; Vanstay 162-B, 246, 6032, 6040, 6053, 6055, 6074, 6078, 6133, 6172, 6191, 6201, RRE, RRZ)

Powder (Nuostabe V-1207, V-1541; Synpron 1343, 1434; Vanstay 3027, HA, HTF)

Color:

White (Vanstay HA)

Cream white (Vanstay 3027, HTF)

Light amber (Mark BB; Therm-Chek 6-V-6A)

Barium-cadmium-zinc compound *(cont'd.)*

    Amber (Vanstay 246, 6032, RRE, RRZ)
    Gardner 8 (Ferro 1288)
    Gardner 8 max. (Interstab BC-4362)
Odor:
    Mild phenolic (Mark BB)
Sp.gr.:
    0.982 (Mark BB)
    1.010–1.020 (Interstab BC-4362)
    1.04 (Therm-Chek 6-V-6A)
Density:
    8.7 lb/gal (Ferro 1288)
    0.94 ± 0.02 mg/m^3 (Vanstay RZ-25)
    0.97 ± 0.02 mg/m^3 (Vanstay RRE, RRZ)
    0.98 ± 0.02 mg/m^3 (Vanstay 246)
    1.02 ± 0.02 mg/m^3 (Vanstay 6032)
Visc.:
    A (Gardner) (Ferro 1288)
    50 cps max. (Interstab BC-4362)
TOXICITY/HANDLING:
    Toxicity normal for barium-cadmium compounds; not to be ingested (Ferro 1288; Therm-Chek 6-V-6A)

Butylated hydroxyanisole

SYNONYMS:
    BHA (CTFA)
    (1,1-Dimethylethyl)-4-Methoxyphenol
    Phenol, (1,1-dimethylethyl)-4-methoxy-
EMPIRICAL FORMULA:
    $C_{11}H_{16}O_2$
STRUCTURE:

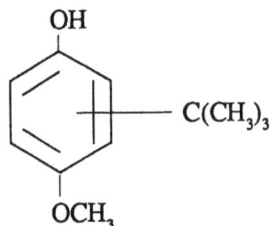

CAS No.:
    25013-16-5

18

TRADENAME EQUIVALENTS:
Sustane BHA, BHA 1-F [UOP Process Div.]
Tenox BHA [Eastman Chem. Prod.]

CATEGORY:
Antioxidant, stabilizer, preservative

APPLICATIONS:
Cosmetic industry preparations: (Sustane BHA 1-F); perfumery/essential oils (Sustane BHA 1-F)

Food applications: (Sustane BHA, BHA 1-F; Tenox BHA); flavors (Sustane BHA 1-F); food packaging (Sustane BHA 1-F; Tenox BHA)

Industrial applications: food-grade plastics (Tenox BHA); food-grade rubber (Tenox BHA)

Pharmaceutical applications: vitamins (Sustane BHA 1-F)

PROPERTIES:
Form:
Flakes (Sustane BHA, BHA 1-F)
Tablets (Sustane BHA, BHA 1-F)
Waxy tablets (Tenox BHA)
Color:
White (Sustane BHA 1-F; Tenox BHA)
Odor:
Slight (Tenox BHA)
Composition:
98.5% min. purity (Sustane BHA 1-F)
Solubility:
Sol. > 25 g/100 g in acetone (Sustane BHA 1-F)
Sol. > 30 g/100 g in cottonseed oil (Sustane BHA 1-F)
Sol. ≥ 50% in diisobutyl adipate (Tenox BHA)
Sol. in edible oils (Tenox BHA)
Sol. ≥ 50% in ethanol (Tenox BHA)
Sol. > 25 g/100 g in ether (Sustane BHA 1-F)
Sol. ≥ 50% in glyceryl monooleate (Tenox BHA)
Sol. ≥ 50% in lard (Tenox BHA)
Sol. > 10 g/100 g in methanol (Sustane BHA 1-F)
Sol. in organic solvents (Tenox BHA)
Sol. ≥ 50% in paraffin (Tenox BHA)
Sol. ≥ 50% in propylene glycol (Tenox BHA)
Sol. ≥ 50% in soya oil (Tenox BHA); > 30 g/100 g in soybean oil (Sustane BHA 1-F)
Negligible sol. in water (Sustane BHA 1-F); insol. (Tenox BHA)
M.W.:
180 (Tenox BHA)

Butylated hydroxyanisole *(cont'd.)*

180.2 (Sustane BHA 1-F)
Visc.:
3.3 cS (99 C) (Sustane BHA 1-F)
B.P.:
264–270 C (733 mm) (Tenox BHA)
270 F (5 mm Hg) (Sustane BHA 1-F)
M.P.:
48–55 C (Tenox BHA)
57 C (Sustane BHA 1-F)
Flash Pt.:
130 C (OC) (Sustane BHA 1-F)
TOXICITY/HANDLING:
Slight irritant of low toxicity (Sustane BHA 1-F)
Wear protective gloves and glasses when handling to avoid irritant effects; use with
adequate ventilation (Tenox BHA)
STORAGE/HANDLING:
Fine dust may create a dust explosion (Tenox BHA)
STD. PKGS.:
100 lb drums (Sustane BHA 1-F tablets)
50 lb drums (Sustane BHA 1-F flakes)
5 lb cartons (Sustane BHA 1-F)
2.27, 11.34, 22.68, or 45.36 kg fiber drums (Tenox BHA)

Butylated hydroxytoluene

SYNONYMS:
BHT (CTFA)
2,6-Bis (1,1-dimethylethyl)-4-methylphenol
DBPC
2,6-Di-*t*-butyl-*p*-cresol
Phenol, 2,6-bis (1,1-dimethylethyl)-4-methyl-
EMPIRICAL FORMULA:
$C_{15}H_{24}O$
STRUCTURE:

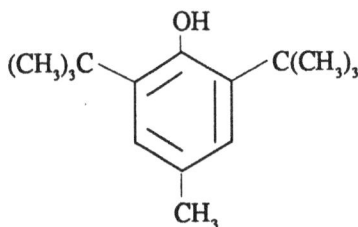

CAS No.:
128-37-0
TRADENAME EQUIVALENTS:
CAO-1 [PMC Specialties; Sherex; Summit]
CAO-3 [PMC Specialties; Sherex; Summit] (food grade)
Ionol, Ionol CP [Shell]
Naugard BHT-Food Grade [Uniroyal]
Naugard BHT-Tech. [Uniroyal]
Sustane BHT [UOP Process] (food grade)
Vanlube PC [R.T. Vanderbilt]
Vulkanox KB [Mobay]
  Food grade generically sold by: [Neville; PMC; Summit]
  Generically sold by: [Aceto Chem.; Biesterfield; Housmex (feed, tech. and food
    grades); Rhodia; Summit; Universal Chem.]

CATEGORY:
Antioxidant, stabilizer, preservative, synergist

APPLICATIONS:
Cleansers: soaps (Sustane BHT)
Cosmetic industry preparations: (Sustane BHT)
Farm products: insecticides/pesticides (CAO-1, -3)
Food applications: food additives (food grade generic; generic—Housemax food
  grade; CAO-3; Sustane BHT); food packaging (Ionol CP; Sustane BHT)
Household products: bathing goods, hoses, toys (Vulkanox KB)
Industrial applications: adhesives (Naugard BHT-Tech.); fabric proofings (Vulkanox
  KB); latex goods (Vulkanox KB); lubricants (CAO-1, -3); oils, fats, greases (CAO-
  1, -3; Sustane BHT; Vanlube PC); packaging materials (CAO-1, -3); petroleum
  industry (CAO-1, -3; Ionol; Vanlube PC); paraffin (Ionol CP); plastics (generic;
  Ionol, CP; Naugard BHT-Food Grade, BHT-Tech.; Sustane BHT; Vulkanox KB);
  polymers (CAO-1, -3); resins (generic); rubber (generic; CAO-1, -3; Ionol, CP;
  Naugard BHT-Tech.; Vulkanox KB); waxes (CAO-1, -3; Vanlube PC)
Pharmaceutical applications: drug packaging (Ionol CP); vitamins (Sustane BHT)

PROPERTIES:
Form:
Crystals (CAO-1, -3; Naugard BHT-Tech.; Sustane BHT)
Crystalline flakes (Vanlube PC)
Crystalline granular (generic—Universal)
Flakes (generic—Rhodia)
Powder (generic—Housmex; Vulkanox KB)
Color:
White (generic—Universal; CAO-1, -3; Naugard BHT-Tech.; Sustane BHT; Vanlube
  PC)

Butylated hydroxytoluene (cont'd.)

White to pink (Vulkanox KB)
APHA 15 max. (generic—Housmex)
Pt-Co 15 max. (30% sol'n. in acetone) (Ionol CP)
Pt-Co 45 max. (30% sol'n. in acetone) (Ionol)

Odor:

Very slight (generic—Universal)
Slight (CAO-1, -3)

Composition:

98% purity min. (generic—Rhodia; Ionol)
99% purity (CAO-3)
99% purity min. (Ionol CP; Sustane BHT)

Solubility:

Sol. 40 g/100 g benzene (Sustane BHT)
Sol. 25% in ethanol (CAO-1, -3)
Sol. 48 g/100 g lard (50 C) (Sustane BHT)
Sol. 28 g/100 g linseed oil (Sustane BHT)
Sol. 20 g/100 g methanol (Sustane BHT)
Sol. 30% in min. oil (CAO-1, -3; Sustane BHT)
Sol. in common organic solvents (Naugard BHT-Tech.)
Sol. in petroleum and synthetic lubricant bases (Vanlube PC)
Sol. 40% in tallow (CAO-1, -3)
Nil sol. in water (Sustane BHT); insol. (CAO-1, -3; Vanlube PC)
Sol. 0.5% without sediment or turbidity in white oil (Ionol)

M.W.:

220.3 (Sustane BHT)

Sp.gr.:

1.03 ± 0.01 (Vulkanox KB)
1.048 (generic—Universal); (20/4 C) (Sustane BHT)

Density:

1.04 mg/m^3 (Vanlube PC)

Bulk Density:

37.5 lb/ft^3 (Sustane BHT)

Visc.:

3.47 cS (80 C) (Sustane BHT)

F.P.:

68 C min. (generic—Rhodia)
69.4 C (Naugard BHT-Tech.)

B.P.:

260–262 C (760 mm Hg) (CAO-1)
265 C (760 mm Hg) (generic—Universal; Sustane BHT)

M.P.:
   69 C (Vanlube PC)
   69–70 C (generic—Universal)
   70 C (Sustane BHT)
Solidification Pt.:
   68.8 C min. (Ionol)
   69 C (Vulkanox KB)
   69.4 C min. (Ionol CP)
Flash Pt.:
   118 C (CC) (Sustane BHT)
   135 C (COC) (Vanlube PC)
Ref. Index:
   1.486 (Sustane BHT)
TOXICITY/HANDLING:
   Slight irritant of low toxicity (Sustane BHT)
STD. PKGS.:
   20-kg paper bags (Vulkanox KB)
   55-lb bags (generic—Housmex)
   100-lb fiber drums, 50-lb bags (CAO-1)
   Bags, drums (generic—Aceto)

Butyl benzyl phthalate (CTFA)

SYNONYMS:
   1,2 Benzenedicarboxylic acid, butyl phenylmethyl ester
EMPIRICAL FORMULA:
   $C_{19}H_{20}O_4$
STRUCTURE:

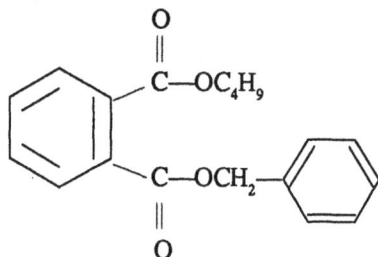

CAS No.:
   85-68-7

23

Butyl benzyl phthalate *(cont'd.)*

TRADENAME EQUIVALENTS:
Santicizer 160 [Monsanto]
CATEGORY:
Plasticizer
APPLICATIONS:
Industrial applications: calendering and extrusion; cellulosics; coatings/lacquers; film
and sheet; flooring industry; packaging; plastics; resins; rubber
PROPERTIES:
Form:
Clear oily liquid
Color:
APHA 40 max.
Odor:
Slight, characteristic
Solubility:
Sol. 0.0003% in water @ 30 C
M.W.:
312
Sp.gr.:
1.115–1.123
Density:
9.3 lb/gal
Visc.:
41.5 ± 1.5 cSt
B.P.:
240 C (10 mm Hg)
Pour Pt.:
–45 C
Crystal. Pt.:
< –35 C
Flash Pt.:
390 F (COC)
Fire Pt.:
450 F (COC)
Ref. Index:
1.535–1.540
Surface Tension:
39.9 dynes/cm
Coeff. of Linear Exp.:
0.00080 cc/cc/C (10–40 C)
TOXICITY/HANDLING:
Low toxicity

SYNONYMS:
1,4-Benzenediol, 2-(1,1-dimethylethyl)-
2-(1,1-Dimethylethyl)-1,4-benzenediol
TBHQ
EMPIRICAL FORMULA:
$C_{10}H_{14}O_2$
STRUCTURE:

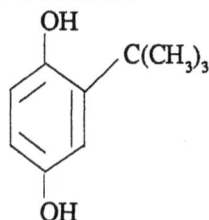

CAS No.:
1948-33-0
TRADENAME EQUIVALENTS:
Sustane TBHQ [UOP]
Tenox TBHQ [Eastman]
CATEGORY:
Antioxidant, stabilizer, preservative
APPLICATIONS:
Food applications: (Sustane TBHQ; Tenox TBHQ)
PROPERTIES:
Form:
Crystals (Sustane TBHQ; Tenox TBHQ)
Color:
White to light tan (Sustane TBHQ; Tenox TBHQ)
Odor:
Very slight (Tenox TBHQ)
Composition:
99.0% min. purity (Sustane TBHQ)
Solubility:
Sol. 10 g/100 g corn oil (Sustane TBHQ)
Sol. ≥ 50% in ethanol (Tenox TBHQ)
Sol. ≥ 50% in ethyl acetate (Tenox TBHQ)
Sol. in edible fats and oils (Tenox TBHQ)
Sol. 5 g/100 g lard (50 C) (Sustane TBHQ)
Sol. 100 g/100 g methanol (Sustane TBHQ)
Sol. in organic solvents (Tenox TBHQ)
Sol. 30 g/100 g propylene glycol (Sustane TBHQ)
Nil sol. in water (Sustane TBHQ); sol. < 1% in water (Tenox TBHQ)

t-Butyl hydroquinone (cont'd.)

M.W.:
166.2 (Sustane TBHQ; Tenox TBHQ)
Bulk Density:
27 lb/ft^3 (Sustane TBHQ)
B.P.:
295 C (Sustane TBHQ)
M.P.:
126.5–128.5 C (Sustane TBHQ; Tenox TBHQ)
Flash Pt.:
171 C (CC) (Sustane TBHQ)
TOXICITY/HANDLING:
May cause eye irritation; prolonged contact may cause moderate skin irritation (Sustane TBHQ)
Wear protective gloves and glasses when handling to avoid irritant effects; use with adequate ventilation (Tenox TBHQ)
STORAGE/HANDLING:
Fine dust may create a dust explosion (Tenox TBHQ)
STD. PKGS.:
100 lb drums; 5 lb cartons (Sustane TBHQ)
2.27, 11.34, 22.68, or 45.36 kg fiber drums (Tenox TBHQ)

Butyl octyl phthalate

STRUCTURE:
$C_4H_9OOCC_6H_4COOC_8H_{17}$
TRADENAME EQUIVALENTS:
Hatcol BOP [Hatco]
PX-914 [USS Chem.]
Staflex BOP [Reichhold]
Generically sold by:
[Ashland; Badische; BASF Canada; Eastman; Nuodex]
CATEGORY:
Plasticizer
APPLICATIONS:
Industrial applications: cellulosics (PX-914; Staflex BOP); floor tile (Hatcol BOP); hose (Hatcol BOP); plastics/plastisols (generic; Hatcol BOP; PX-914; Staflex BOP); rubber (PX-914)

PROPERTIES:
Form:
  Liquid (generic)
Color:
  Water-white (generic)
  APHA 30 (Staflex BOP)
Odor:
  Mild characteristic (generic)
Solubility:
  Miscible with most organic solvents (generic)
M.W.:
  334 (Hatcol BOP)
  368 (Staflex BOP)
  377 (PX-914)
Sp.gr.:
  0.991–0.997 (20 C) (generic)
  0.992 (20/20 C) (generic—Ashland)
  0.996 (23 C) (Staflex BOP)
Density:
  8.3 lb/gal (Staflex BOP)
Visc.:
  287 cSt (kinematic, 0 C) (Staflex BOP)
M.P.:
  < –50 C (generic)
Pour Pt.:
  –52 F (Staflex BOP)
Flash Pt.:
  199 C (COC) (generic)
  415 F (Staflex BOP)
Cloud Pt.:
  –22 F (Staflex BOP)
Saponification No.:
  298–308 (generic)
Ref. Index:
  1.485 (generic)
  1.4862 (23 C) (Staflex BOP)
STORAGE/HANDLING:
  Combustible (generic)

Butyl oleate (CTFA)

SYNONYMS:
Butyl 9-octadecenoate
9-Octadecenoic acid, butyl ester
EMPIRICAL FORMULA:
$C_{22}H_{42}O_2$
STRUCTURE:

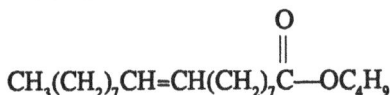

$$CH_3(CH_2)_7CH=CH(CH_2)_7\overset{\overset{\textstyle O}{\|}}{C}\text{—}OC_4H_9$$

CAS No.:
142-77-8
TRADENAME EQUIVALENTS:
Butyl Oleate C-914 [C.P. Hall]
Butyl Oleate DLC [Harwick]
Diamond Shamrock Butyl Oleate, light [Diamond Shamrock]
Emerest 2328 [Emery]
Graden Butyl Oleate [Graden]
Grocor 4000 [A. Gross]
Kemester 4000 [Humko]
Kessco Butyl Oleate [Armak]
Plasthall 503 [C.P. Hall]
Polycizer Butyl Oleate [Harwick]
Radia 7040 [Oleofina S.A.]
Uniflex BYO [Union Camp]

CATEGORY:
Plasticizer, softener, heat stabilizer, lubricant, emollient, coupling agent, solubilizer, solvent, defoamer, wetting agent, chemical intermediate, antistat, processing aid, cure retarder
APPLICATIONS:
Cosmetic industry preparations: creams and lotions (Kessco Butyl Oleate); hair sprays (Kessco Butyl Oleate); makeup (Kessco Butyl Oleate)
Farm products: agricultural defoamer (Diamond Shamrock Butyl Oleate, light)
Industrial applications: cellulosics (Uniflex BYO); chemical synthesis (Radia 7040); lubricating/cutting oils (Radia 7040; Uniflex BYO); metalworking (Uniflex BYO); mold release (Uniflex BYO); plastics (Croda Butyl Oleate; Grocor 4000; Kessco Butyl Oleate; Polycizer Butyl Oleate; Uniflex BYO); polishes and waxes (Uniflex BYO); polymers (Polycizer Butyl Oleate; Uniflex BYO); rubber (Butyl Oleate C-914; Croda Butyl Oleate; Diamond Shamrock Butyl Oleate, light; Graden Butyl Oleate; Grocor 4000; Kemester 4000; Kessco Butyl Oleate; Plasthall 503; Uniflex BYO); textile/leather processing (Grocor 4000; Kemester 4000; Kessco Butyl Oleate; Plasthall 503; Radia 7040; Uniflex BYO); waterproofing agents (Uniflex BYO)

PROPERTIES:

Form:
Liquid (Croda Butyl Oleate; Diamond Shamrock Butyl Oleate, light; Emerest 2328; Grocor 4000; Kemester 4000)
Clear liquid (Graden Butyl Oleate; Polycizer Butyl Oleate)
Clear oily liquid (Plasthall 503)
Oily liquid (Butyl Oleate C-914)
Powder (Butyl Oleate DLC)

Color:
Light (Butyl Oleate C-914)
Slightly yellow (Grocor 4000)
Light yellow (Polycizer Butyl Oleate)
APHA 70 (Uniflex BYO)
Gardner 2 max. (Graden Butyl Oleate)
Gardner 11 (Plasthall 503)

Odor:
Mild (Butyl Oleate C-914; Plasthall 503)
Mild, fatty (Graden Butyl Oleate)

Composition:
72% active on pressure-sensitive silicate (Butyl Oleate DLC)
97% active min. (Grocor 4000)
100% conc. (Diamond Shamrock Butyl Oleate, light; Emerest 2328)

Solubility:
Sol. in acetone (Graden Butyl Oleate; Plasthall 503)
Sol. in benzol (Graden Butyl Oleate)
Sol. in carbon tetrachloride (Graden Butyl Oleate)
Sol. in castor oil (Graden Butyl Oleate)
Sol. in ethanol (Plasthall 503)
Insol. in glycerol (Graden Butyl Oleate)
Sol. in hexane (Plasthall 503)
Sol. in hydrocarbons (Grocor 4000)
Sol. in isopropanol (Graden Butyl Oleate; Grocor 4000)
Sol. in kerosene (Plasthall 503)
Sol. in min. oil (Graden Butyl Oleate; Grocor 4000; Plasthall 503; Radia 7040)
Sol. in peanut oil (Graden Butyl Oleate)
Insol. in propylene glycol (Graden Butyl Oleate)
Sol. in most solvents (Radia 7040)
Sol. in toluene (Plasthall 503)
Sol. in veg. oil (Radia 7040)
Insol. in water (Graden Butyl Oleate)

Butyl oleate *(cont'd.)*

Ionic Nature:
Nonionic (Diamond Shamrock Butyl Oleate, light; Emerest 2328; Kessco Butyl Oleate)

M.W.:
327 (avg.) (Radia 7040)
388 (avg.) (Plasthall 503)

Sp.gr.:
0.854 (37.8 C) (Radia 7040)
0.86 (Grocor 4000)
0.864–0.877 (Graden Butyl Oleate)
0.868 (20/20 C) (Uniflex BYO)
0.87 (Polycizer Butyl Oleate)
0.873 (Butyl Oleate C-914)
0.875 (Plasthall 503)

Density:
7.62 lb/gal (Butyl Oleate C-914; Plasthall 503)
7.2 lb/gal (Graden Butyl Oleate)

Visc.:
5.30 cps (37.8 C) (Radia 7040)
14 cps (Plasthall 503)
8 cSt (Uniflex BYO)

F.P.:
–55 C (Plasthall 503)
–20 C (Grocor 4000)
–10 C (Graden Butyl Oleate)

B.P.:
170 C (Grocor 4000)
173–227 C (2 mm) (Butyl Oleate C-914; Plasthall 503)
190–220 C (Graden Butyl Oleate)

M.P.:
10 C (Uniflex BYO)

Flash Pt.:
179 C (Plasthall 503)
180 C (Butyl Oleate C-914)
195 C (COC) (Radia 7040)
199 C (COC) (Uniflex BYO)
375 F (COC) (Graden Butyl Oleate)

Fire Pt.:
218 C (COC) (Uniflex BYO)

Cloud Pt.:
–15 C (Radia 7040)

Acid No.:
   0.5 (Uniflex BYO)
   1.5 (Plasthall 503)
Iodine No.:
   70–80 (Graden Butyl Oleate)
   75 (Polycizer Butyl Oleate; Uniflex BYO)
   127 (Plasthall 503)
Saponification No.:
   165 (Plasthall 503)
   170 (Uniflex BYO)
Ref. Index:
   1.45 (Grocor 4000)
   1.451 (Graden Butyl Oleate; Uniflex BYO)
   1.4518 (Radia 7040)
   1.457 (Plasthall 503)
pH:
   Neutral (Grocor 4000)
Surface Tension:
   31.5 dynes/cm (Radia 7040)
STD. PKGS.:
   Drums (Butyl Oleate C-914; Plasthall 503)
   Bulk or drums (Grocor 4000)

Butyltin mercaptide

TRADENAME EQUIVALENTS:
   Advastab TM-180 [Cincinnati Milacron]
   Stanclere T-94 C, T-126, T-801 [Interstab]
CATEGORY:
   Heat and light stabilizer
APPLICATIONS:
   Industrial applications: film and sheet (Stanclere T-94 C, T-801); pipe and profile
      (Stanclere T-801); plastics/plastisols (Advastab TM-180; Stanclere T-94 C, T-
      126); powder coatings (Advastab TM-180); siding (Stanclere T-801)
PROPERTIES:
Form:
   Liquid (Stanclere T-94 C, T-126, T-801)
   Clear liquid (Advastab TM-180)

Butyltin mercaptide *(cont'd.)*

Color:
  Gardner 4 max. (Advastab TM-180)
Sp.gr.:
  1.110–1.130 (75 F) (Advastab TM-180)
Density:
  9.24–9.48 lb/gal (Advastab TM-180)
Visc.:
  50 cps max. (75 F) (Advastab TM-180)
STD. PKGS.:
  55-gal (450 lb net) drums (Advastab TM-180)

C_{12-15} alcohols lactate (CTFA)

STRUCTURE:

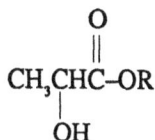

$$CH_3CHC-OR$$

where R represents the C_{12-15} alkyl group

CAS No.:
RD No. 977064-17-7

TRADENAME EQUIVALENTS:
Ceraphyl 41 [Van Dyk]

CATEGORY:
Plasticizer, gloss agent

APPLICATIONS:
Cosmetic industry preparations: hair resin preparations

PROPERTIES:

Solubility:
Sol. in 95% ethanol
Sol. in isopropyl myristate
Sol. in min. oil
Insol. in water

Capryl hydroxyethyl imidazoline (CTFA)

SYNONYMS:
3H-Imidazole-1-ethanol, 4,5-dihydro-2-nonyl-
2-Nonyl-4,5-dihydro-1H-imidazole-1-ethanol

EMPIRICAL FORMULA:
$C_{14}H_{26}N_2O$

STRUCTURE:

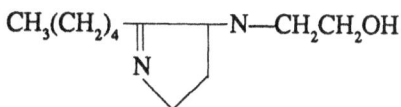

Capryl hydroxyethyl imidazoline (cont'd.)

CAS No.:
37478-68-5
TRADENAME EQUIVALENTS:
Monazoline CY [Mona]
CATEGORY:
Thickener, wetting agent, emulsifier, detergent, corrosion inhibitor, antistat, softener, bactericide
APPLICATIONS:
Farm products: agricultural oils/sprays; herbicides/fungicides
Industrial applications: dyes and pigments; flotation agents; paint mfg.; petroleum industry; plastic films; textile/leather processing
PROPERTIES:
Form:
Liquid (may crystallize on aging)
Color:
Amber
Composition:
90% min. imidazoline
Solubility:
Sol. in chlorinated hydrocarbons
Sol. in ethanol
Sol. in toluene
Sol. in veg. oil
Ionic Nature:
Cationic
M.W.:
212
Sp.gr.:
0.99
Density:
8.25 lb/gal
Acid No.:
1 max.
pH:
10.5–12.0 (10% disp.)
Biodegradable:
TOXICITY/HANDLING:
Fairly strong organic base—handle with caution; wear protective goggles and gloves
STD. PKGS.:
55-gal open-head steel drums

SYNONYMS:
1H-Isoindole-1,3(2H)-dione, 3a,4,7,7a-tetrahydro-2-[(trichloromethyl)thio]-
3a,4,7,7a-Tetrahydro-2-[(trichloromethyl) thio]-1H-isoindole-1,3(2H)-dione
N-Trichloromethylthio-4-cyclohexene-1,2-dicarboximide
N-Trichloromethylthiotetrahydro-phthalimide
EMPIRICAL FORMULA:
$C_9H_8Cl_3NO_2S$
STRUCTURE:

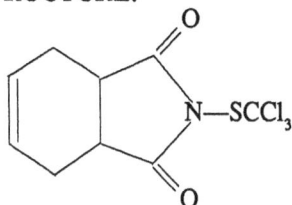

CAS No.:
133-06-2
TRADENAME EQUIVALENTS:
Vancide 89, 89RE [R.T. Vanderbilt]
CATEGORY:
Antimicrobial, preservative, fungicide
APPLICATIONS:
Cleansers: hand soaps (Vancide 89RE)
Cosmetic industry preparations: (Vancide 89RE); creams and lotions (Vancide 89RE);
shampoos (Vancide 89RE)
Industrial applications: paint/lacquer mfg. (Vancide 89); rubber (Vancide 89); soaps
(Vancide 89); wallpaper paste (Vancide 89)
Pharmaceutical applications: (Vancide 89RE); antidandruff preparations (Vancide
89RE); ointments (Vancide 89RE)
Veterinary products (Vancide 89RE)
PROPERTIES:
Form:
Powder (Vancide 89)
Fine powder (Vancide 89RE)
Color:
White to off-white (Vancide 89RE)
Light tan (Vancide 89)
Composition:
90% min. assay (Vancide 89)
97% assay (Vancide 89RE)
Solubility:
Moderately sol. in acetone (Vancide 89); sol. 4.2 g/100 g (Vancide 89RE)
Sol. in chloroform (Vancide 89); sol. 5.3 g/100 g (Vancide 89RE)

35

Captan *(cont'd.)*

Sol. in cyclohexanone (Vancide 89); sol. 5.8 g/100 g (Vancide 89RE)
Sol. 5.0 g/100 g ethyl acetate (Vancide 89RE)
Sol. in ethylene dichloride (Vancide 89)
Sol. in tetrachloroethane (Vancide 89); sol. 5.2 g/100 g (Vancide 89RE)
Moderately sol. in toluene (Vancide 89)
Insol. in water (Vancide 89, 89RE)
Sol. in xylene (Vancide 89); sol. 7.5 g/100 g (Vancide 89RE)
M.W.:
300.6 (Vancide 89)
Sp.gr.:
1.69 ± 0.03 (Vancide 89)
Density:
1.7 mg/m³ (Vancide 89RE)
M.P.:
168–174 C (Vancide 89)
171–176 C (Vancide 89RE)
Stability:
Avoid use with alkaline materials and water systems in general (Vancide 89)
Very stable in dry condition at ordinary temps.; stable to uv light, in w/o and o/w emulsions; optimum stability in nonaq. systems and in systems with acid pH (Vancide 89RE)
Storage Stability:
May be stored indefinitely in closed containers (Vancide 89)
pH:
5.0–6.0 (1% disp.) (Vancide 89RE)
TOXICITY/HANDLING:
Avoid inhalation of dust; avoid prolonged/repeated contact with skin; toxic to fish (Vancide 89)
STD. PKGS.:
50-lb Leverpak drum (Vancide 89)

Carrageenan *(CTFA)*

SYNONYMS:
Chondrus
Irish moss extract
CAS No.:
9000-07-1

Carrageenan *(cont'd.)*

TRADENAME EQUIVALENTS:
Genugel Series [Hercules]
Genuvisco Series [Hercules]
Genulacta Series [Hercules]
CATEGORY:
Thickener, gelling agent, stabilizer, suspending agent
APPLICATIONS:
Cosmetic industry preparations: (Genugel Series; Genuvisco Series; Genulacta Series)
Food applications: (Genugel Series; Genuvisco Series; Genulacta Series)
Pharmaceutical applications: (Genugel Series; Genuvisco Series; Genulacta Series)

Castor oil *(CTFA)*

SYNONYMS:
Ricinus oil
CAS No.:
8001-79-4
TRADENAME EQUIVALENTS:
Crystal Crown [CasChem Inc.] (deodorized, refined)
Crystal O [CasChem Inc.] (deodorized, refined)
Surfactol-13 [NL Industries] (modified)
    Generically sold by: [Acme Hardesty]
CATEGORY:
Plasticizer, solvent, lubricant, release agent, antifoam agent, wetting agent, thickener
APPLICATIONS:
Cosmetic industry preparations: (Surfactol-13)
Household products: (Surfactol-13)
Industrial applications: dyes and pigments (Surfactol-13); lubricating/cutting oils (Surfactol-13); paint mfg. (Surfactol-13); rubber (generic—Acme Hardesty); textile/leather processing (Surfactol-13); waxes (Surfactol-13)
PROPERTIES:
Form:
Liquid (Surfactol-13)
Viscous liquid (generic—Acme Hardesty)
Color:
Gardner 4 (Surfactol-13)
Odor:
Faint (Surfactol-13)
Faint, mild (Crystal O)

Castor oil *(cont'd.)*

Taste:
Bland, mildly characteristic (Crystal O)
Composition:
100% conc. (Surfactol-13)
Solubility:
Sol. in alcohols (Crystal Crown, O)
Partly sol. in aliphatic hydrocarbons (Crystal Crown, O)
Sol. in aromatic solvents (Crystal Crown, O)
Sol. in esters (Crystal Crown, O)
Sol. in ethers (Crystal Crown, O)
Sol. in ketones (Crystal Crown, O)
80% sol. in water (Surfactol-13)
Ionic Nature:
Nonionic (Surfactol-13)
Sp.gr.:
1.005 (Surfactol-13)
Density:
8.36 lb/gal (Surfactol-13)
Visc.:
17 stokes (Surfactol-13)
STORAGE/HANDLING:
Keep containers closed (Surfactol-13)
STD. PKGS.:
55-gal drums (Surfactol-13)

Castor oil, sulfated

SYNONYMS:
Alizarin assistant
Alizarin oil
Castor oil, soluble
Castor oil, sulfonated
Sulfated castor oil (CTFA)
Turkey red oil
CAS No.
8002-33-3
TRADENAME EQUIVALENTS:
Actrasol C50, C75, C85, PSR [Southland]
Ahco AJ-110 [ICI]

TRADENAME EQUIVALENTS *(cont'd.):*
Chemax SCO [Chemax]
Consos Castor Oil [Consos]
Cordon NU 890/50, 890/75 [Finetex]
Eureka 102 [Atlas Refinery]
Haroil SCO-50, -65, -75, -7525 [Graden]
Hartenol V-63 [Hart]
Hartex V63, V64 [Hart]
Hipochem Dispersol SCO [High Point]
Marvanol SCO 75% [Marlowe-Van Loan]
Monopol Oil 75 [GAF]
Nopcocastor, Nopcocastor L [Diamond Shamrock]
Nopcosulf CA-60, CA-70 [Diamond Shamrock]
Standapol SCO [Henkel]
Supratol VF [Hart]

CATEGORY:
Wetting agent, lubricant, penetrant, dispersant, emulsifier, surfactant, softener, retarder, leveling agent, plasticizer, detergent, superfatting agent, antisagging agent,
antisettling agent

APPLICATIONS:
Bath products: bubble bath (Standapol SCO)
Cleansers: cleansing creams (Haroil SCO-50, SCO-65); face cleanser (Haroil SCO-
75); hand cleanser (Haroil SCO-65)
Cosmetic industry preparations: cosmetic base (Haroil SCO-50, SCO-75; Nopco-
castor); personal care products (Haroil SCO-50); shampoo base (Standapol SCO);
shampoos (Standapol SCO)
Industrial applications: adhesives (Haroil SCO-65); dyes and pigments (Actrasol C50,
C75, C85, PSR; Ahco AJ-110; Eureka 102; Hartenol V-63; Hipochem Dispersol
SCO; Marvanol SCO 75%; Monopol 75; Supratol VF); industrial processing
(Haroil SCO-65, SCO-75; Marvanol SCO 75%); lubricating/cutting oils (Cordon
NU 890/75); metalworking (Chemax SCO); paint mfg. (Haroil SCO-75, SCO-
7525); polishes and waxes (Nopcosulf CA-60, CA-70); textile/leather processing
(Ahco AJ-110; Chemax SCO; Consos Castor Oil; Eureka 102; Monopol Oil 75;
Nopcocastor L; Supratol VF)
Industrial cleaners: drycleaning compositions (Nopcosulf CA-60, CA-70)

PROPERTIES:
Form:
Liquid (Actrasol C50, C75, C85, PSR; Ahco AJ-110; Chemax SCO; Consos Castor
Oil; Cordon NU 890/75; Eureka 102; Haroil SCO-50, SCO-65, SCO-75, SCO-
7525; Hartenol V-63; Hartex V63, V64; Hipochem Dispersol SCO; Marvanol SCO
75%; Nopcosulf CA-60, CA-70; Supratol VF)

Castor oil, sulfated (cont'd.)

    Clear liquid (Monopol Oil 75)
    Viscous liquid (Standapol SCO)
Color:
    Light (Eureka 102)
    Amber (Ahco AJ-110; Haroil SCO-65, SCO-7525; Hartex V63, V64; Marvanol SCO
       75%; Standapol SCO; Supratol VF)
    Dark amber (Hipochem Dispersol SCO)
    Yellow (Monopol Oil 75)
Odor:
    Castor oil (Ahco AJ-110)
    Mild sulfate (Hipochem Dispersol SCO)

Composition:
    30% active (Supratol VF)
    45% active (Hartex V64)
    50% active (Actrasol C-50; Haroil SCO-50, SCO-65; Nopcocastor L)
    50% active in water (Ahco AJ-110)
    63% active (Nopcosulf CA-70)
    68% active (Haroil SCO-75; Nopcosulf CA-60)
    70% active (Actrasol C-75; Hartex V63; Hipochem Dispersol SCO)
    72% active (Eureka 102)
    75% active (Actrasol C-85, PSR; Cordon NU 890/75; Haroil SCO-7525; Hartenol V-
       63; Marvanol SCO 75%; Monopol Oil 75; Nopcocastor)

Solubility:
    Sol. in water @ 25 C (Ahco AJ-110); disp. at all temps. (Hipochem Dispersol SCO)

Ionic Nature:
    Anionic
Sp.gr.:
    1.0 (Hipochem Dispersol SCO)
    1.02 (Hartex V64)
    1.025 @ 68 F (Haroil SCO-7525)
    1.05 (Ahco AJ-110; Monopol Oil 75)
    1.056 (Hartex V63)
    1.112 (Supratol VF)
Density:
    8.33 lb/gal (Hipochem Dispersol SCO)
Visc.:
    90 cps (Supratol VF)
    250 cps (Hartex V64)
    520 cps (Hartex V63)
Hydroxyl No.:
    16.0 max. (Standapol SCO)

Stability:
Excellent (Hipochem Dispersol SCO)
Stable to acids, alkalies and electrolytes (Monopol Oil 75)
Ref. Index:
1.4022 (Supratol VF)
1.4448 (Hartex V63)
pH:
7.0 ± 0.5 (2% aq. sol'n.) (Haroil SCO-65)
7.6 (Eureka 102)
TOXICITY/HANDLING:
Irritating to skin and eyes in conc. form (Standapol SCO)
Eye irritant (Ahco AJ-110)
Skin and eye irritant (Supratol VF)
Avoid contact with skin and eyes (Hartex V63, V64; Supratol VF)
STORAGE/HANDLING:
Store in cool, dry place; protect from moisture (Standapol SCO)
STD. PKGS.:
55-gal drums (Haroil SCO-65, SCO-7525)
Bulk or drums (Hipochem Dispersol SCO)
480 lb net fiber drums (Standapol SCO)
Drums, T/L (Hartex V63, V64)
Drums, T/T (Supratol VF)

Cetyl dimethyl amine oxide

SYNONYMS:
Cetamine oxide
N,N-Dimethyl-1-hexadecanamine-N-oxide
1-Hexadecanamine, N,N-dimethyl-, N-oxide
Palmitamine oxide (CTFA)
Palmityl dimethylamine oxide
EMPIRICAL FORMULA:
$C_{18}H_{39}NO$
STRUCTURE:

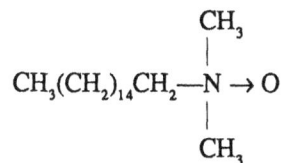

$$CH_3$$
$$|$$
$$CH_3(CH_2)_{14}CH_2\!-\!N \rightarrow O$$
$$|$$
$$CH_3$$

Cetyl dimethyl amine oxide *(cont'd.)*

CAS No.:
7128-91-8

TRADENAME EQUIVALENTS:
Ammonyx CO [Onyx]
Aromox DM16 [Armak]
Conco XA-C [Continental]
Jordamox CDA-40 [Jordan]

CATEGORY:
Wetting agent, emulsifier, stabilizer, foam stabilizer, foaming agent, antistat, softener, conditioner, viscosity builder

APPLICATIONS:
Cosmetic industry preparations: (Ammonyx CO; Aromox DM16; Conco XA-C); hair conditioning (Jordamox CDA-40); shampoos (Aromox DM16); skin preparations (Jordamox CDA-40)
Household detergents: (Ammonyx CO; Aromox DM16; Conco XA-C)
Industrial applications: chemical specialties (Aromox DM16); electroplating/metal plating (Aromox DM16; Conco XA-C); paper mfg. (Aromox DM16); petroleum industry (Aromox DM16); photography (Conco XA-C); plastics (Aromox DM16); polymers/polymerization (Conco XA-C); rubber (Aromox DM16); textile/leather processing (Aromox DM16; Conco XA-C)
Industrial cleaners: (Conco XA-C); janitorial cleaners (Ammonyx CO)

PROPERTIES:
Form:
Liquid (Ammonyx CO; Conco XA-C; Jordamox CDA-40)
Clear liquid (Aromox DM16)
Color:
Gardner 1 max. (Aromox DM16)
Composition:
20% conc. (Conco XA-C)
29–31% amine oxide (Ammonyx CO)
39–41% active (Jordamox CDA-40)
40% active in aq. isopropanol (Aromox DM16)
Solubility:
Insol. in carbon tetrachloride (Conco XA-C)
Sol. in ethanol (Conco XA-C)
Sol. in hexylene glycol (Conco XA-C)
Sol. in isopropanol (Conco XA-C)
Insol. in kerosene (Conco XA-C)
Sol. in water (Conco XA-C)
Insol. in xylene (Conco XA-C)

Cetyl dimethyl amine oxide *(cont'd.)*

Ionic Nature:
　Nonionic; mildly cationic under acidic conditions (Ammonyx CO)
　Cationic (Conco XA-C)
Sp.gr.:
　0.885 (Aromox DM16)
　0.96 (25/20 C) (Ammonyx CO)
Visc.:
　19 cps (Brookfield, #1 spindle) (Aromox DM16)
Pour Pt.:
　0 F (Aromox DM16)
Flash Pt.:
　50 F (PM) (Aromox DM16)
　> 200 F (Ammonyx CO)
Fire Pt.:
　183 F (Aromox DM16)
Cloud Pt.:
　44 F (Aromox DM16)
Surface Tension:
　31.6 dynes/cm (0.1%) (Aromox DM16)
Biodegradable: (Aromox DM16; Conco XA-C)
TOXICITY/HANDLING:
　Protective clothing, gloves, goggles should be worn (Aromox DM16)
STORAGE/HANDLING:
　Store in SS, glass, or fiberglass reinforced polyester tanks (Aromox DM16)

Chlorinated paraffin

SYNONYMS:
　Chlorinated alkane
TRADENAME EQUIVALENTS:
　Cereclor 42, 42P, 50LV, 51-L, 52P, 70L, AP45, AP52, LP4446, LP4985, S45, S-52 [ICI
　　Americas]
　CPF-0001, -0003, -0008, -0019, -0022 [Pearsall/Witco]
　Flexchlor 0001, 0002, 0008, 0009, 0010, 0012, 0018, 0023 [Pearsall/Witco]
　Paroil 1160 [Dover]
CATEGORY:
　Plasticizer, extender, extreme pressure additive, flame retardant

Chlorinated paraffin (cont'd.)

APPLICATIONS:

Industrial applications: adhesives/sealants/caulks (Cereclor 42, 42P, 52P, LP4985, S45, S-52; CPF-0001, -0003, -0008, -0019, -0022; Flexchlor 0001, 0002, 0008, 0009, 0012); carpet backings (Paroil 1160); flame-retardant applications (Cereclor 42, 42P, 52P, 70L, AP45, AP52, LP4446, LP4985, S45, S-52; CPF-0001, -0003, –0008, -0019, -0022; Flexchlor 0001, 0002, 0008, 0009, 0023); metalworking lubricants (Cereclor 42, 42P, 50LV, 51-L, 70L, LP4446, LP4985); paint mfg. (Cereclor 42, 42P, 52P, LP4985, S45, S-52; CPF-0001, -0003, -0008, -0019, -0022; Flexchlor 0001, 0002, 0008, 0009); plastics (Cereclor 42, 42P, S45, S-52; Flexchlor 0010, 0018, 0023; Paroil 1160); resins (Paroil 1160); rubber (Cereclor 42, 42P, S45, S-52; CPF-0001, -0003, -0008, -0019, -0022; Flexchlor 0001, 0002, 0008, 0009; Paroil 1160); textile processing (Cereclor 42, 42P; Paroil 1160)

PROPERTIES:

Form:

Liquid (CPF-0001, -0003, -0008, -0019, -0022; Flexchlor 0001, 0002, 0008, 0009, 0010, 0012, 0018, 0023)

Clear viscous liquid (Cereclor 42, 42P, 52LV, 51-L, 52P, 70L, AP45, AP52, LP4446, LP4985, S45, S-52)

Color:

Gardner < 1 (Cereclor AP45, AP52)

Gardner 1 (Cereclor 42, 52LV, S45, S-52; Flexchlor 0001, 0002, 0012; Paroil 1160)

Gardner 2 (Flexchlor 0010, 0018)

Gardner 3 (CPF-0001, -0022; Flexchlor 0008, 0009, 0023)

Gardner 4 (Cereclor 42P, 52P, LP4446; CPF-0008)

Gardner 5 (CPF-0019)

Gardner 6 (Cereclor 51-L; CPF-0003)

Gardner 8 (Cereclor LP4985)

Gardner 11 (Cereclor 70L)

Composition:

40% chlorine (CPF-0001)

41% chlorine (CPF-0022; Flexchlor 0001)

42% chlorine (Cereclor 42, 42P)

44% chlorine (Cereclor LP4446)

45% chlorine (Cereclor AP45, S45)

46% chlorine (CPF-0003)

50% chlorine (Cereclor 50LV, LP4985; CPF-0008, -0019; Flexchlor 0002)

51% chlorine (Cereclor 51-L)

52% chlorine (Cereclor AP52, S-52; Flexchlor 0008)

55% chlorine (Flexchlor 0010, 0018, 0023)

57% chlorine (Flexchlor 0009)

58% chlorine (Flexchlor 0012)

60% chlorine (Paroil 1160)

70% chlorine (Cereclor 70L)

Solubility:
Miscible with benzene (Cereclor 42, 42P, 50LV, 51-L, 52P, 70L, AP45, AP52, LP4446, LP4985, S45, S-52)
Miscible with cyclohexanol (Cereclor 42, 42P, 50LV, 51-L, 52P, 70L, AP45, AP52, LP4446, LP4985, S45, S-52)
Miscible with esters (Cereclor 42, 42P, 50LV, 51-L, 52P, 70L, AP45, AP52, LP4446, LP4985, S45, S-52)
Miscible with ethers (Cereclor 42, 42P, 50LV, 51-L, 52P, 70L, AP45, AP52, LP4446, LP4985, S45, S-52)
Miscible with ketones (Cereclor 42, 42P, 50LV, 51-L, 52P, 70L, AP45, AP52, LP4446, LP4985, S45, S-52)
Miscible with petroleum ether (Cereclor 42, 42P, 50LV, 51-L, 52P, 70L, AP45, AP52, LP4446, LP4985, S45, S-52)
Miscible with trichlorethylene (Cereclor 42, 42P, 50LV, 51-L, 52P, 70L, AP45, AP52, LP4446, LP4985, S45, S-52)
Miscible with veg. oils (Cereclor 42, 42P, 50LV, 51-L, 52P, 70L, AP45, AP52, LP4446, LP4985, S45, S-52)

Sp.gr.:
1.10 (Flexchlor 0001)
1.14 (CPF-0001)
1.16 (CPF-0022)
1.22 (CPF-0003; Flexchlor 0002)
1.26 (CPF-0008)
1.27 (Flexchlor 0008)
1.28 (CPF-0019)
1.30 (Flexchlor 0010)
1.33 (Flexchlor 0023)
1.335–1.355 (50/25 C) (Paroil 1160)
1.35 (Flexchlor 0012, 0018)
1.41 (Flexchlor 0009)

Density:
1.16 g/ml (Cereclor 42P, AP45)
1.17 g/ml (Cereclor 42, S45)
1.19 g/ml (Cereclor 50LV, LP4446)
1.25 g/ml (Cereclor 51-L, 52P, AP52, S-52)
1.28 g/ml (Cereclor LP4985)
1.55 g/ml (Cereclor 70L)
11.1–11.3 lb/gal (Paroil 1160)

Visc.:
0.5 poise (Flexchlor 0001)
1 poise (Cereclor 50LV, AP45)
2 poise (Cereclor S45)

Chlorinated paraffin (cont'd.)

3 poise (Flexchlor 0002)
11 poise (Cereclor AP52; CPF-0001)
12 poise (Cereclor 52P, S-52)
13 poise (Cereclor 51-L)
15 poise (Flexchlor 0008)
20 poise (Flexchlor 0012)
24 poise (Cereclor 42)
25 poise (Flexchlor 0010)
28 poise (Cereclor 42P)
35 poise (CPF-0022)
45 poise (Cereclor LP4446)
55 poise (Flexchlor 0023)
115 poise (CPF-0003)
145 poise (CPF-0008)
169 poise (Cereclor LP4985)
500 poise (CPF-0019; Flexchlor 0018)
800 poise (Flexchlor 0009)
20,000 poise (Cereclor 70L)
1800–2200 cps (Paroil 1160)

Stability:

Resistant to acids and alkalis; stable at ordinary temps.; unaffected by sunlight except on prolonged exposure when slight darkening takes place (Cereclor 42, 42P, 50LV, 51-L, 52P, 70L, AP45, AP52, LP4446, LP4985, S45, S-52)

Ref. Index:

1.4825–1.4925 (105 C) (Paroil 1160)

Cocamidopropylamine oxide (CTFA)

SYNONYMS:

Amides, coco, N-[3-(dimethylamino) propyl], N-oxide
Coco amides, N-[3-(dimethylamino) propyl], N-oxide
N-[3-(Dimethylamino) propyl] coco amides-N-oxide
Coco amido propyl dimethyl amine oxide

STRUCTURE:

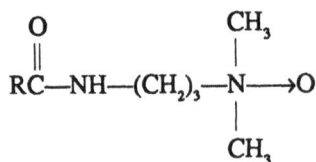

46

Cocamidopropylamine oxide (*cont'd.*)

CAS No.:
68155-09-9
TRADENAME EQUIVALENTS:
Alkamox CAPO [Alkaril]
Aminoxid WS35 [Goldschmidt AG]
Ammonyx CDO [Onyx]
Barlox C [Lonza]
Cyclomox CO [Cyclo]
Jordamox CAPA [Jordan]
Ninox CA [Stepan]
Ninox FCA [Stepan Europe]
Rewominoxid B 204 [Dutton & Reinisch]
Schercamox C-AA [Scher]
Standamox CAW [Henkel]
Varox 1770 [Sherex]
CATEGORY:
Foam stabilizer, foaming agent, foam booster, thickener, viscosity builder, wetting agent, antistat, detergent, emollient, conditioner, emulsifier, dispersant, softener, surfactant
APPLICATIONS:
Bath products: bubble bath (Alkamox CAPO; Ammonyx CDO; Ninox CA, FCA; Standamox CAW); bath oils (Ammonyx CDO)
Cosmetic industry preparations: (Aminoxid WS35); conditioners (Ninox CA; Schercamox C-AA); hair preparations (Ammonyx CDO); personal care products (Ninox CA); shampoos (Alkamox CAPO; Barlox C; Jordamox CAPA; Ninox CA, FCA; Rewominoxid B 204; Schercamox C-AA; Standamox CAW); shaving preparations (Ninox CA; Schercamox C-AA); skin preparations (Standamox CAW)
Household detergents: (Aminoxid WS35; Jordamox CAPA; Ninox FCA; Rewominoxid B 204); carpet & upholstery shampoos (Alkamox CAPO); dishwashing (Alkamox CAPO; Ammonyx CDO); laundry detergent (Alkamox CAPO); light-duty cleaners (Ninox CA; Schercamox C-AA)
Industrial applications: electroplating (Alkamox CAPO); foam rubber (Alkamox CAPO); paper coatings (Alkamox CAPO)
Pharmaceutical applications: (Aminoxid WS35)

PROPERTIES:
Form:
Liquid (Aminoxid WS35; Ammonyx CDO; Cyclomox CO; Jordamox CAPA; Varox 1770)
Clear liquid (Alkamox CAPO; Ninox CA; Rewominoxid B 204; Standamox CAW)
Clear to slightly hazy liquid (Schercamox C-AA)
Color:
Straw (Ninox CA)

Cocamidopropylamine oxide *(cont'd.)*

Pale yellow (Rewominoxid B 204)
Gardner 4.0 max. (Schercamox C-AA)
Odor:
Almost odorless (Rewominoxid B 204)
Composition:
29–31% amine oxide (Alkamox CAPO)
29.5–31.5% active (Jordamox CAPA)
30% amine oxide (Ammonyx CDO)
30% amine oxide min. in water (Schercamox C-AA)
30% active (Ninox CA)
30% conc. (Cyclomox CO; Standamox CAW)
35% conc. (Aminoxid WS35; Varox 1770)
35% solids min. (Rewominoxid B 204)
40% conc. (Barlox C)

Solubility:
Sol. in alcohols (Schercamox C-AA)
Sol. in glycol ethers (Schercamox C-AA)
Sol. in glycols (Schercamox C-AA)
Sol. in most hydrophilic solvents (Schercamox C-AA)
Disp. in min. oil (@ 10%) (Alkamox CAPO)
Sol. in polyols (Schercamox C-AA)
Sol. in triols (Schercamox C-AA)
Sol. in water (Schercamox C-AA); (@ 10%) (Alkamox CAPO)

Ionic Nature:
Nonionic (Aminoxid WS35; Ammonyx CDO; Rewominoxid B 204; Schercamox C-AA; Varox 1770)
Nonionic at pH > 7; exhibits mild cationic properties at acidic pH (Ninox CA)
Cationic (Cyclomox CO)
M.W.:
320 (Schercamox C-AA)
Sp.gr.:
0.986 ± 0.01 (Schercamox C-AA)
1.02 (25/20 C) (Ammonyx CDO)
Density:
1.0 g/ml (Alkamox CAPO)
Flash Pt.:
> 200 F (Ammonyx CDO)
Stability:
Stable over a wide acid and alkaline pH range (Aminoxid WS35)
Storage Stability:
1 yr min. shelf life; free amine and free peroxide decreases slightly with aging which does not affect product performance (Schercamox C-AA)

48

pH:
 5.0–7.0 (10% solids) (Rewominoxid B 204)
 7.0 ± 0.5 (Schercamox C-AA)
Biodegradable: (Ninox CA; Rewominoxid B 204)
STD. PKGS.:
 55-gal (450 lb net) open-head Liquipak drums (Schercamox C-AA)

Cocamidopropyl hydroxysultaine (CTFA)

SYNONYMS:
 (3-Cocamidopropyl)(2-hydroxy-3-sulfopropyl) dimethyl quaternary ammonium compounds, hydroxide, inner salt
 Quaternary ammonium compounds, (3-cocamidopropyl) (2-hydroxy-3-sulfopropyl) dimethyl, hydroxide, inner salt
 Coco amido sulfobetaine
STRUCTURE:

$$
\begin{array}{c c}
O & CH_3 \\
\| & | \\
RC{-}NH{-}(CH_2)_3{-}N^+{-}CH_2CHCH_2SO_3^- \\
 & | \quad\;\; | \\
 & CH_3 \quad OH
\end{array}
$$

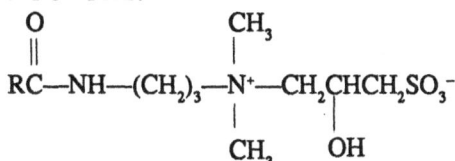

 where RCO⁻ represents the coconut acid radical
CAS No.:
 68139-30-0
TRADENAME EQUIVALENTS:
 Cycloteric BET-CS [Cyclo]
 Jortaine COSB, CSB, CSB-50 [Jordan]
 Lexaine CSB-35, CSB-50 [Inolex]
 Lonzaine CS [Lonza]
 Mirataine CBS [Miranol]
 Schercotaine SCAB [Scher]
CATEGORY:
 Viscosity builder, foam booster, emulsifier, dispersant, detergent, wetting agent, cloud pt. depressant, counter-irritant, hydrotrope
APPLICATIONS:
 Bath products: bubble bath (Lexaine CSB-35, CSB-50; Lonzaine CS; Schercotaine SCAB)

49

Cocamidopropyl hydroxysultaine (cont'd.)

Cleansers: skin cleansers (Lexaine CSB-35, CSB-50; Schercotaine SCAB)

Cosmetic industry preparations: (Jortaine CSB-50; Lonzaine CS); conditioners (Lonzaine CS); creams and lotions (Lonzaine CS); hair preparations (Lexaine CSB-35, CSB-50); shampoos (Cycloteric BET-CS; Lexaine CSB-35, CSB-50; Lonzaine CS; Mirataine CBS; Schercotaine SCAB); skin preparations (Lexaine CSB-35, CSB-50); toiletries (Lonzaine CS)

Household detergents: (Jortaine COSB, CSB); dishwashing (Mirataine CBS); liquid detergents (Jortaine COSB, CSB, CSB-50)

Industrial applications: (Lonzaine CS)

PROPERTIES:
Form:

Liquid (Cycloteric BET-CS; Jortaine COSB, CSB, CSB-50; Lexaine CSB-35, CSB-50; Lonzaine CS; Mirataine CBS)

Clear liquid (Schercotaine SCAB)

Color:

Light lemon (Schercotaine SCAB)

Odor:

Slight, typical (Schercotaine SCAB)

Composition:

35% active (Jortaine CSB; Lexaine CSB-35; Schercotaine SCAB)

48% conc. (Lexaine CSB-50)

49–51% conc. (Cycloteric BET-CS)

50% active (Jortaine COSB, CSB-50; Lonzaine CS; Mirataine CBS)

Solubility:

Sol. in water (Mirataine CBS; Schercotaine SCAB)

Ionic Nature:

Amphoteric (Cycloteric BET-CS; Jortaine COSB, CSB, CSB-50; Lonzaine CS; Mirataine CBS; Schercotaine SCAB)

M.W.:

500 (avg.) (Schercotaine SCAB)

Sp.gr.:

1.07 ± 0.01 (Schercotaine SCAB)

Density:

8.9 lb/gal (Schercotaine SCAB)

9.1 lb/gal (Mirataine CBS)

Visc.:

100 cps max. (Schercotaine SCAB)

Cloud Pt.:

≤ -12 C (Mirataine CBS)

−7 C (Schercotaine SCAB)

Stability:

Stable over wide pH range; effective in acid or alkaline media (Schercotaine SCAB)

Cocamidopropyl hydroxysultaine *(cont'd.)*

Stable in systems containing acids, alkalis, and electrolytes (Lexaine CSB-35, CSB-50)

pH:

6.0 ± 1.0 (Schercotaine SCAB)

8.2 (Mirataine CBS)

Biodegradable: (Lonzaine CS; Schercotaine SCAB)

STD. PKGS.:

55-gal poly-lined drums (Schercotaine SCAB)

Coconut diethanolamide

SYNONYMS:

Amides, coco, N,N-bis (2-hydroxyethyl)-

N,N-Bis (2-hydroxyethyl) coco amides

N,N-Bis (2-hydroxyethyl) coco fatty acid amide

Cocamide DEA (CTFA)

Coco amides, N,N-bis (2-hydroxyethyl)-

Coco diethanolamide

Coco fatty acid diethanolamide

Coconut fatty acid diethanolamide

Coconut oil diethanolamide

Coconut oil fatty acid diethanolamide

Coco oil diethanolamide

Cocoyl diethanolamide

Diethanolamine coconut fatty acid condensate

STRUCTURE:

$$
\begin{array}{c} O \\ \parallel \\ RC\!-\!N(CH_2CH_2OH)_2 \end{array}
$$

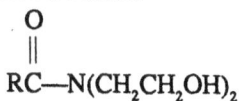

where RCO^- represents the coconut acid radical

CAS No.:

61791-31-9; 68603-42-9

TRADENAME EQUIVALENTS:

Accomid C [Armstrong]

Alkamide 1002, 2104, 2106, 2204 [Alkaril]

Alkamide CDE, CDO [Alkaril]

Alrosol B [Ciba-Geigy]

Aminol HCA [Finetex]

Ardet CB, DC, DCAE, DCX, DMA, DMC, DMF, WO [Ardmore]

51

Coconut diethanolamide (cont'd.)

TRADENAME EQUIVALENTS *(cont'd.):*

Carsamide 7644, C-3, CA, LE, SAC [Carson]
Clindrol 101CG, 200CG, 200CGN, 200HC, 200RC, 202CGN, 203CG [Clintwood]
Clindrol 206CGN, 207CGN, 209CGN [Clintwood]
Clindrol Superamide 100C, 100CG [Clintwood]
Comperlan KD, KDO [Henkel Canada]
Conco Emulsifier K [Continental]
Condensate CO, NP [Continental]
Condensate PA, PN, PO, PS [Continental]
Cyclomide DC212 [Cyclo]
Cyclomide DC212/S, DC212/SE, KD [Cyclo]
Cyclomide DC212M [Cyclo]
Emid 6514, 6515 [Emery]
Emid 6531, 6533, 6534 [Emery]
Empilan 2502 [Albright & Wilson/Detergents]
Empilan CDE [Albright & Wilson/Marchon]
Empilan CDEY [Albright & Wilson/Australia]
Empilan CDX [Albright&Wilson/Detergents]
Empilan FD, FD20, FE [Albright & Wilson/Australia]
ESI-Terge 10, B-15, C-5, S-10 [Emulsion Systems]
Ethylan A15, LD, LDG, LDS [Lankro]
Gafamide CDD-518 [GAF]
Hartamide OD [Hart]
Hetamide MC, RC [Heterene]
Hymolon CWC, K-90 [Hart]
Incromide CA [Croda Surfactants]
Jordamide JT128, WC Conc. [Jordan]
Lakeway 100-CA [Bofors Lakeway]
Lauridit KD, KDG [Akzo Chemie]
Loramine DC212/S, DC212/SE, DC220/SE [Dutton & Reinisch]
Mackamide 100-A, C, MC [McIntyre]
Manro CD, CDS [Manro]
Marlamid D1218 [Chem.Werke Huls]
Marsamid 10, 40, 50 [Mars]
Mazamide 70, 80, CA-20, CS-148 [Mazer]
Monamid 7-100, 7-153CS, 150-AD, 150-ADD, 150-DR, 759 [Mona]
Monamine AA-100, AD-100, ADD-100, ADD-100LE, ALX-80SS, ALX-100S, ARA-100, I-76 [Mona]
Ninol 128 Extra, 2012 Extra [Stepan]
Nitrene 11120, 11230, 13026, A-309, A-567, C, C-Extra [Henkel]
Onyxol 345 [Onyx]
Rewomid DC212/S, DC220/SE [Dutton & Reinisch]

Coconut diethanolamide *(cont'd.)*

TRADENAME EQUIVALENTS *(cont'd.):*
Rewomid DC212/SE, DC220/LS [Rewo Chem. Werke]
Richamide 6404, M-3 [Richardson]
Schercomid CCD, CDA, CDA-H, SCE, SCO-Extra [Scher]
Sipomide 1500 [Alcolac]
Stafoam DF-1, DF-4, F [Nippon Oil & Fats]
Standamid KD, PD, SD [Henkel]
Steinamid DC212/S, DC220/SE [Dutton & Reinisch]
Sterling DEA [Canada Packers]
Super Amide GR [Onyx]
Surco 128-T, SR-200, WC Conc. [Onyx]
Surco Coco Condensate [Onyx]
Synotol 119N, CN60, CN80, CN90 [PVO]
Teric CDE [ICI Australia]
Unamide CDX, JJ-35, LDL, N-72-3 [Lonza]
Varamide A-2, A-10, A-12, A-83 [Sherex]
Witcamide 82, 1017, 5130, 5133 [Witco/Organics]

CATEGORY:
Detergent, surfactant, scouring aid, emulsifier, wetting agent, stabilizer, foam stabilizer, foam booster, thickener, viscosity builder, dispersant, suspending agent, emollient, conditioner, lubricant, superfatting agent, softener, solubilizer, penetrant, rust preventive, corrosion inhibitor, intermediate, antistat, coupling agent

APPLICATIONS:
Automobile cleaners: (Mazamid 70, CA-20); car shampoo (Alrosol B; Carsamide C-3; Clindrol 206CGN, 207CGN; Condensate CO, NP; ESI-Terge S-10; Hetamide MC, RC; Marsamid 10, 50; Nitrene 11120, C-Extra; Schercomid CDA); car waxes (Schercomid CCD); tire cleaners (Marsamid 10, 50)
Bath products: (Loramine DC212/SE); bubble bath (Accomid C; Calamide C, CW-100; Carsamide SAC; Clindrol Superamide 100C, 100CG; Comperlan KD, KDO; Condensate PA, PO; Empilan CDE, FD, FE; Gafamide CDD-518; Lakeway 100-CA; Lauridit KD, KDG; Manro CD, CDS; Marsamid 50; Monamid 7-100, 7-153CS, 150-AD, 150-ADD, 150-DR, 759; Monamine AA-100, AD-100, ADD-100, ADD-100LE, ALX-80SS, ALX-100S, ARA-100, I-76; Ninol 128 Extra, 2012 Extra; Schercomid SCE; Sipomide 1500; Standamid KD, PD, SD; Surco WC Conc.; Synotol 119N, CN60, CN80; Unamide CDX, JJ-35); bath gels (Standamid SD)
Cleansers: (Carsamide CA, SAC; Synotol 119N, CN60, CN80); body cleansers (Empilan FE; Monamid 759); cleansing creams (Empilan FE); germicidal skin cleanser (Synotol 119N, CN60, CN80); hand cleanser (Clindrol 200CGN, 202CGN; Clindrol Superamide 100C, 100CG; Empilan FD; Ethylan LD; Gafamide CDD-518; Hetamide MC, RC; Marsamid 50; Mazamid 70, CA-20; Monamid 7-100, 7-153CS, 150-AD, 150-ADD, 150-DR; Monamine AA-100, AD-100, ADD-

100, ADD-100LE, ALX-80SS, ALX-100S, ARA-100, I-76; Nitrene C-Extra; Schercomid CCD, CDA, CDA-H; Sipomide 1500); scrub soap base (Witcamide 82, 5133)

Cosmetic industry preparations: (Alkamide CDE, CDO; Clindrol 200HC, 200RC, 203CG, 206CGN, 207CGN; Condensate PA, PO; Cyclomide DC212; Emid 6531, 6534; ESI-Terge 10; Hetamide MC, RC; Incromide CA; Jordamide JT128, WC Conc.; Lauridit KD, KDG; Marsamid 50; Mazamid 80, CS-148; Monamid 7-100, 7-153CS, 150-AD, 150-ADD, 150-DR; Monamine AA-100, AD-100, ADD-100, ADD-100LE, ALX-80SS, ALX-100S, ARA-100, I-76; Onyxol 345; Schercomid CDA-H, SCE; Super Amide GR; Surco 128-T; Synotol CN90; Unamide CDX, JJ-35; Witcamide 82, 5133); conditioners (Monamid 7-100, 7-153CS, 150-AD, 150-ADD, 150-DR; Monamine AA-100, AD-100, ADD-100, ADD-100LE, ALX-80SS, ALX-100S, ARA-100, I-76); creams and lotions (Schercomid CDA-H); personal care products (Loramine DC212/SE; Sipomide 1500); shampoos (Accomid C; Aminol HCA; Calamide C, CW-100; Carsamide C-3; Carsamide SAC; Clindrol 200HC, 200RC; Clindrol Superamide 100C, 100CG; Comperlan KD, KDO; Condensate PA, PO; Cyclomide DC212/SE; Emid 6531, 6534; Empilan CDE, CDEY, CDX, FE; Ethylan LD, LDG, LDS; Gafamide CDD-518; Lakeway 100-CA; Lauridit KD, KDG; Loramine DC212/S, DC212/SE, DC220/SE; Mackamide C; Manro CD, CDS; Marsamid 50; Monamid 7-100, 7-153CS, 150-AD, 150-ADD, 150-DR, 759; Monamine AA-100, AD-100, ADD-100, ADD-100LE, ALX-80SS, ALX-100S, ARA-100, I-76; Ninol 128 Extra, 2012 Extra; Rewomid DC212/S; Richamide 6404, M-3; Schercomid CDA-H, SCE; Sipomide 1500; Stafoam DF-1, DF-4; Standamid KD, PD, SD; Steinamid DC212/S, DC220/SE; Surco WC Conc.; Synotol 119N, CN60, CN80; Unamide CDX, JJ-35); shaving preparations (Synotol 119N, CN60, CN80); toiletries (Empilan CDX, FD, FE; Loramine DC212/SE; Mazamid 80, CS-148; Monamid 759; Rewomid DC212/SE; Unamide CDX, JJ-35; Witcamide 82, 5133)

Degreasers: (Calamide C; Clindrol 200CGN, 202CGN, 209CGN; Cyclomide DC212M; Empilan FD; Ethylan A15; Hetamide MC, RC; Marsamid 40; Nitrene 11230; Schercomid CDA; Sipomide 1500)

Farm products: agricultural oils/sprays (Clindrol 209CGN; Monamid 7-100, 7-153CS, 150-AD, 150-ADD, 150-DR; Monamine AA-100, AD-100, ADD-100, ADD-100LE, ALX-80SS, ALX-100S, ARA-100, I-76; Schercomid CCD); herbicides/fungicides (Clindrol 206CGN, 207CGN); insecticides/pesticides (Clindrol 200RC, 206CGN, 207CGN, 209CGN; Condensate CO, NP; Marsamid 10)

Food applications: food machinery cleaner (Clindrol 206CGN, 207CGN; Condensate PA, PO); bottle cleaning (Hetamide MC, RC)

Household detergents: (Alkamide 1002, CDE, CDO; Aminol HCA; Ardet DC; Carsamide 7644, CA, SAC; Clindrol 101CG, 200CGN, 200HC, 200RC, 202CGN, 203CG, 206CGN, 207CGN, 209CGN; Condensate CO, NP, PA, PO, PS; Cyclomide DC212/SE; Emid 6533; ESI-Terge 10; Incromide CA; Jordamide JT128, WC

Coconut diethanolamide (cont'd.)

Conc.; Lauridit KD, KDG; Loramine DC212/S, DC212/SE, DC220/SE; Nitrene 11120, A-309, A-567, C; Rewomid DC212/S; Schercomid SCO-Extra; Sipomide 1500; Steinamid DC212/S, DC220/SE; Super Amide GR; Surco 128-T; Synotol 119N, CN60, CN80, CN90; Unamide LDL, N-72-3; Varamide A-2, A-12); all-purpose cleaner (Clindrol 200CG, 203CG; ESI-Terge B-15, C-5; Ethylan LD, LDG, LDS; Hymolon CWC, K-90; Marsamid 10, 40; Schercomid CCD, CDA; Varamide A-83); built detergents (Clindrol 200CGN, 202CGN; Condensate PN); carpet & upholstery shampoos (Monamid 7-100, 7-153CS, 150-AD, 150-ADD, 150-DR; Monamine AA-100, AD-100, ADD-100, ADD-100LE, ALX-80SS, ALX-100S, ARA-100, I-76; Schercomid SCE; Sipomide 1500); detergent base (Alkamide 2104; Schercomid CCD); dishwashing (Accomid C; Alrosol B; Ca-lamide C; Clindrol 206CGN, 207CGN; Clindrol Superamide 100C, 100CG; Condensate PA, PO; ESI-Terge S-10; Gafamide CDD-518; Lauridit KD, KDG; Marsamid 50; Mazamid 70, CA-20; Monamid 7-100, 7-153CS, 150-AD, 150-ADD, 150-DR; Monamine AA-100, AD-100, ADD-100, ADD-100LE, ALX-80SS, ALX-100S, ARA-100, I-76; Ninol 128 Extra, 2012 Extra; Richamide 6404, M-3; Schercomid CCD, CDA, SCE; Stafoam DF-1, DF-4; Standamid SD; Surco WC Conc.; Teric CDE); hard surface cleaner (Alkamide 2106; Alrosol B; Clindrol 200CG, 200CGN, 202CGN, 203CG, 206CGN, 207CGN; Cyclomide DC212M; Empilan FD; Ethylan LD, LDS; Mazamid 70, 80, CA-20; Nitrene 11230; Richa-mide 6404, M-3; Sipomide 1500; Surco SR-200; Unamide LDL, N-72-3); heavy-duty cleaner (Alkamide 2104, 2106, 2204; Calamide C, CW-100; Condensate CO, NP, PN; Marsamid 10; Stafoam F; Surco Coco Condensate); laundry detergent (Alkamide CDE, CDO; Calamide C; Clindrol 200CGN, 202CGN, 206CGN, 207CGN; Cyclomide DC212; Empilan FD; Gafamide CDD-518; Stafoam DF-1, DF-4, F; Standamid SD); light-duty cleaners (Empilan FD; Lakeway 100-CA; Marsamid 50; Surco Coco Condensate; Unamide LDL, N-72-3); liquid detergents (Calamide C, CW-100; Carsamide C-3; Emid 6514, 6531; Empilan CDE, CDX, FD, FD20; Ethylan LD, LDS; Hartamide OD; Lauridit KD, KDG; Manro CD, CDS; Nitrene C-Extra; Stafoam F; Unamide LDL, N-72-3); powdered detergents (Emid 6531; Empilan FD; Hartamide OD)

Industrial applications: (ESI-Terge 10; Jordamide JT128, WC Conc.); aerosols (Clin-drol 101CG; Witcamide 5130); dyes and pigments (Alkamide CDO; Monamid 7-100, 7-153CS, 150-AD, 150-ADD, 150-DR; Monamine AA-100, AD-100, ADD-100, ADD-100LE, ALX-80SS, ALX-100S, ARA-100, I-76; Stafoam F); lubricat-ing/cutting oils (Alkamide CDE, CDO; Clindrol 200CGN, 202CGN, 206CGN, 207CGN; Monamid 7-100, 7-153CS, 150-AD, 150-ADD, 150-DR; Monamine AA-100, AD-100, ADD-100, ADD-100LE, ALX-80SS, ALX-100S, ARA-100, I-76; Nitrene A-309; Schercomid CCD, CDA-H); metalworking (Alkamide 2204; Ardet DMA, DMC, DMF; Clindrol 200CGN, 202CGN; Mazamid 70, CA-20; Monamid 7-100, 7-153CS, 150-AD, 150-ADD, 150-DR; Monamine AA-100, AD-100, ADD-100, ADD-100LE, ALX-80SS, ALX-100S, ARA-100, I-76); petroleum

Coconut diethanolamide (cont'd.)

industry (Monamid 7-100, 7-153CS, 150-AD, 150-ADD, 150-DR; Monamine AA-100, AD-100, ADD-100, ADD-100LE, ALX-80SS, ALX-100S, ARA-100, I-76; Witcamide 1017); plastics (Empilan CDE, CDX; Ethylan LD); polishes and waxes (Condensate PA, PO; Ethylan A15; Monamid 7-100, 7-153CS, 150-AD, 150-ADD, 150-DR; Monamine AA-100, AD-100, ADD-100, ADD-100LE, ALX-80SS, ALX-100S, ARA-100, I-76); printing inks (Clindrol 200RC); textile/leather processing (Alkamide CDE, CDO; Carsamide 7644; Condensate CO, NP, PA, PN, PO, PS; Hymolon CWC; Monamid 7-100, 7-153CS, 150-AD, 150-ADD, 150-DR; Monamine AA-100, AD-100, ADD-100, ADD-100LE, ALX-80SS, ALX-100S, ARA-100, I-76; Stafoam DF-1, DF-4, F)

Industrial cleaners: (Alkamide 2104, CDE, CDO; Calamide C; Carsamide CA, SAC; Clindrol 200CGN, 202CGN, 206CGN, 207CGN; Condensate CO, NP, PA, PN, PO, PS; Cyclomide DC212M; Empilan FD; Ethylan A15; Gafamide CDD-518; Hetamide MC, RC; Marsamid 10, 40, 50; Nitrene 11120, 11230, 13026, A-567, C; Schercomid CCD, CDA, SCO-Extra; Sipomide 1500; Super Amide GR; Surco 128-T; Teric CDE; Unamide LDL, N-72-3; Varamide A-10; Witcamide 82, 5130, 5133); all-purpose cleaners (Alkamide 1002, 2104, 2106, 2204; Ardet CB, DC; Incromide CA; Schercomid CCD); drycleaning compositions (Carsamide C-3; Hetamide MC, RC; Monamid 7-100, 7-153CS, 150-AD, 150-ADD, 150-DR; Monamine AA-100, AD-100, ADD-100, ADD-100LE, ALX-80SS, ALX-100S, ARA-100, I-76; Nitrene C-Extra; Stafoam DF-1, DF-4, F); engine cleaners (Schercomid CCD, CDA); institutional cleaners (Alkamide 1002, 2104, 2106, 2204, CDE, CDO; Ardet DB, DC; Calamide C; Carsamide CA, SAC; Empilan FD); janitorial/maintenance products (Condensate CO, NP, PN; Emid 6531); metal processing surfactants (Alrosol B; Calamide C; Clindrol 200CG, 200CGN, 200HC, 202CGN, 203CG; Condensate CO, NP, PN; Ethylan A15; Monamid 7-100, 7-153CS, 150-AD, 150-ADD, 150-DR; Monamine AA-100, AD-100, ADD-100, ADD-100LE, ALX-80SS, ALX-100S, ARA-100, I-76; Stafoam DF-1, DF-4, F; Witcamide 82, 5130, 5133); sanitizers/germicides (Clindrol 203CG); solvent cleaners (Carsamide C-3; Nitrene C-Extra; Schercomid CDA-H); specialty cleaners (Nitrene 11230, 13026); textile cleaning (Alkamide CDE, CDO; Carsamide 7644; Clindrol 200CG, 200CGN, 202CGN, 203CG; Emid 6533; Hymolon K-90; Monamid 7-100, 7-153CS, 150-AD, 150-ADD, 150-DR; Monamine AA-100, AD-100, ADD-100, ADD-100LE, ALX-80SS, ALX-100S, ARA-100, I-76; Witcamide 82, 5130, 5133)

PROPERTIES:
Form:

Liquid (Accomid C; Alkamide 1002, 2104, 2106, 2204, CDE, CDO; Ardet CB, DC, DCX, DMA, DMC, DMF, WO; Carsamide 7644, LE; Clindrol 101CG, 200CG, 200CGN, 200HC, 200RC, 202CGN, 203CG, 206CGN, 207CGN, 209CGN; Clindrol Superamide 100C, 100CG; Conco Emulsifier K; Condensate CO, NP, PN, PS; Cyclomide DC212, DC212M, DC212/S, DC212/SE, KD; Emid 6514, 6515, 6531,

Coconut diethanolamide *(cont'd.)*

6533, 6534; Empilan 2502; ESI-Terge 10, B-15, C-5, S-10; Hartamide OD; Hymolon CWC, K-90; Incromide CA; Jordamide JT128, WC Conc.; Lakeway 100-CA; Lauridit KD, KDG; Loramine DC212/S, DC212/SE; Mackamide 100A, C, MC; Marlamid D1218; Mazamid 70, 80, CA-20, CS-148; Monamid 7-100, 7-153CS, 150-AD, 150-ADD, 150-DR, 759; Monamine AA-100, AD-100, ADD-100, ADD-100LE, ALX-80SS, ALX-100S, ARA-100, I-76; Ninol 128 Extra, 2012 Extra; Nitrene 11120, 11230, 13026, A-567, C, C-Extra; Onyxol 345; Rewomid DC212/S; Richamide 6404, M-3; Schercomid CDA; Sipomide 1500; Stafoam DF-1, DF-4, F; Standamid PD, SD; Steinamid DC212/S; Sterling DEA; Super Amide GR; Surco 128-T, SR-200, WC Conc.; Surco Coco Condensate; Synotol 119N, CN60, CN80, CN90; Teric CDE; Unamide CDX, JJ-35, LDL, N-72-3; Varamide A-10, A-12, A-83; Witcamide 82, 1017, 5130, 5133); (@ 20 C) (Hetamide MC, RC)

Clear liquid (Calamide C, CW-100; Ethylan A15, LD, LDG, LDS; Marsamid 10, 40; Schercomid CDA-H, SCO-Extra)

Clear, viscous liquid (Empilan CDEY, FD, FD20, FE; Gafamide CDD-518; Manro CD, CDS; Schercomid SCE)

Viscous liquid (Alrosol B; Carsamide SAC; Comperlan KD, KDO; Condensate PA, PO; Marsamid 50; Schercomid CCD; Standamid PD); (@ 20 C) (Empilan CDE, CDX)

Med. visc. liquid (Rewomid DC212/SE, DC220/LS)

Soft paste (Loramine DC220/SE; Rewomid DC220/SE; Steinamid DC220/SE)

Paste (Ardet DCAE; Standamid KD)

Color:

Light straw (ESI-Terge S-10; Lakeway 100-CA; Unamide JJ-35, LDL)

Straw (Clindrol 200HC; Clindrol Superamide 100C; Ethylan A15, LD, LDS; Unamide CDX)

Straw to light amber (Clindrol 203CG)

Straw to amber (Clindrol Superamide 100CG)

Light/pale amber (Alkamide CDE; Carsamide LE, SAC; Condensate PS; Nitrene C-Extra; Schercomid SCE, SCO-Extra)

Amber (Alkamide CDO; Alrosol B; Calamide C; Clindrol 200RC, 207CGN, 209CGN; Comperlan KD, KDO; Condensate CO, NP, PN; Cyclomide DC212, DC212M; ESI-Terge B-15, C-5; Marsamid 40; Monamid 759; Schercomid CDA; Standamid KD, PD)

Dark amber (Clindrol 101CG, 200CG, 200CGN, 202CGN, 206CGN; Marsamid 10; Schercomid CCD)

Light/pale yellow (Condensate PO; Cyclomide KD; Empilan CDE, CDEY, CDX, FD, FD20, FE; Lauridit KDG; Loramine DC212/S; Marsamid 50; Rewomid DC212/S; Stafoam DF-1, DF-4; Steinamid DC212/S)

Golden yellow (Manro CD, CDS)

Yellow (Calamide CW-100; Condensate PA; Hartamide OD; Hymolon K-90; Jordamide JT128, WC Conc.; Lauridit KD; Ninol 128 Extra, 2012 Extra; Schercomid CDA-H; Unamide N-72-3)

Coconut diethanolamide (cont'd.)

Gardner 3 (Emid 6514, 6515, 6531)
Gardner 4 (Accomid C; Emid 6533, 6534)
Gardner 6 (Carsamide 7644)
GVCS-33: 3 max. (Monamid 7-153CS; Monamine ADD-100, ADD-100LE)
GVCS-33: 4 max. (Monamid 150-ADD)
GVCS-33: 5 max. (Monamid 150-DR)
GVCS-33: 6 max. (Monamid 7-100; Monamine ALX-80SS)
GVCS-33: 7 max. (Monamine AA-100, ARA-100)
GVCS-33: 8 max. (Monamine ALX-100S)
GVCS-33: 11 max. (Monamid 150-AD; Monamine AD-100)
GVCS-33: 12 max. (Monamine I-76)
Hazen > 250 (Teric CDE)
Odor:
Low (Alkamide CDE, CDO; Schercomid CDA, CDA-H)
Characteristic (Condensate PS)
Mild (Accomid C; Ethylan LD, LDS; Manro CD, CDS; Schercomid CCD, SCE)
Mild, fatty (Carsamide SAC)
Typical (Condensate CO, NP, PA, PN, PO)
Typical, slight (Loramine DC212/S; Rewomid DC212/S; Steinamid DC212/S)
Typical alkanolamide (Clindrol 101CG, 200CG, 200HC, 202CGN, 203CG)
Composition:
20.0 ± 2.0% amine (Empilan FD20)
25 ± 2.0% free amine (Empilan CDX)
40% conc. (Ardet CB)
50% amide (Synotol 119N, CN60)
75% amide (Synotol CN80)
78.0% amide content (Empilan CDEY)
80% active (Monamine ALX-80SS)
80% amide min. (Schercomid SCO-Extra)
80% conc. (Super Amide GR)
82% active (Alkamide CDO)
82% amide content (Empilan FD)
82% conc. (Ethylan LDG)
84% active (Gafamide CDD-518)
85% active (Lauridit KDG; Manro CDS)
85% amide content min. (Condensate PO)
85% conc. (Synotol CN90)
85–90% amide content (Standamid KD)
87% active (Ethylan LDS)
90% active (Alkamide CDE; Ethylan LD; Lauridit KD; Manro CD)
90% conc. (Lakeway 100-CA)
90–95% amide content (Condensate PA)

58

91% active (Hymolon K-90)
97% active min. (Condensate CO, NP, PN, PS)
97% conc. (Onyxol 345)
98% active (Accomid C)
98% conc. (Surco SR-200; Witcamide 5130)
99% active (Ethylan A15; Schercomid CDA-H)
99+% solids (Marsamid 40)
100% active (Carsamide CA, SAC; Clindrol 101CG, 200CG, 200CGN, 200HC, 200RC, 202CGN, 203CG, 206CGN, 207CGN, 209CGN; Clindrol Superamide 100C, 100CG; Cyclomide DC212, DC212M, DC212/S, DC212/SE, KD; Emid 6514, 6515, 6533, 6534; Empilan CDE; ESI-Terge B-15, C-5, S-10; Hartamide OD; Hetamide MC, RC; Loramine DC212/S, DC212/SE, DC220/SE; Marlamid D1218; Monamid 7-100, 7-153CS, 150-AD, 150-ADD, 150-DR, 759; Monamine AA-100, AD-100, ADD-100, ADD-100LE, ALX-100S, ARA-100, I-76; Nitrene 13026; Rewomid DC212/S, DC220/SE; Schercomid CCD, CDA, SCE; Sipomide 1500; Steinamid DC212/S, DC220/SE; Sterling DEA; Teric CDE; Unamide CDX, JJ-35, LDL, N-72-3; Varamide A-2, A-10, A-12, A-83)
100% conc. (Alkamide 1002, 2104, 2106, 2204; Ardet DC, DCAE, DCX, DMA, DMC, DMF, WO; Conco Emulsifier K; Empilan 2502; ESI-Terge 10; Incromide CA; Mackamide 100A, C, MC; Ninol 128 Extra, 2012 Extra; Nitrene 11120, 11230, A-309, A-567, C, C-Extra; Rewomid DC212/SE, DC220/LS; Standamid PD, SD; Surco 128-T, WC Conc.; Witcamide 82, 1017, 5133)
100% solids (Marsamid 10, 50)
Solubility:
Sol. in acetone (Nitrene C-Extra)
Sol. in alcohol (Carsamide 7644, CA, SAC; Clindrol 101CG, 200CG, 200CGN, 200HC, 200RC, 202CGN, 203CG, 206CGN, 207CGN, 209CGN; Clindrol Superamide 100C, 100CG; Condensate PS; Marsamid 50; Schercomid CCD, CDA, CDA-H, SCE); lower alcohols (Nitrene C-Extra)
Sol. in aliphatic hydrocarbons (Clindrol 101CG; Schercomid SCE); disp. (Schercomid CDA-H)
Sol. in aromatic hydrocarbons (Alkamide CDE, CDO; Carsamide 7644, CA, SAC; Clindrol 101CG, 200CG, 200CGN, 200HC, 200RC, 202CGN, 203CG, 206CGN, 207CGN, 209CGN; Clindrol Superamide 100C, 100CG; Schercomid CCD, CDA, CDA-H, SCE); (@ 10%) (Monamid 7-100, 7-153CS, 150-AD, 150-ADD; Monamine AA-100, ADD-100LE, ALX-100S, ARA-100, I-76); disp. (Nitrene C-Extra); disp. (@ 10%) (Monamid 150-DR; Monamine AD-100, ADD-100)
Sol. in benzene (Stafoam DF-1, DF-4, F; Teric CDE)
Sol. in butyl ether (Gafamide CDD-518)
Sol. in butyl stearate (Emid 6514, 6515); (@ 5%) (Emid 6531); disp. (Emid 6533, 6534)
Sol. in Carbitols (Carsamide CA, SAC)

Coconut diethanolamide (cont'd.)

Sol. in carbon tetrachloride (Condensate PS; Marsamid 50)

Sol. in Cellosolve (Carsamide CA, SAC); butyl Cellosolve (Gafamide CDD-518); Cellosolve-type ethers (Marsamid 50)

Sol. in chlorinated aliphatic solvents (Alkamide CDE, CDO)

Sol. in chlorinated hydrocarbons (Carsamide 7644, CA, SAC; Clindrol 200CG, 200CGN, 200HC, 200RC, 202CGN, 203CG, 206CGN, 207CGN, 209CGN; Clindrol Superamide 100C, 100CG; Schercomid CCD, CDA, CDA-H, SCE); sol. in many chlorinated organic compounds (Marsamid 40); (@ 10%) (Monamid 7-100, 7-153CS, 150-AD, 150-ADD; Monamine AA-100, AD-100, ADD-100, ADD-100LE, ALX-100S, ARA-100, I-76); disp. (@ 10%) (Monamid 150-DR)

Sol. in detergent systems (Alkamide CDE)

Sol. in diethylene glycol (Stafoam DF-1, DF-4, F)

Sol. in dimethyl sulfoxide (Marsamid 50)

Sol. in diols (Schercomid CCD, CDA, SCE)

Sol. in esters (Clindrol 101CG, 206CGN, 207CGN); disp. (Nitrene C-Extra)

Sol. in ethanol (Gafamide CDD-518; Stafoam DF-1, DF-4, F; Teric CDE)

Disp. in ethyl acetate (Teric CDE)

Sol. in ethyl alcohol (@ 10%) (Monamid 7-100, 7-153CS, 150-AD, 150-ADD, 150-DR; Monamine AA-100, AD-100, ADD-100, ADD-100LE, ALX-80SS, ALX-100S, ARA-100, I-76)

Sol. in ethylene glycol (Gafamide CDD-518)

Sol. in natural fats and oils (@ 10%) (Monamid 7-100, 150-AD, 150-ADD, 150-DR); disp. (Carsamide 7644, CA, SAC; Schercomid CDA-H); (@ 10%) (Monamid 7-153CS)

Sol. in glycerol trioleate (Emid 6514, 6515); disp. (Emid 6533, 6534); disp. (@ 5%) (Emid 6531)

Sol. in glycols (Clindrol 101CG, 200CG, 200RC, 203CG, 206CGN, 207CGN, 209CGN; Clindrol Superamide 100C, 100CG; Marsamid 50; Schercomid CDA-H)

Sol. in glycol ethers (Clindrol 209CGN; Schercomid CCD, CDA, CDA-H, SCE)

Sol. in many hydrocarbons (Marsamid 40)

Sol. in isopropyl alcohol (Stafoam DF-1, DF-4, F)

Sol. in kerosene (@ 10%) (Monamid 7-100, 7-153CS, 150-DR); disp. (Carsamide 7644)

Sol. in ketones (Clindrol 101CG, 200CG, 200CGN, 200HC, 200RC, 202CGN, 203CG, 206CGN, 207CGN; Clindrol Superamide 100C); disp. in higher ketones (Nitrene C-Extra)

Sol. in min. oils (Alkamide CDE, CDO); disp. (Carsamide 7644; Emid 6514, 6515, 6533, 6534; Schercomid CDA-H); disp. (@ 5%) (Emid 6531)

Sol. in min. spirits (@ 10%) (Monamid 7-100, 7-153CS, 150-DR); disp. (Carsamide 7644)

Sol. in olein (Teric CDE)

Sol. in most organic solvents (Schercomid CCD, CDA, CDA-H, SCE)

Sol. in perchloroethylene (Emid 6514, 6515, 6533, 6534; Gafamide CDD-518; Teric CDE)

Sol. in petroleum ether (Stafoam DF-1, DF-4)

Sol. in petroleum solvents (Alkamide CDE); most petroleum solvents (Alkamide CDO)

Sol. in pine oil (Marsamid 50)

Sol. in polar organic solvents (Marsamid 40)

Sol. in polyethylene glycols and derivatives (Nitrene C-Extra)

Sol. in polyols (Schercomid CCD, CDA, CDA-H, SCE)

Sol. in propylene glycol (Gafamide CDD-518; Stafoam DF-1, DF-4, F)

Sol. in Stoddard solvent (Emid 6514, 6515, 6533, 6534; Marsamid 50); (@ 5%) (Emid 6531)

Sol. in tetrachloroethylene (Stafoam DF-1, DF-4, F)

Sol. in toluene (Nitrene C-Extra)

Sol. in triols (Schercomid CCD, CDA, SCE)

Sol. in water (Calamide C, CW-100; Carsamide 7644, CA, SAC; Clindrol 200CG, 200CGN, 200HC, 200RC, 202CGN, 203CG, 206CGN, 207CGN, 209CGN; Clindrol Superamide 100CG; Condensate CO, NP, PN, PS; Emid 6514, 6515, 6534; ESI-Terge B-15, C-5; Gafamide CDD-518; Nitrene C-Extra; Schercomid CCD, CDA, CDA-H; Stafoam F; Witcamide 5130); sol. in water in all concs. (Marsamid 40); completely sol. in water at low concs. (Accomid C); 800 g/l (@ 20 C) (Lauridit KDG); 40 g/l (@ 20 C) (Lauridit KD); (@ 10%) (Monamid 150-AD, 150-ADD; Monamine AA-100, AD-100, ADD-100, ADD-100LE, ALX-80SS, ALX-100S, ARA-100, I-76); hazy (@ < 10%) (Unamide CDX); (@ 5%) (Emid 6531, 6533); disp. in water in low concs. and forms gels at higher levels; completely sol. when blended with anionic or nonionic surfactants (Condensate PA, PO); disp. (Ethylan LD, LDS; Marsamid 50; Schercomid SCE); (@ 10%) (Monamid 7-100, 7-153CS, 150-DR); disp. up to 1% (Teric CDE)

Sol. in white min. oil (@ 10%) (Monamid 7-100, 7-153CS, 150-DR)

Sol. in xylene (Nitrene C-Extra; Stafoam DF-1, DF-4, F); (@ 5%) (Emid 6531)

Ionic Nature:

Nonionic (Accomid C; Alkamide CDE, CDO; Carsamide CA, SAC; Clindrol 101CG, 200CG, 200HC, 200RC; Clindrol Superamide 100C, 100CG; Condensate CO, NP, PN; Emid 6514, 6415, 6534; Empilan CDE, FD, FE; ESI-Terge 10, B-15, C-5, S-10; Ethylan LD, LDG, LDS; Hartamide OD; Hymolon K-90; Lakeway 100-CA; Lauridit KD, KDG; Loramine DC212/S, DC212/SE, DC220/SE; Manro CD; Marlamid D1218; Marsamid 40, 50; Mazamid 70, 80; Monamid 7-100, 150-AD, 150-ADD, 150-DR; Monamine AD-100, ADD-100, ADD-100LE; Nitrene 13026, C-Extra; Rewomid DC212/S, DC212/SE, DC220/LS, DC220/SE; Schercomid SCE; Standamid KD, SD; Steinamid DC212/S, DC220/SE; Sterling DEA; Super Amide GR; Surco 128-T, WC Conc.; Surco Coco Condensate; Synotol 119N, CN60, CN80; Teric CDE; Unamide CDX, JJ-35, LDL, N-72-3; Varamide A-2, A-

Coconut diethanolamide (cont'd.)

10, A-12, A-83; Witcamide 82, 1017, 5133)
Nonionic/anionic (Carsamide 7644; Clindrol 200CGN, 202CGN, 209CGN; Marsamid 10; Mazamide CA-20; Monamid 7-153CS; Monamine AA-100, ALX-80SS, ALX-100S, ARA-100; Schercomid CCD, CDA, CDA-H; Witcamide 5130)
Anionic (Clindrol 206CGN, 207CGN; Gafamide CDD-518; Hymolon CWC; Surco SR-200)

Sp.gr.:
0.96 (20 C) (Monamid 7-100)
0.97 (Condensate PA; Empilan FE; ESI-Terge C-5; Schercomid SCE)
0.974 (30 C) (Stafoam DF-4)
0.98 (Condensate PO; ESI-Terge S-10; Unamide LDL); (@ 20 C) (Hetamide MC; Monamid 150-DR)
0.98–1.00 (Mazamid 80, CS-148)
0.981 (30 C) (Stafoam DF-1); (20 C) (Ethylan LD)
0.982 (20 C) (Manro CD)
0.99 (Accomid C; Clindrol 101CG, 200CGN, 202CGN; Clindrol Superamide 100C, 100CG; Monamid 150-AD; Unamide CDX); (@ 25/20 C) (Super Amide GR; Surco 128-T, WC Conc.); (@ 20 C) (Ethylan LDS; Manro CDS)
0.99 ± 0.01 (Schercomid CCD)
0.99–1.02 (Mazamid 70, CA-20)
0.995 (30 C0 (Stafoam F)
1.00 (Carsamide CA; Clindrol 200CG, 203CG; Gafamide CDD-518; Monamid 150-ADD); (25/20 C) (Onyxol 345); (20 C) (Monamine AA-100, ARA-100, I-76)
1.00–1.01 (20 C) (Empilan FD)
1.01 (ESI-Terge B-15; Ethylan A15; Marsamid 10); (@ 25/20 C) (Surco Coco Condensate); (@ 20 C) (Hetamide RC; Teric CDE)
1.01 ± 0.01 (Schercomid CDA, CDA-H)
1.02 (Clindrol 200HC, 200RC, 206CGN, 207CGN; Condensate PS; Hartamide OD; Lauridit KD, KDG); (20 C) (Monamine AD-100, ADD-100, ADD-100LE)
1.025 (Marsamid 40)
1.03 (@ 25.20 C) (Surco SR-200)
1.04 (Hymolon K-90); (20 C) (Alkamide CDE, CDO); (60F) (Condensate PN)
1.05 (20 C) (Monamid 7-153CS; Monamine ALX-80SS); (60 F) (Condensate CO, NP)
1.06 (Clindrol 209CGN); (20 C) (Monamine ALX-100S)

Density:
0.99 g/cc (20 C) (Loramine DC212/S; Rewomid DC212/S; Steinamid DC212/S)
1.0 g/cc (20 C) (Empilan CDE, CDX)
8.0 lb/gal (Monamid 7-100)
8.1 lb/gal (ESI-Terge C-5; Schercomid SCE)
8.15 lb/gal (ESI-Terge S-10)
8.2 lb/gal (Carsamide SAC; Emid 6514; Monamid 150-DR; Ninol 2012 Extra; Unamide CDX, LDL)

8.21 lb/gal (Marsamid 50)

8.25 lb/gal (Monamid 150-AD)

8.3 lb/gal (Accomid C; Carsamide 7644, CA; Emid 6515, 6533; Lakeway 100-CA; Monamid 150-ADD; Monamine AA-100, ARA-100, I-76; Ninol 128 Extra; Nitrene 11120, 11230, A-567, C; Richamide 6404; Schercomid CCD)

8.33 lb/gal (Schercomid CDA)

8.4 lb/gal (Emid 6531; ESI-Terge B-15; Marsamid 10; Nitrene A-309; Richamide M-3; Schercomid CDA-H)

8.41 lb/gal (Nitrene C-Extra)

8.5 lb/gal (Calamide CW-100; Emid 6534; Monamine AD-100, ADD-100, ADD-100LE)

8.55 lb/gal (Marsamid 40)

8.75 lb/gal (Monamid 7-153CS; Monamine ALX-80SS)

8.85 lb/gal (Monamine ALX-100S)

Visc.:

250 cP (Brookfield #1 spindle, 12 rpm) (Empilan FE)

360 cSt (100 F) (Emid 6531)

600 cps (20 C) (Teric CDE)

625 cps (30 C) (Stafoam DF-1)

675 cps (30 C) (Stafoam DF-4)

755 cps (30 C) (Stafoam F)

800 cps (20 C) (Manro CDS)

810 cs (20 C) (Ethylan LDS)

880 cs (20 C) (Ethylan A15)

900 cps (Hartamide OD)

950 cps (20 C) (Manro CD)

960 cs (20 C) (Ethylan LD)

1000 cps min. (Schercomid CDA)

1050 cst (Ostwald-Fenske) (Gafamide CDD-518)

1100 cps min. (Schercomid CDA-H)

1450 cps (Marsamid 40)

1500 cps (Marsamid 50)

1900 cps (Marsamid 10)

F.P.:

–8 C (Clindrol 200CG, 203CG)

–7 to –3 C (Clindrol 209CGN)

–5 C (Clindrol 206CGN, 207CGN)

–3 C (Clindrol Superamide 100CG)

0 C (Clindrol 200CGN, 202CGN)

6 C (Clindrol 101CG)

8 C (Clindrol Superamide 100C)

14–15 C (Clindrol 200HC)

Coconut diethanolamide (cont'd.)

M.P.:
  < 0 C (Teric CDE)
  14–15 C (Lauridit KD)
  25 F (Standamid SD)
  100 F (Standamid KD)
Congeal Pt.:
  1.0 C (Schercomid SCE)
  6.0 C max. (Schercomid CDA-H)
Set Pt.:
  10 C max. (Manro CDS)
  20 C max. (Manro CD)
Pour Pt.:
  –10 C (Emid 6531)
  10 C (Ethylan LDS)
  10 C min. (Marsamid 10)
  15 C (Ethylan LD)
Flash Pt.:
  28 C (PMCC) (Empilan FE)
  104 C (PMCC) (Lauridit KD)
  114 C (PMCC) (Lauridit KDG)
  > 170 C (OC) (Schercomid CCD, CDA, CDA-H)
  > 180 C (OC) (Schercomid SCE)
  146 F (Super Amide GR; Surco 128-T)
  178 F (Surco WC Conc.)
  > 200 F (Onyxol 345; Surco Coco Condensate; Surco SR-200)
  338 F (Emid 6531)
  > 350 F (COC) (Ethylan LD, LDS)
Cloud Pt.:
  –3 C (Lakeway 100-CA)
  0 C (Monamid 759)
Clear Pt.:
  33 C (Monamid 759)
HLB:
  16.6 (Teric CDE)
Acid No.:
  0–2 (Monamid 7-100, 7-153CS; Monamine ADD-100LE; Schercomid SCE; Surco
    128-T, WC Conc.)
  0–3 (Monamid 150-AD, 150-ADD; Schercomid SCO-Extra; Unamide CDX)
  0–5 (Monamid 150-DR)
  0–10 (Unamide JJ-35, LDL, N-72-3)
  1 max. (Monamid 759; Nitrene C-Extra; Standamid KD, SD)
  1.5 max. (Lauridit KDG)

2–6 (Monamine ADD-100)
2–8 (Monamine AD-100)
4 max. (Lauridit KD)
5 max. (Surco Coco Condensate)
7–14 (Schercomid CDA-H)
8–13 (Onyxol 345)
< 10 (Marsamid 40, 50; Super Amide GR)
10–15 (Condensate NP)
15–20 (Schercomid CCD, CDA)
20–25 (Condensate CO)
28–32 (Monamine AA-100, ARA-100)
35–45 (Marsamid 10)
45–50 (Condensate PN)
45–55 (Monamine I-76)
50–75 (Surco SR-200)
52–60 (Monamine ALX-80SS)
62–70 (Monamine ALX-100S)

Alkali No.:
20–40 (Schercomid SCE, SCO-Extra)
21–46 (Surco WC Conc.)
27 max. (Standamid KD)
30–40 (Standamid SD)
30–50 (Condensate PA)
32 max. (Super Amide GR; Surco 128-T)
35 ± 5 (Monamid 759)
38–50 (Condensate PO)
45 max. (Nitrene C-Extra)
105–150 (Surco SR-200)
130–165 (Onyxol 345; Schercomid CDA-H)
140–160 (Surco Coco Condensate)

Amine Value:
20–35 (Unamide JJ-35, LDL)
27–35 (Unamide CDX)
128–138 (Unamide N-72-3)

Stability:
Stable (Clindrol Superamide 100C)
Good hard water, alkali, soap, and organic acid stability; fair to poor inorganic acid stability (Condensate CO, NP, PN)
Good hard water stability (Condensate PS)
Good in presence of hard water and electrolytes (Ethylan LD, LDS)
Excellent in highly alkaline systems (Marsamid 10)
Good in alkaline systems (Marsamid 40)

Coconut diethanolamide (cont'd.)

Alkali value may decrease slightly with aging which does not affect product performance (Schercomid CCD, CDA, CDA-H, SCE)

Chemically stable at pH 5–12 (Standamid KD, SD)

Stable to mild acids and alkalis, but tends to hydrolyze under extreme conditions (Teric CDE)

Storage Stability:

1 yr min. shelf life (Schercomid CCD, CDA, CDA-H, SCE)

May become cloudy and solidify under prolonged cold storage conditions—reliquefies on warming (Empilan CDEY)

Product solidifies on prolonged cold storage and is not readily reconstituted (Empilan FD)

May gel if stored for prolonged periods at low temps. ($\approx$ 10 C); may form a soft, waxy opaque solid at temps. near 4 C (Empilan FE)

pH:

8.0–9.0 (10% sol'n.) (Monamid 7-100)

8.0–9.5 (1% sol'n.) (Condensate PN)

8.0–10.0 (Teric CDE)

8.1–9.5 (1% sol'n.) (Condensate NP)

8.5–9.5 (10% sol'n.) (Monamid 7-153CS, 150-DR; Monamine ALX-80SS, ALX-100S, I-76); (5% sol'n.) (Marsamid 10); (1% aq.) (Condensate PS); (1% sol'n.) (ESI-Terge C-5)

8.5–10.5 (1% aq.) (Ethylan LD, LDS)

9.0 (Lauridit KD)

9.0–9.5 (10% sol'n.) (Standamid KD); (2% sol'n.) (Marsamid 50)

9.0–9.7 (1% disp.) (Condensate PO)

9.0–9.8 (1% disp.) (Condensate PA)

9.0–10.0 (ESI-Terge S-10; Mazamid 70, CA-20); (1% sol'n.) (Condensate CO); (1% aq.) (Hetamide RC)

9.0–10.5 (Mazamid 80, CS-148); (5% sol'n.) (Unamide CDX, JJ-35, LDL, N-72-3)

9.0–11.0 (10% sol'n.) (Nitrene C-Extra)

9.5 (Lauridit KDG); (1% aq.) (Calamide C)

9.5–10.5 (10% sol'n.) (Monamine AA-100, AD-100, ADD-100, ARA-100; Standamid SD); (2% sol'n.) (Marsamid 40); (1% aq.) (Hetamide MC; Stafoam F)

9.6 ± 0.5 (101% sol'n.) (Schercomid CDA)

9.6–10.6 (1% aq.) (Stafoam DF-1, DF-4)

9.8–10.8 (10% sol'n.) (Monamid 150-AD); (1% sol'n.) (ESI-Terge B-15)

9.9 ± 0.5 (10% sol'n.) (Schercomid CCD)

10.0 (10%) (Lakeway 100-CA); (1% sol'n.) (Gafamide CDD-518)

10.0 ± 0.5 (10% aq. disp.) (Schercomid CDA-H)

10.0–11.0 (10% sol'n.) (Monamid 150-ADD)

10.5 (1% aq.) (Calamide CW-100)

10.5 ± 0.5 (10% sol'n.) (Monamid 759)

10.5–11.5 (10% sol'n.) (Monamine ADD-100LE)

Foam (Ross Miles):
 175 cc (1%) (Condensate PS)
Surface Tension:
 7 dynes/cm (0.1%) (Stafoam F)
 14 dynes/cm (0.1%) (Stafoam DF-1)
 16 dynes/cm (0.1%) (Stafoam DF-4)
Biodegradable: (Carsamide CA; ESI-Terge B-15, C-5, S-10; Manro CD, CDS; Marsamid 10, 40, 50; Mazamid 70, 80, CA-20, CS-148; Monamid 7-100, 7-153CS, 150-AD, 150-ADD, 150-DR; Monamine AA-100, AD-100, ADD-100, ADD-100LE, ALX-80SS, ALX-100S, ARA-100, I-76; Teric CDE; Unamide CDX, LDL); > 80% (Empilan FD)

TOXICITY/HANDLING:
 Skin and eye irritant (Clindrol 101CG, 200CG, 200CGN, 200HC, 200RC, 202CGN, 203CG, 206CGN, 207CGN, 209CGN; Clindrol Superamide 100C, 100CG; Teric CDE)
 Skin and eye irritant in conc. form; avoid ingestion (Nitrene C-Extra)
 Avoid contact with skin and eyes (Hartamide OD)
 Avoid prolonged contact with conc. form (Ethylan LD, LDS)
 Avoid prolonged contact with skin (Empilan CDEY, FD, FD20, FE)
 Defatting to skin (Mazamid 70, CA-20)
 Spillages are slippery (Empilan CDEY, FD, FD20, FE; Ethylan LD, LDS; Teric CDE)
 Contains small amt. of residual methanol—avoid inhaling vapor and using in a confined space (Empilan FE)

STORAGE/HANDLING:
 Store under cover on pallets in winter months (Empilan FD)
 Store in closed containers between 20–40 C (Nitrene C-Extra)
 Flammable—store and handle away from naked flames (Empilan FE)
 Do not mix with strong oxidizing or reducing agents (potentially explosive) (Ethylan LD, LDS)

STD. PKGS.:
 Drums/T/L (Hartamide OD)
 Mild steel drums (Loramine DC212/S; Rewomid DC212/S; Steinamid DC212/S)
 23-kg net plastic containers, 200-kg net closed-head mild steel drums (Empilan CDEY)
 125-kg drums (Lauridit KD, KDG)
 190-kg D/M, 17 kg C/N, bulk (Stafoam DF-1, DF-4)
 200-kg D/M, 17 kg C/N, bulk (Stafoam F)
 200-kg net closed-head mild steel drums (Empilan FD, FD20)
 200-kg net open-head mild steel drums (Empilan FE)
 45-gal drums or road tankers (Manro CD, CDS)
 55-gal drums (ESI-Terge B-15, C-5, S-10)
 55-gal steel drums (Carsamide C-3)

Coconut diethanolamide *(cont'd.)*

55-gal (425 lb net) open-head drums (Clindrol Superamide 100C)
55-gal (450 lb net) lined steel drums (Calamide C, CW-100)
55-gal (450 lb net) lined open-head steel drums (Schercomid CCD, CDA)
55-gal (450 lb net) open-head Liquipak drums (Schercomid SCE)
55-gal (450 lb net) tight-head steel drums (Clindrol 101CG, 200CG, 200CGN, 200HC, 200RC, 202CGN, 203CG, 206CGN, 207CGN, 209CGN; Clindrol Superamide 100CG)
450 lb net fiber drums, bulk, tank wagons, rail cars (Nitrene C-Extra)
450 lb net open-head steel drums (Monamid 150-DR)
450 lb net tight-head steel drums (Monamid 7-100, 150-AD, 150-ADD; Monamine AA-100, AD-100, ADD-100, ADD-100LE, ARA-100, I-76)
480 lb net open-head steel drums (Monamine ALX-80SS, ALX-100S)
480 lb net Liquipak (Monamid 7-153CS)

Coconut monoethanolamide

SYNONYMS:
Amides, coco, N-(2-hydroxyethyl)-
Cocamide MEA (CTFA)
Coco monoethanolamide
Coconut fatty acid, monoethanolamide
Cocoyl monoethanolamine
N-(2-hydroxyethyl) coco fatty acid amide
Monoethanolamine coconut acid amide

STRUCTURE:

$$RC\underset{\displaystyle \|}{\overset{\displaystyle O}{}}-NHCH_2CH_2OH$$

where RCO⁻ represents the coconut acid radical

CAS No.
68140-00-1

TRADENAME EQUIVALENTS:
Alkamide CME, CMO [Alkaril]
Aminol CM [Finetex]
Ardet CEA [Ardmore]
Chimipal MC [La Tassilchimka]
Clindrol 100MC, 100MCG [Clintwood]
Cyclomide C212 [Cyclo]
Emid 6500 [Emery Ind.]

Coconut monoethanolamide (cont'd.)

TRADENAME EQUIVALENTS *(cont'd.):*
  Empilan CM [Albright & Wilson/Australia]
  Empilan CME [Albright &Wilson/Marchon]
  Incromide CM, CME [Croda Surfactants]
  Jordamide CMEA, CMEA Extra [Jordan]
  Loramine C212 [Dutton & Reinisch]
  Mackamide CMA [McIntyre]
  Manro CMEA [Manro]
  Marlamid M1218 [Chem.Werke Huls]
  Monamid CMA [Mona]
  Monamide [Zohar Detergent]
  Ninol CNR [Stepan Europe]
  Nissan Stafoam MF [Nippon Oil & Fat]
  Rewomid C212 [Dutton & Reinisch]
  Richamide MX [Richardson]
  Schercomid CME [Scher]
  Standamid 100 [Henkel Argentina]
  Standamid CM, CMG, KM, SM [Henkel]
  Steinamid C212 [Dutton & Reinisch]
  Sterling Granulated Wax [Canada Packers]
  Surco CMEA, CMEA Flake [Onyx]
  Swanic 51 [Swastik]
  Unamide CMX [Lonza]

CATEGORY:
  Detergent, foam booster, viscosity builder, opacifier, stabilizer, superfatting agent

APPLICATIONS:
  Bath products: (Standamid KM, SM); bubble bath (Schercomid CME; Standamid 100, CM, CMG; Unamide CMX)
  Cleansers: liquid hand cleanser (Standamid KM, SM)
  Cosmetic industry preparations: (Alkamide CME, CMO; Incromide CM; Jordamide CMEA, CMEA Extra; Surco CMEA, CMEA Flake; Unamide CMX); creams and lotions (Schercomid CME); hair preparations (Standamid 100); shampoos (Chimipal MC; Clindrol 100MC; Emid 6500; Empilan CME; Nissan Stafoam MF; Richamide MX; Schercomid CME; Standamid 100, CMG, KM, SM; Unamide CMX); stick-form cosmetics (Standamid 100); toiletries (Aminol CM; Empilan CM; Unamide CMX)
  Household detergents: (Alkamide CME, CMO; Empilan CM; Incromide CM; Jordamide CMEA, CMEA Extra; Loramine C212; Marlamid M1218; Rewomid C212; Standamid 100; Steinamid C212; Sterling Granulated Wax; Surco CMEA, CMEA Flake); carpet & upholstery shampoos (Emid 6500; Empilan CME); dishwashing (Clindrol 100MC; Empilan CME; Richamide MX; Standamid CMG); hard surface cleaner (Richamide MX); laundry detergent (Alkamide CME, CMO); liquid

Coconut monoethanolamide (cont'd.)

    detergents (Emid 6500; Empilan CME; Manro CMEA; Monamid CMA); powdered detergents (Cyclomide C212; Empilan CME; Manro CMEA; Monamid CMA; Schercomid CME; Swanic 51)

Industrial applications: industrial processing (Ardet CEA; Ninol CNR)

Industrial cleaners: (Incromide CM; Jordamide CMEA, CMEA Extra; Surco CMEA, CMEA Flake); institutional cleaners (Alkamide CME)

Pharmaceutical applications: (Alkamide CME, CMO); medicated shampoos (Standamid 100); ointments (Standamid 100)

PROPERTIES:
Form:
Paste (Richamide MX)

Solid paste (Swanic 51)

Solid (Aminol CM; Mackamide CMA; Marlamid M1218; Ninol CNR; Standamid 100, CMG; Unamide CMX)

Beads (Standamid KM, SM)

Flakes (Ardet CEA; Chimipal MC; Clindrol 100MC, 100MCG; Cyclomide C212; Emid 6500; Incromide CM, CME; Jordamide CMEA, CMEA Extra; Loramine C212; Nissan Stafoam MF; Rewomid C212; Standamid KM, SM; Steinamid C212; Sterling Granulated Wax; Surco CMEA Flake)

Granular (Monamid CMA; Standamid CM)

Wax (Schercomid CME; Surco CMEA)

Waxy flake (Empilan CME; Manro CMEA)

Waxy solid (Alkamide CME, CMO; Empilan CM)

Color:
Light (Unamide CMX)

White (Alkamide CME; Swanic 51)

Off-white (Loramine C212; Rewomid C212; Steinamid C212)

Cream (Alkamide CMO; Cyclomide C212; Empilan CM, CME; Manro CMEA)

Light yellow (Jordamide CMEA, CMEA Extra)

Tan (Clindrol 100MC, 100MCG; Schercomid CME)

Gardner 4 (Emid 6500)

Gardner 4 max. (Incromide CM)

Gardner 5 max. (Incromide CME)

Odor:
Low (Alkamide CME)

Slight amidic (Loramine C212; Rewomid C212; Steinamid C212)

Slight amine (Incromide CM)

Ammoniacal (Clindrol 100MC; Schercomid CME)

Mild (Manro CMEA)

Composition:
84% min. amide content (Clindrol 100MCG)

90% active (Alkamide CMO; Monamide)

90% amide content (Standamid 100)

94% min. active (Manro CMEA)

94.2% active (Empilan CME)

95% active (Alkamide CME)

100% active (Clindrol 100MC; Cyclomide C212; Emid 6500; Incromide CM; Loramine C212; Marlamid M1218; Rewomid C212; Schercomid CME; Steinamid C212; Sterling Granulated Wax; Unamide CMX)

100% conc. (Aminol CM; Ardet CEA; Chimipal MC; Mackamide CMA; Monamid CMA; Ninol CNR; Nissan Stafoam MF; Standamid CM, CMG; Surco CMEA)

Solubility:

Sol. in alcohols (Clindrol 100MCG; Schercomid CME); lower alcohols (Standamid KM, SM)

Sol. in aliphatic hydrocarbons (Clindrol 100MC; Schercomid CME); at elevated temps. (Clindrol 100MCG)

Sol. in aromatic hydrocarbons (Alkamide CME, CMO; Clindrol 100MC; Schercomid CME); at elevated temps. (Clindrol 100MCG)

Sol. in butyl stearate (Emid 6500)

Sol. in chlorinated aliphatic solvents (Alkamide CME, CMO)

Sol. in chlorinated hydrocarbons (Schercomid CME); at elevated temps. (Clindrol 100MCG)

Sol. in detergent systems (Alkamide CME, CMO)

Sol. in ethanol (Clindrol 100MC)

Sol. in glycerol trioleate (Emid 6500)

Partly sol. in glycerols (Schercomid CME)

Sol. in glycols (Clindrol 100MCG; Schercomid CME); insol. in propylene glycol, polyethylene glycols (Standamid KM, SM)

Sol. in glycol ethers (Schercomid CME)

Sol. in isopropanol (Clindrol 100MC)

Sol. in min. oils (Alkamide CME, CMO; Emid 6500)

Sol. in perchloroethylene (Emid 6500)

Sol. in petroleum solvents (Alkamide CME, CMO)

Sol. in Stoddard solvent (Emid 6500)

Partly sol. in triols (Schercomid CME)

Disp. in water (Emid 6500; Schercomid CME); disp. in warm water (Clindrol 100MCG); disp. in hot water (Empilan CME); insol. (Standamid KM, SM)

Ionic Nature:

Nonionic

Sp.gr.:

0.884 (Standamid KM)

0.92 (25/20 C) (Surco CMEA, CMEA Flake)

0.934 (Standamid SM)

1.0 (Clindrol 100MC)

Coconut monoethanolamide *(cont'd.)*

Density:
0.4 g/cc (20 C) (Empilan CME)
9.0 lb/gal (Richamide MX)

M.P.:
50 C (Alkamide CMO)
56 C (Alkamide CME)
60–66 C (Clindrol 100MCG)
62 C (Standamid SM)
63 C (Clindrol 100MC)
63 ± 3 C (Schercomid CME)
68 C (Clindrol 100MCG; Incromide CME)
71 C (Emid 6500)
≈ 73 C (Loramine C212; Rewomid C212; Steinamid C212)
80 C (Standamid KM)
140–148 F (Standamid CM)

Flash Pt.:
> 180 C (OC) (Schercomid CME)
146 F (Surco CMEA, CMEA Flake)

Acid No.:
0–1.0 (Standamid CM)
0–3 (Unamide CMX)
1 max. (Incromide CM, CME; Schercomid CME; Standamid SM)
2 max. (Standamid KM; Surco CMEA, CMEA Flake)

Amine Value:
0–18 (Unamide CMX)

Alkali No.:
8.0–12.0 (Standamid CM)
12 max. (Standamid KM, SM)
15 max. (Schercomid CME)
19 max. (Surco CMEA, CMEA Flake)

Stability:
Stable to acids and alkalies under certain conditions (Empilan CME)
Chemically stable at pH 5–14 (Standamid CM)
Excellent stability characteristics (Standamid CMG)

Storage Stability:
Storable over a wide temp. range (0-45 C) with no product degradation (Standamid KM, SM)

pH:
8–10.5 (1% disp.) (Clindrol 100MCG)
9–10.5 (5% sol'n.) (Unamide CMX)
9–11 (1.0% sol'n.) (Standamid KM, SM)
9.5–10.5 (10% sol'n.) (Standamid CM)

Coconut monoethanolamide (cont'd.)

Biodegradable: (Manro CMEA)
TOXICITY/HANDLING:
   Skin and eye irritant (Clindrol 100MC)
   Avoid prolonged contact with skin (Empilan CM; Incromide CM, CME)
STORAGE/HANDLING:
   Store in a dry place at 50–120 F (Incromide CM, CME)
STD. PKGS.:
   Paper sacks (Loramine C212; Rewomid C212; Steinamid C212)
   25 kg polypropylene bags (Manro CMEA)
   180 kg net open-head mild steel drums, 18 kg net plastic pails (Empilan CM)
   44-gal (150 lb net) Leverpak with polyethylene (Incromide CM, CME)
   55-gal (200 lb net) open-head fiber drums with polyethylene bag liner (Clindrol 100MC, 100MCG)
   200 lb fiber drums, 50 lb boxes (Standamid KM, SM)

Coconut monoisopropanolamide

SYNONYMS:
   Amides, coco, N-(2-hydroxypropyl)-
   Cocamide MIPA (CTFA)
   Coconut fatty acid isopropanolamide
   Coconut isopropanolamide
   Monoisopropanolamine coconut acid amide
STRUCTURE:

$$RC—NH—CH_2CHOH$$

with O double-bonded to the RC carbon and CH_3 attached to the CHOH carbon.

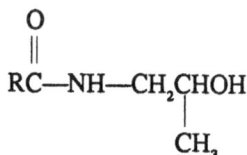

   where RCO⁻ represents the coconut acid radical
CAS No.:
   68333-82-4
TRADENAME EQUIVALENTS:
   Empilan CIS [Albright & Wilson/Detergents]
   Loramine IPP 240 [Dutton & Reinisch]
   Rewomid IPP 240 [Dutton & Reinisch]
   Schercomid CMI [Scher]
   Steinamid IPP 240 [Dutton & Reinisch]

73

Coconut monoisopropanolamide *(cont'd.)*

CATEGORY:
   Stabilizer, foam stabilizer, foam booster, viscosity modifier, antidefatting agent, detergent
APPLICATIONS:
   Cosmetic industry preparations: shampoos (Empilan CIS)
   Household detergents: (Schercomid CMI); carpet & upholstery shampoos (Empilan CIS); liquid detergents (Empilan CIS); powdered detergents (Empilan CIS)
PROPERTIES:
Form:
   Flake (Loramine IPP 240; Rewomid IPP 240; Steinamid IPP 240)
   Wax (Schercomid CMI)
   Waxy flake (Empilan CIS)
Color:
   White (Loramine IPP 240; Rewomid IPP 240; Steinamid IPP 240)
   Cream (Empilan CIS)
   Light tan (Schercomid CMI)
Odor:
   Typical, slight (Loramine IPP 240; Rewomid IPP 240; Steinamid IPP 240)
Composition:
   90% amide min. (Schercomid CMI)
   100% active (Loramine IPP 240; Rewomid IPP 240; Steinamid IPP 240)
Ionic Nature:
   Nonionic (Empilan CIS; Loramine IPP 240; Rewomid IPP 240; Schercomid CMI; Steinamid IPP 240)
Density:
   0.4 g/cm³ (20 C) (Empilan CIS)
M.P.:
   46 C (Empilan CIS)
   ≈ 50 C (Loramine IPP 240; Rewomid IPP 240; Steinamid IPP 240)
Acid No.:
   2.0 max. (Schercomid CMI)
Alkali No.:
   20 max. (Schercomid CMI)
pH:
   Alkaline (Loramine IPP 240; Rewomid IPP 240; Steinamid IPP 240)
STD. PKGS.:
   Mild steel drums (Loramine IPP 240; Rewomid IPP 240; Steinamid IPP 240)

SYNONYMS:
Amides, coconut oil
Cocamide (CTFA)
Cocoamide
Coconut acid amide

STRUCTURE:

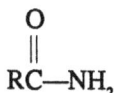

O
‖
RC—NH₂
where RCO⁻ represents the coconut acid radical

CAS No.:
61789-19-3

TRADENAME EQUIVALENTS:
Armid C [Armak]
Amides of Cocout Oil [Murphy-Phoenix]
Ecconol 628 [Essential]

CATEGORY:
Lubricant, slip agent, foam stabilizer, intermediate, dispersant, thickener, detergent, wetting agent, emulsifier

APPLICATIONS:
Household detergents: all-purpose cleaner (Ecconol 628)
Industrial applications: dyes and pigments (Armid C); paint and coatings mfg. (Armid C); paper mfg. (Armid C); plastics (Armid C); printing inks (Armid C); textile/leather processing (Armid C)

PROPERTIES:
Form:
Liquid (Ecconol 628)
Flake (Armid C)
Color:
Gardner 10 max. (Armid C)
Composition:
90% amide min. (Armid C)
100% active (Ecconol 628)
Solubility:
Insol. in water (Armid C)
Ionic Nature:
Nonionic (Amides of Cocout Oil; Ecconol 628)
M.W.:
214 (Armid C)
Sp.gr.:
0.845 (100 C) (Armid C)

Coconut oil amide *(cont'd.)*

M.P.:
85 C min. (Armid C)
Flash Pt.:
≈ 347 F (COC) (Armid C)
Fire Pt.:
≈ 365 F (Armid C)
Iodine No.:
10 max. (Armid C)
Biodegradable: (Ecconol 628)
STD. PKGS.:
60-lb cardboard containers (Armid C)

Cocoyl sarcosine (CTFA)

SYNONYMS:
N-Cocoyl-N-methyl glycine
N-Cocoyl sarcosine
Glycine, N-methyl-, N-coco acyl derivs.
Glycine, N-methyl-N-(1-oxococonut alkyl)-
N-Methylglycine, N-coco acyl derivs.
N-Methyl-N-(1-oxococonut alkyl) glycine
STRUCTURE:

$$R-\overset{\overset{\displaystyle O}{\|}}{C}-\overset{\overset{\displaystyle CH_3}{|}}{N}-CH_2COOH$$

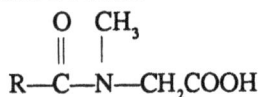

where RCO⁻ represents the coconut acid radical
CAS No.:
68411-97-2
TRADENAME EQUIVALENTS:
Hamposyl C [W.R. Grace]
Sarkosyl LC [Ciba-Geigy]
CATEGORY:
Foam stabilizer, foam booster, foaming agent, detergent, corrosion inhibitor, wetting
agent, lubricant, emulsifier, shampoo base
APPLICATIONS:
Bath products: bath oils (Sarkosyl LC)
Cleansers: hand cleanser (Sarkosyl LC); skin cleanser (Hamposyl C)
Cosmetic industry preparations: (Hamposyl C); aerosol products (Sarkosyl LC);
shampoos (Hamposyl C; Sarkosyl LC); toilet soaps (Sarkosyl LC)

76

Degreasers: solvent degreasing (Sarkosyl LC)
Household detergents: carpet & upholstery shampoos (Hamposyl C; Sarkosyl LC); dishwashing (Sarkosyl LC); laundry detergent (Sarkosyl LC); window cleaners (Sarkosyl LC)
Industrial applications: lubricating/cutting oils (Sarkosyl LC); metalworking (Sarkosyl LC)
Pharmaceutical applications: (Sarkosyl LC); antiperspirant/deodorant (Sarkosyl LC); antidandruff preparations (Sarkosyl LC); balms/lotions (Sarkosyl LC); dental preparations (Sarkosyl LC); mouthwash (Sarkosyl LC)
PROPERTIES:
Form:
Liquid (Hamposyl C)
Powder (Sarkosyl LC)
Color:
Gardner 3 (Hamposyl C)
Composition:
94% active (Hamposyl C; Sarkosyl LC)
Solubility:
Sol. in aliphatic hydrocarbons (Sarkosyl LC)
Sol. in glycerin (Sarkosyl LC)
Sol. in glycols (Sarkosyl LC)
Sol. in most organic solvents (Hamposyl C; Sarkosyl LC)
Sol. in phosphate esters (Sarkosyl LC)
Sol. in silicones (Sarkosyl LC)
Insol. in water (Sarkosyl LC)
Ionic Nature:
Anionic (Hamposyl C; Sarkosyl LC)
M.W.:
285–300 (Sarkosyl LC)
Sp.gr.:
0.969 (Sarkosyl LC)
0.97–0.99 (Hamposyl C)
M.P.:
23–26 C (Hamposyl C)
23–28 C (Sarkosyl LC)
Biodegradable: (Hamposyl C)
TOXICITY/HANDLING:
Skin irritant on prolonged contact with conc. acid (Sarkosyl LC)
STD. PKGS.:
55-gal drums (Hamposyl C)

Decaglycerol tetraoleate

SYNONYMS:
Decaglyceryl tetraoleate
9-Octadecenoic acid, tetraester with decaglycerol
Polyglyceryl-10 tetraoleate (CTFA)
EMPIRICAL FORMULA:
$C_{102}H_{190}O_{25}$
CAS No.
34424-98-1
TRADENAME EQUIVALENTS:
Caprol 10G40 [Capital City]
Drewmulse 10-4-O [PVO Int'l.]
Drewpol 10-4-O [PVO Int'l.]
Hodag SVO-1047 [Hodag]
Mazol PGO-104 [Mazer]
CATEGORY:
Emulsifier, stabilizer, solubilizer, dispersant
APPLICATIONS:
Cosmetic industry preparations: (Drewmulse 10-4-O; Mazol PGO-104); emollient creams and lotions (Drewmulse 10-4-O); personal care products (Mazol PGO-104)
Food applications: food emulsifying (Drewpol 10-4-O; Hodag SVO-1047; Mazol PGO-104)
Industrial applications: industrial processing (Caprol 10G40); lubricating/cutting oils (Mazol PGO-104); plastics (Mazol PGO-104); textile/leather processing (Mazol PGO-104)
Pharmaceutical applications: (Drewmulse 10-4-O; Mazol PGO-104)
PROPERTIES:
Form:
Liquid (Caprol 10G40; Drewmulse 10-4-O; Drewpol 10-4-O; Hodag SVO-1047; Mazol PGO-104)
Color:
Amber (Drewpol 10-4-O)
Gardner 8 (Drewmulse 10-4-O; Mazol PGO-104)
Composition:
100% conc. (Caprol 10G40; Hodag SVO-1047)
Solubility:
Sol. in isopropanol (Drewmulse 10-4-O)
Sol. in min. oils (Drewmulse 10-4-O)

Sol. in peanut oil (Drewmulse 10-4-O)
Disp. in water (Drewmulse 10-4-O)
Ionic Nature:
Nonionic
HLB:
6.0 (Drewmulse 10-4-O; Drewpol 10-4-O; Hodag SVO-1047)
6.2 (Caprol 10G40; Mazol PGO-104)
Acid No.:
8 max. (Mazol PGO-104)
Iodine No.:
60 max. (Mazol PGO-104)
Saponification No.:
125–145 (Drewmulse 10-4-O; Drewpol 10-4-O)
125–150 (Mazol PGO-104)

Dibasic lead phosphite

STRUCTURE:
$2PbO \cdot PbHPO_3 \cdot {}^1/_2 H_2O$
CAS No.:
1344-40-7
TRADENAME EQUIVALENTS:
Dyphos [Associated Lead]
Dyphos XL [Associated Lead] (coated)
CATEGORY:
Heat, light, and outdoor weathering stabilizer, uv absorber, antioxidant, activator, anticorrosion agent
APPLICATIONS:
Industrial applications: chlorinated hydrocarbons/rubber (Dyphos, XL); electrical insulation (Dyphos, XL); foam mfg. (Dyphos, XL); plastics/plastisols (Dyphos); protective coatings (Dyphos)
PROPERTIES:
Form:
Fine acicular crystals (Dyphos, XL)
Color:
White (Dyphos, XL)
Composition:
81.2% PbO (Dyphos XL)
90% PbO (Dyphos)

Dibasic lead phosphite *(cont'd.)*

Sp.gr.:
4.5 (25/4 C) (Dyphos XL)
6.1 (25/4 C) (Dyphos)

Stability:
Excellent heat stability, outdoor weathering properties, and resistance to color change in use (Dyphos, XL)

Dibasic lead phthalate

STRUCTURE:
$C_6H_4(COO)_2Pb \cdot PbO$

TRADENAME EQUIVALENTS:
Cyastab 908 [Amer. Cyanamid]
Cyastab 988 [Amer. Cyanamid] (coated)
Dibasic Lead Phthalate Conc. 9-73-A, 13-191, 14-81, 17-239 [Santech]
Dythal [Associated Lead]
Dythal XL [Associated Lead] (coated)
Epithal 120 [Eagle-Picher]
Halthal [Halstab]
Halthal-EP [Halstab] (coated)
Generically sold by: [Akzo]

CATEGORY:
Heat stabilizer, activator, vulcanizing agent

APPLICATIONS:
Industrial applications: calendered sheet (Halthal, -EP); chlorinated rubber (Dythal, XL); elastomers (Dythal, XL); electrical insulation (Cyastab 908, 988; Dythal, XL; Epithal 120; Halthal, -EP); foam mfg. (Dythal, XL); plastics/plastisols (Cyastab 908, 988; Dibasic Lead Phthalate Conc. 9-73-A, 13-191, 14-81, 17-239; Dythal, XL; Halthal, -EP); profile extrusion (Halthal, -EP)

PROPERTIES:

Form:
Fine particle size (Cyastab 988)
Fluffy powder (Dythal, XL)
Paste (Dibasic Lead Phthalate Conc. 14-81, 17-239)
Pellets (Dibasic Lead Phthalate Conc. 9-73-A)

Color:
White (Dythal, XL)

Composition:
71.5% PbO (Dythal XL)
79.8% PbO (Dythal)

Dibasic lead phthalate (cont'd.)

Sp.gr.:
 3.5 (25/4 C) (Dythal XL)
 4.5 (25/4 C) (generic; Dythal)
Ref. Index:
 1.99 (generic)

Dibutyl maleate

SYNONYMS:
 DBM
STRUCTURE:
 $C_4H_9OOCCH:CHCOOC_4H_9$
CAS No.:
 105-76-0
TRADENAME EQUIVALENTS:
 PX-504 [USS Chem.]
 Staflex DBM [Reichhold]
  Generically sold by:
   [BASF Canada; Nuodex; Unitex]
CATEGORY:
 Plasticizer, comonomer, chemical intermediate
APPLICATIONS:
 Industrial applications: adhesives (PX-504); coatings (PX-504); lubricating/cutting
  oils (PX-504); paint mfg. (PX-504); plastics/plastisols (PX-504; Staflex DBM)
PROPERTIES:
Form:
 Clear liquid (PX-504)
 Oily liquid (generic)
Color:
 Colorless (generic)
 APHA 50 (Staflex DBM)
Solubility:
 Insol. in water (generic)
M.W.:
 228 (Staflex DBM)
Sp.gr.:
 0.993 (23 C) (Staflex DBM)
 0.9964 (20/20 C) (generic)

Dibutyl maleate *(cont'd.)*

Density:
   8.3 lb/gal (Staflex DBM); (20 C) (generic)
Visc.:
   3.7 cSt (kinematic, 100 F) (Staflex DBM)
B.P.:
   280.6 C (generic)
M.P.:
   < −80 to < −65 C (generic)
Pour Pt.:
   < −76 F (Staflex DBM)
Flash Pt.:
   285 F (Staflex DBM); (OC) (generic)
Cloud Pt.:
   < −76 F (Staflex DBM)
Ref. Index:
   1.4444 (generic)
   1.4545 (23 C) (Staflex DBM)

Dibutyl phthalate *(CTFA)*

SYNONYMS:
   1,2-Benzenedicarboxylic acid, dibutyl ester
   DBP
   Dibutyl 1,2-benzenedicarboxylate
EMPIRICAL FORMULA:
   $C_{16}H_{22}O_4$
STRUCTURE:

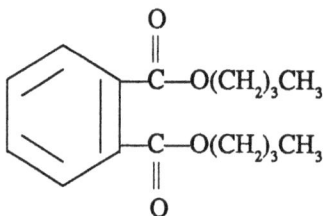

CAS No.:
   84-74-2
TRADENAME EQUIVALENTS:
   Hatcol DBP [Hatco]
   Hexaflex DBP [Hexagon]

Dibutyl phthalate (cont'd.)

TRADENAME EQUIVALENTS *(cont'd.):*
Kodaflex DBP [Eastman]
Palatinol DBP [Badische]
Polycizer DBP [Harwick]
PX-104 [USS Chem.]
Staflex DBP [Reichhold]
Unimoll DB [Bayer AG]
  Generically sold by:
    [Ashland; FMC Corp.; C.P. Hall; Monsanto; Unocal]

CATEGORY:
Plasticizer, softener, solvent, antifoam agent, lubricant, carrier, fixative

APPLICATIONS:
Cosmetic industry preparations: perfumes (Palatinol DBP)
Farm products: insecticides (Kodaflex DBP)
Household products: detergent solutions (generic—FMC)
Industrial applications: adhesives (generic—FMC, Monsanto; Hexaflex DBP; Palatinol DBP; PX-104; Unimoll DB); cellulosics (generic—Monsanto; Hatcol DBP; Hexaflex DBP; Kodaflex DBP; Palatinol DBP; Polycizer DBP; PX-104; Staflex DBP); coatings and lacquers (generic—Monsanto; Hatcol DBP; Hexaflex DBP; Kodaflex DBP; Palatinol DBP; PX-104); dyes and pigments (Kodaflex DBP; Palatinol DBP; Staflex DBP); laminates (generic—Monsanto; Palatinol DBP); military smokeless powders (Palatinol DBP); packaging films (PX-104); paper coating (generic—FMC); plastics (generic—Monsanto; Hatcol DBP; Hexaflex DBP; Kodaflex DBP; Palatinol DBP; Polycizer DBP; PX-104; Unimoll DB); polishes and waxes (Palatinol DBP); printing inks (generic—FMC; Palatinol DBP; Staflex DBP); rubber (generic—Hall, Monsanto; Hatcol DBP; Kodaflex DBP; Palatinol DBP; Polycizer DBP; Staflex DBP; Unimoll DB); textile processing (generic—FMC, Monsanto; Kodaflex DBP)

PROPERTIES:
Form:
Liquid (Kodaflex DBP)
Oily liquid (generic—Hall; Hatcol DBP; Polycizer DBP)
Color:
Essentially colorless (generic—Hall; Hatcol DBP; Polycizer DBP)
APHA 20 (Staflex DBP)
APHA 35 max. (Palatinol DBP)
Hazen ≤ 30 (Unimoll DB)
Pt-Co 15 ppm (Kodaflex DBP)
Odor:
Odorless (generic—Hall; Hatcol DBP; Kodaflex DBP; Polycizer DBP)
Mild, characteristic (Palatinol DBP)

Dibutyl phthalate *(cont'd.)*

Composition:
99% min. ester content (Kodaflex DBP; Palatinol DBP)

Solubility:
Misc. with coal tar solvents and diluents (PX-104)
Misc. with petroleum solvents and diluents (PX-104)
Sol. < 0.01% in water @ 20 C (Palatinol DBP)

M.W.:
278 (Hatcol DBP; Hexaflex DBP; Kodaflex DBP; PX-104; Staflex DBP)
278.35 (calc.) (Palatinol DBP)

Sp.gr.:
1.04 (Hatcol DBP; Polycizer DBP)
1.045–1.051 (20/20 C) (Palatinol DBP)
1.046 (generic—Unocal)
1.047 (generic—Hall)
1.048 (20/20 C) (Kodaflex DBP)
1.049 (20/20 C) (generic—Ashland)
1.050 (23 C) (Staflex DBP)

Density:
1.04 kg/l (Kodaflex DBP)
1.043–1.049 g/cc (Unimoll DB)
8.7 lb/gal (Hatcol DBP; Palatinol DBP; Polycizer DBP; Staflex DBP)
8.74 lb/gal (generic—Hall)

Visc.:
15 cP (Kodaflex DBP)
19–21 MPa•s (20 C) (Unimoll DB)
20.1 cps (20 C) (Palatinol DBP)
75 cSt (0 C, kinematic) (Staflex DBP)
102 SUS (generic—Hall; Hatcol DBP; Polycizer DBP)

F.P.:
–35 C (Kodaflex DBP)

B.P.:
182 C (5 mm Hg) (Palatinol DBP)
340 C (760 mm) (Kodaflex DBP)

Pour Pt.:
–40 C (Palatinol DBP)
< –80 F (Staflex DBP)

Flash Pt.:
171 C (COC) (Palatinol DBP)
173–177 C (OC) (Unimoll DB)
190 C (COC) (Kodaflex DBP)
360 F (Staflex DBP)

Fire Pt.:
202 C (COC) (Kodaflex DBP)
Cloud Pt.:
< −80 F (Staflex DBP)
Acid No.:
0.07 (Palatinol DBP)
≤ 0.1 (Unimoll DB)
Saponification No.:
400–410 (Unimoll DB)
Storage Stability:
Can be stored indefinitely (Palatinol DBP)
Ref. Index:
1.4910 (23 C) (Staflex DBP)
1.4915–1.4930 (20 C) (Unimoll DB)
1.4920 (Kodaflex DBP)
1.4926 (20 C) (Palatinol DBP)
Dissipation Factor:
0.08×10^{-2} (1 MHz) (Kodaflex DBP)
Dielectric Constant:
5.8 (1 MHz) (Kodaflex DBP)
Volume Resistivity:
3.0×10^9 ohm-cm (Kodaflex DBP)
TOXICITY/HANDLING:
Low toxicity (generic—Hall; Kodaflex DBP; Polycizer DBP)
Avoid repeated/prolonged skin and eye contact; avoid breathing vapors; use with adequate ventilation (Palatinol DBP)
STD. PKGS.:
Bulk (Palatinol DBP)

Dibutyl sebacate (CTFA)

SYNONYMS:
DBS
Decanedioic acid, dibutyl ester
Dibutyl decanedioate
EMPIRICAL FORMULA:
$C_{18}H_{34}O_4$

Dibutyl sebacate (cont'd.)

STRUCTURE:

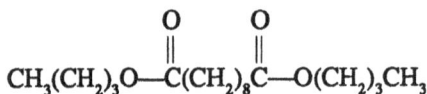

$$CH_3(CH_2)_3O-\overset{\overset{O}{\|}}{C}(CH_2)_8\overset{\overset{O}{\|}}{C}-O(CH_2)_3CH_3$$

CAS No.:

109-43-3

TRADENAME EQUIVALENTS:

Hallco DBS [C.P. Hall]

Polycizer DBS [Harwick]

Rilanit DBS [Henkel KGaA]

Uniflex DBS [Union Camp]

CATEGORY:

Plasticizer, softener, lubricant, fatting agent, emulsifier, dispersant

APPLICATIONS:

Food applications: food packaging (Hallco DBS; Polycizer DBS; Uniflex DBS)

Industrial applications: cellulosics (Uniflex DBS); ceramics (Rilanit DBS); latexes (Hallco DBS; Polycizer DBS); lubricating/cutting oils (Rilanit DBS); metalworking fluids (Rilanit DBS); petroleum products (Rilanit DBS); plastics (Polycizer DBS; Uniflex DBS); rubber (Hallco DBS; Polycizer DBS; Uniflex DBS); textile/ leather processing (Rilanit DBS)

PROPERTIES:

Form:

Liquid (Polycizer DBS; Rilanit DBS)

Clear liquid (Hallco DBS)

Color:

Colorless (Hallco DBS; Polycizer DBS)

APHA 5 (Uniflex DBS)

Odor:

Slight residual odor (Hallco DBS; Polycizer DBS)

M.W.:

316 (Hallco DBS)

Sp.gr.:

0.93–0.94 (Polycizer DBS)

0.935 (Hallco DBS)

0.937 (20/20 C) (Uniflex DBS)

Density:

7.8 lb/gal (Hallco DBS; Polycizer DBS)

Visc.:

7.9 cps (Uniflex DBS)

14 cps (Hallco DBS)

F.P.:

–11 C (Uniflex DBS)

B.P.:
344–349 C (Hallco DBS; Polycizer DBS)
M.P.:
–11 C (Hallco DBS; Polycizer DBS)
Flash Pt.:
180 C (Polycizer DBS)
185 C (Hallco DBS)
193 C (COC) (Uniflex DBS)
Fire Pt.:
213 C (COC) (Uniflex DBS)
Acid No.:
< 0.1 (Uniflex DBS)
TOXICITY/HANDLING:
Nontoxic (Hallco DBS; Polycizer DBS; Uniflex DBS)
STD. PKGS.:
Drum, tankcar, tanktruck (Hallco DBS; Polycizer DBS)

Dicapryl adipate (CTFA)

SYNONYMS:
Didecyl hexanedioate
Hexanedioic acid, didecyl ester
EMPIRICAL FORMULA:
$C_{26}H_{50}O_4$
STRUCTURE:

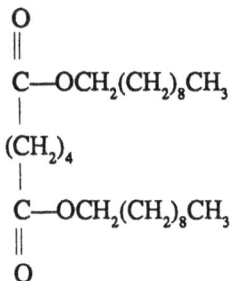

CAS No.:
105-97-5
TRADENAME EQUIVALENTS:
Uniflex DCA [Union Camp]

Dicapryl adipate *(cont'd.)*

CATEGORY:
Plasticizer
APPLICATIONS:
Industrial applications: plastics/plastisols; rubber
PROPERTIES:
Color:
APHA 10
Sp.gr.:
0.914 (20/20 C)
Visc.:
13 cSt
F.P.:
−32 C
Flash Pt.:
207 C (COC)
Fire Pt.:
224 C (COC)
Acid No.:
< 0.1
Ref. Index:
1.439

Dicetyl thiodipropionate (CTFA)

SYNONYMS:
Propanoic acid, 3,3′-thiobis-, dihexadecyl ester
3,3′-Thiobispropanoic acid, dihexadecyl ester
EMPIRICAL FORMULA:
$C_{38}H_{74}O_4S$
STRUCTURE:

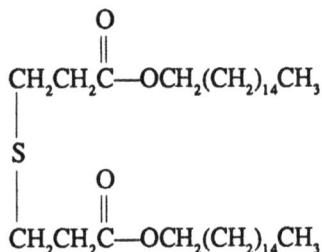

Dicetyl thiodipropionate (cont'd.)

CAS No.:
3287-12-5
TRADENAME EQUIVALENTS:
Evanstab 16 [Evans Chemetics]
CATEGORY:
Antioxidant
APPLICATIONS:
Food applications: food packaging
Industrial applications: polymers
PROPERTIES:
Form:
Free-flowing crystalline flakes
Color:
White
Composition:
98% assay min.
M.W.:
627
M.P.:
59–62 C
Acid No.:
1.0 max.
Saponification No.:
176–183
Stability:
Good color stability during aging; improved melt stability during polymer fabrication; improved retention of physical and electrical properties during polymer fabrication and during environmental aging of the polymer

Diethylaminoethyl stearate (CTFA)

SYNONYMS:
2-(Diethylamino) ethyl octadecanoate
Octadecanoic acid, 2-(diethylamino) ethyl ester
EMPIRICAL FORMULA:
$C_{24}H_{50}NO_2$

Diethylaminoethyl stearate (cont'd.)

STRUCTURE:

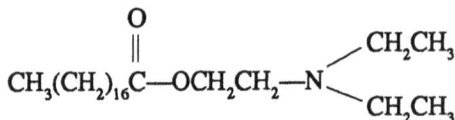

CAS No.:
 3179-81-5
TRADENAME EQUIVALENTS:
 Cerasynt 303 [Van Dyk]
CATEGORY:
 Viscosity builder, emulsifier
APPLICATIONS:
 Cosmetic industry preparations: hair dyes
 Pharmaceutical applications
PROPERTIES:
Form:
 Liquid dispersion in water
Solubility:
 Sol. in 95% ethanol
 Sol. in isopropyl myristate
 Partly sol. in min. oil
 Insol. in propylene glycol
Sp.gr.:
 0.860–0.880
Saponification No.:
 150–160

Diethylene glycol dibenzoate (CTFA)

SYNONYMS:
 PEG 100 dibenzoate
 POE (2) dibenzoate
STRUCTURE:

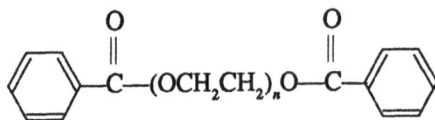

where avg. $n = 2$

Diethylene glycol dibenzoate *(cont'd.)*

TRADENAME EQUIVALENTS:
Benzoflex 2-45 [Velsicol]
  Generically sold by:
    [Nuodex; Unitex]
CATEGORY:
Plasticizer

APPLICATIONS:
Industrial applications: adhesives (Benzoflex 2-45); cellulosics (generic); plastics (generic; Benzoflex 2-45)

PROPERTIES:
Form:
Liquid (Benzoflex 2-45)
Color:
APHA 100 max. (Benzoflex 2-45)
Odor:
Mild ester (Benzoflex 2-45)
Composition:
98% purity (Benzoflex 2-45)
Solubility:
Sol. in aliphatic hydrocarbons (Benzoflex 2-45)
Sol. in aromatic hydrocarbons (Benzoflex 2-45)
Sol. < 0.01% in water (Benzoflex 2-45)
M.W.:
314.3 (Benzoflex 2-45)
Sp.gr.:
1.178 (Benzoflex 2-45)
Density:
9.8 lb/gal (Benzoflex 2-45)
Visc.:
110 cps (Benzoflex 2-45)
F.P.:
16 and 28 C (two crystal forms) (Benzoflex 2-45)
B.P.:
240 C (5 mm Hg) (Benzoflex 2-45)
Flash Pt.:
199 C (COC) (Benzoflex 2-45)
Hydroxyl No.:
15 max. (Benzoflex 2-45)
Ref. Index:
1.5424 (Benzoflex 2-45)
Dielectric Constant:
7.16 (10 kHz) (Benzoflex 2-45)

Diethylene glycol dibenzoate *(cont'd.)*

TOXICITY/HANDLING:
Repeated exposure may cause defatting of skin and mild irritation to the eyes and nose; avoid skin and eye contact and inhalation of mist; 5.44 g/kg body weight (Oral LD_{50} rats) (Benzoflex 2-45)

STORAGE/HANDLING:
Due to potential freezing at R.T., bulk shipping and storage must be in heated tanks; combustible (Benzoflex 2-45)

STD. PKGS.:
55-gal (525 lb net) epoxy-phenolic-lined steel drums and bulk (Benzoflex 2-45)

Diethyl phthalate *(CTFA)*

SYNONYMS:
1,2-Benzenedicarboxylic acid, diethyl ester
DEP
Diethyl, 1,2-benzenedicarboxylate
Ethyl phthalate

EMPIRICAL FORMULA:
$C_{12}H_{14}O_4$

STRUCTURE:

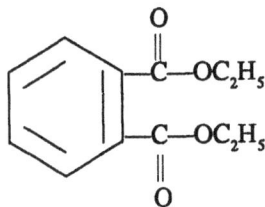

CAS No.:
84-66-2

TRADENAME EQUIVALENTS:
Kodaflex DEP [Eastman]
Generically sold by:
[Ashland; Dynamit-Nobel; Kay-Fries; Mobay; Morflex; Unitex]

CATEGORY:
Plasticizer, solvent, wetting agent, dispersant, fixative

APPLICATIONS:
Cosmetic industry preparations: perfumery (generic)
Farm products: insecticides/pesticides (generic)

Industrial applications: cellulosics (generic); dyes and pigments (Kodaflex DEP); plastics (generic); polymers/polymerization (Kodaflex DEP); resins (Kodaflex DEP); rocket propellants (generic)

PROPERTIES:
Form:
   Liquid (generic; Kodaflex DEP)
Color:
   Water-white (generic)
   Pt-Co 10 ppm (Kodaflex DEP)
Odor:
   Odorless (generic)
Taste:
   Bitter (generic)
Composition:
   99.0% assay min. (Kodaflex DEP)
Solubility:
   Miscible with alcohols (generic)
   Partly miscible with aliphatic solvents (generic)
   Miscible with aromatic hydrocarbons (generic)
   Miscible with esters (generic)
   Miscible with ketones (generic)
   Sol. in water: 0.12 g/l @ 20 C (Kodaflex DEP); insol. (generic)
M.W.:
   222 (theoret.) (Kodaflex DEP)
Sp.gr.:
   1.120 (20/20 C) (Kodaflex DEP); (25/25 C) (generic)
Density:
   1.12 kg/l (Kodaflex DEP)
   9.31 lb/gal (20 C) (generic)
Visc.:
   9.5 cP (D445) (Kodaflex DEP)
   31.3 cstk (0 C) (generic)
F.P.:
   < −50 C (Kodaflex DEP)
   −40.5 C (generic)
B.P.:
   298 C (generic); (760 mm) (Kodaflex DEP)
Flash Pt.:
   161 C (COC) (Kodaflex DEP)
   162.7 C (OC) (generic)
Fire Pt.:
   171 C (Kodaflex DEP)

Diethyl phthalate *(cont'd.)*

Stability:
Good stability to heat and uv light (Kodaflex DEP)
Ref. Index:
1.4990 (Kodaflex DEP)
1.5002 (25 C) (generic)
Surface Tension:
37.5 dynes/cm (20 C) (generic)
ELECTRICAL PROPERTIES:
Dissip. Factor:
0.10×10^{-2} (1 MHz, D150) (Kodaflex DEP)
Dielec. Constant:
6.7 (1 MHz, D150) (Kodaflex DEP)
Volume Resistivity:
1.45×10^{9} ohm-cm (D257) (Kodaflex DEP)
TOXICITY/HANDLING:
May cause possible transient irritation of nose and throat from inhalation of heated vapor; use with adequate ventilation (Kodaflex DEP)
Toxic by ingestion and inhalation; strong irritant to eyes and mucous membranes (generic)
STORAGE/HANDLING:
Combustible (generic)

Dihydroabietyl alcohol *(CTFA)*

SYNONYMS:
Dodecahydro-1,4a-dimethyl-7-(1-methylethyl)-1-phenanthrenemethanol
Hydroabietyl alcohol
1-Phenanthrenemethanol, dodecahydro-1,4a-dimethyl-7-(1-methylethyl)-
EMPIRICAL FORMULA:
$C_{20}H_{34}O$
STRUCTURE:

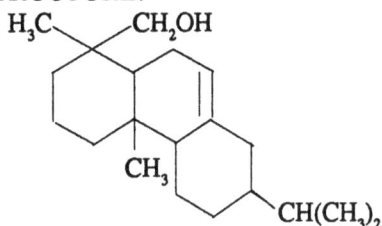

CAS No.:
26266-77-3

TRADENAME EQUIVALENTS:
Abitol [Hercules]

CATEGORY:
Plasticizer, tackifier, chemical intermediate, softener

APPLICATIONS:
Industrial applications: adhesives; lacquers; plastics; polishes and waxes; polymers/ polymerization; printing inks; resins; rubber

PROPERTIES:

Form:
Viscous liquid
Color:
Water-white
Odor:
Low
Composition:
90% solids in toluene or xylene
100% conc.
Solubility:
Sol. in alcohols
Sol. in aliphatic hydrocarbons
Sol. in aromatic hydrocarbons
Sol. in chlorinated solvents
Sol. in esters
Sol. in ketones
Sol. in terpene hydrocarbons
Sp.gr.:
1.008
Density:
8.4 lb/gal
Visc.:
400 poises (40 C)
Flash Pt.:
185 C (COC)
Acid No.:
0.3
Saponification No.:
15

Dihydroabietyl alcohol *(cont'd.)*

Stability:
   Resistant to oxidation
Ref. Index:
   1.5262 (20 C)
STD. PKGS.:
   208-1 (195-kg net) metal drums

2,4-Dihydroxybenzophenone

SYNONYMS:
   Benzophenone-1 (CTFA)
   Benzoresorcinol
   4-Benzoyl resorcinol
   (2,4-Dihydroxyphenyl) phenylmethanone
   Methanone, (2,4-dihydroxyphenyl) phenyl-
EMPIRICAL FORMULA:
   $C_{13}H_{10}O_3$
STRUCTURE:

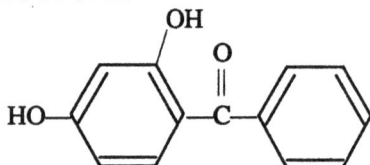

CAS No.:
   131-56-6
TRADENAME EQUIVALENTS:
   Mixxim LS-24 [Fairmount]
   Syntase 100 [Neville]
   Uvasorb 20H [3-V Chem.]
   Uvinul 400 [BASF Wyandotte]
CATEGORY:
   UV stabilizer, UV absorber, stabilizer
APPLICATIONS:
   Cleansers: (Uvinul 400)
   Cosmetic industry preparations: (Uvinul 400); alcohol-based cosmetics (Syntase 100);
      toiletries (Uvinul 400)
   Industrial applications: adhesives (Syntase 100; Uvinul 400); cellulosics (Mixxim LS-
      24; Uvasorb 20H); latexes (Syntase 100); paint/lacquer/varnish mfg. (Syntase 100;

96

2,4-Dihydroxybenzophenone *(cont'd.)*

Uvinul 400); photography (Uvinul 400); plastics (Mixxim LS-24; Syntase 100; Uvasorb 20H); resins (Syntase 100); rubber (Syntase 100)

PROPERTIES:

Form:

Powder (Mixxim LS-24; Syntase 100; Uvasorb 20H; Uvinul 400)

Color:

Pale yellow (Syntase 100)

Composition:

99.5% min. purity (Syntase 100)

100% active (Uvinul 400)

Solubility:

Sol. in alcohols (Syntase 100; Uvinul 400)

Sol. in aliphatic hydrocarbons (Syntase 100)

Sol. in aromatic hydrocarbons (Syntase 100)

Sol. in esters (Syntase 100; Uvinul 400)

Sol. in ethers (Uvinul 400)

Sol. in ketones (Syntase 100; Uvinul 400)

Sol. in most organic solvents (Syntase 100)

Insol. in water (Syntase 100; Uvinul 400)

M.W.:

214 (Syntase 100)

M.P.:

136–149 C (Mixxim LS-24; Uvasorb 20H)

140–142 C (Syntase 100)

Stability:

Stable for a reasonable period of time at temps. of 230–240 C (Syntase 100)

Dihydroxyethyl tallow glycinate *(CTFA)*

SYNONYMS:

Tallow dihydroxyethyl betaine

STRUCTURE:

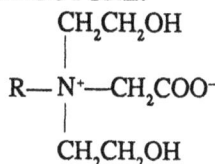

97

Dihydroxyethyl tallow glycinate *(cont'd.)*

TRADENAME EQUIVALENTS:
  Jortaine TM [Jordan]
  Mirataine TM [Miranol]
CATEGORY:
  Thickener; conditioner
APPLICATIONS:
  Cosmetic industry preparations: shampoos (Jortaine TM)
  Industrial applications: (Mirataine TM)
PROPERTIES:
Form:
  Viscous liquid (Mirataine TM)
  Viscous liquid to gel (Jortaine TM)
Composition:
  35% active (Mirataine TM)
  39% active (Jortaine TM)
Solubility:
  Sol. in water (Mirataine TM)
Ionic Nature:
  Amphoteric (Mirataine TM)

2,2´-Dihydroxy-4-methoxybenzophenone

SYNONYMS:
  Benzophenone-8 (CTFA)
  Dioxybenzone
  (2-Hydroxy-4-methoxyphenyl) (2-hydroxyphenyl) methanone
  Methanone, (2-Hydroxy-4-methoxyphenyl) (2-hydroxyphenyl)-
EMPIRICAL FORMULA:
  $C_{14}H_{12}O_4$
STRUCTURE:

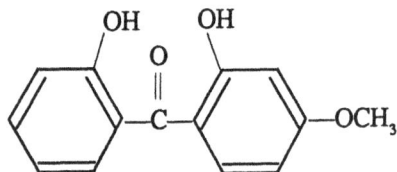

CAS No.:
  131-53-3

2,2´-Dihydroxy-4-methoxybenzophenone *(cont'd.)*

TRADENAME EQUIVALENTS:
Cyasorb UV 24 [Amer. Cyanamid]
CATEGORY:
Stabilizer, uv absorber
APPLICATIONS:
Industrial applications: coatings; plastics; resins
PROPERTIES:
Form:
Powder
Color:
Pale yellow
Solubility:
Sol. 46.6 g/100 g in benzene
Sol. 22.2 g/100 g in carbon tetrachloride
Sol. 31.1 g/100 g in di-2-ethylhexyl phthalate
Sol. 21.4 g/100 g in 95% ethanol
Sol. 2.3 g/100 g in n-hexane
Sol. 55.3 g/100 g in MEK
M.W.:
244.2
Sp.gr.:
1.382
Bulk Density:
3.5 lb/gal
B.P.:
160–170 C (0.5–1.0 mm Hg)
M.P.:
68–70 C
TOXICITY/HANDLING:
Considered practically nontoxic in single doses by mouth; not irritating to rabbit eyes
or skin

Diisobutyl adipate (CTFA)

SYNONYMS:
Bis (2-methylpropyl) hexanedioate
DIBA
Diisobutyl hexanedioate
Hexanedioic acid, bis (2-methylpropyl) ester
Hexanedioic acid, diisobutyl ester

Diisobutyl adipate *(cont'd.)*

EMPIRICAL FORMULA:
$C_{14}H_{26}O_4$

STRUCTURE:

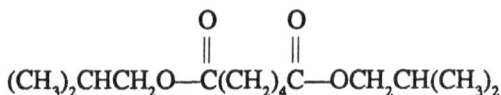

$$(CH_3)_2CHCH_2O-\overset{\overset{O}{\|}}{C}(CH_2)_4\overset{\overset{O}{\|}}{C}-OCH_2CH(CH_3)_2$$

CAS No.:
141-04-8

TRADENAME EQUIVALENTS:
Hatcol DIBA [Hatco]
Plasthall DIBA [C.P. Hall]
Radia 7197 [Oleofina]
    Generically sold by: [Croda]

CATEGORY:
Plasticizer, processing aid, softener, chemical intermediate, lubricant

APPLICATIONS:
Food applications: food packaging (Hatcol DIBA; Plasthall DIBA)
Industrial applications: adhesives (Hatcol DIBA); chemical synthesis (Radia 7197);
    lamination (Radia 7197); lubricating/cutting oils (Radia 7197); plastics (generic—
    Croda; Hatcol DIBA; Plasthall DIBA); rubber (generic—Croda; Hatcol DIBA;
    Plasthall DIBA); rust inhibitors (Radia 7197); textile/leather processing (Radia
    7197)

PROPERTIES:
Form:
Liquid (generic—Croda; Hatcol DIBA)
Clear liquid (Plasthall DIBA)
Color:
APHA 20 max. (Hatcol DIBA)
APHA 30 (Plasthall DIBA)
Odor:
Mild, characteristic (Hatcol DIBA; Plasthall DIBA)
Solubility:
Sol. in min. oils (Radia 7197)
Sol. in most solvents (Radia 7197)
Sol. in veg. oils (Radia 7197)
M.W.:
258 (Hatcol DIBA; Radia 7197)
Sp.gr.:
0.939 (37.8 C) (Radia 7197)
0.950 ± 0.002 (Hatcol DIBA; Plasthall DIBA)
Visc.:
3.9 cps (37.8 C) (Radia 7197)

7 cps (Hatcol DIBA)
13 cps (Plasthall DIBA)
B.P.:
145–163 C (4 mm) (Hatcol DIBA; Plasthall DIBA)
M.P.:
–20 C (Hatcol DIBA; Plasthall DIBA)
Flash Pt.:
153 C (COC) (Radia 7197)
Cloud Pt.:
–30 C (Radia 7197)
Ref. Index:
1.4300 (Radia 7197)
Surface Tension:
30.70 dynes/cm (Radia 7197)
TOXICITY/HANDLING:
Nontoxic (Hatcol DIBA; Plasthall DIBA)

Diisodecyl adipate

SYNONYMS:
DIDA
STRUCTURE:
$C_{10}H_{21}OOC(CH_2)_4COOC_{10}H_{21}$
TRADENAME EQUIVALENTS:
Hatcol DIDA [Hatco]
Hexaflex DIDA [Hexagon]
Monoplex DDA [Rohm & Haas]
Morflex 330 [Morflex]
Plasthall DIDA [C.P. Hall]
Staflex DIDA [Reichhold]
CATEGORY:
Plasticizer, softener
APPLICATIONS:
Industrial applications: coated fabrics (Plasthall DIDA); foam applications (Plasthall DIDA); lubricating/cutting oils (Hexaflex DIDA); plastics/plastisols (Hatcol DIDA; Hexaflex DIDA; Monoplex DDA; Plasthall DIDA); rubber (Monoplex DDA; Plasthall DIDA)

Diisodecyl adipate *(cont'd.)*

PROPERTIES:
Form:
  Clear liquid (Monoplex DDA; Plasthall DIDA)
Color:
  Water-white (Plasthall DIDA)
  APHA 10 (Staflex DIDA)
  APHA 15 (Monoplex DDA)
Odor:
  Mild (Plasthall DIDA)
M.W.:
  426 (Hatcol DIDA; Hexaflex DIDA; Monoplex DDA)
  427 (Plasthall DIDA; Staflex DIDA)
Sp.gr.:
  0.914 (25/15 C) (Monoplex DDA)
  0.918 (Plasthall DIDA); (23 C) (Staflex DIDA)
Density:
  7.6 lb/gal (Monoplex DDA; Staflex DIDA)
Visc.:
  0.23 poise (Monoplex DDA)
  27 cps (Plasthall DIDA)
  50 cSt (kinematic, 0 C) (Staflex DIDA)
F.P.:
  < −60 F (Monoplex DDA)
B.P.:
  239 C (5 mm) (Plasthall DIDA)
M.P.:
  −50 C (Monoplex DDA)
  > −50 C (Plasthall DIDA)
Pour Pt.:
  < −80 F (Staflex DIDA)
Flash Pt.:
  227 C (Monoplex DDA)
  445 F (Staflex DIDA)
Cloud Pt.:
  < −80 F (Staflex DIDA)
Acid No.:
  0.1 (Monoplex DDA)
Saponification No.:
  261 (Monoplex DDA)
Ref. Index:
  1.449 (Monoplex DDA)
  1.4513 (23 C) (Staflex DIDA

SYNONYMS:
DIDP

STRUCTURE:
$C_6H_4(COOC_{10}H_{21})_2$

TRADENAME EQUIVALENTS:
Hatcol DIDP [Hatco]
Hatcol DIDP-EG [Hatco] (electrical grade)
Hexaflex DIDP [Hexagon]
Jayflex DIDP, DIDP-E [Exxon]
Palatinol DIDP [Badische]
Palatinol DIDP-E [Badische] (with 0.25% Bisphenol-A)
Polycizer DIDP [Harwick]
PX-120 [USS Chem.]
Staflex DIDP [Reichhold]
    Generically sold by:
        [Ashland; C.P. Hall; Monsanto; Unocal]

CATEGORY:
Plasticizer, softener

APPLICATIONS:
Automotive applications: (Jayflex DIDP, DIDP-E; Palatinol DIDP, DIDP-E; Staflex DIDP)
Cosmetics applications (Staflex DIDP)
Industrial applications: adhesives (Staflex DIDP); coated fabrics (PX-120); construction (Staflex DIDP); consumer goods (Staflex DIDP); electrical insulation (generic, generic—C.P. Hall; Hatcol DIDP-EG; Hexaflex DIDP; Jayflex DIDP, DIDP-E; Palatinol DIDP, DIDP-E; Polycizer DIDP; Staflex DIDP); extruded goods (PX-120); flooring (PX-120); foam applications (generic—C.P. Hall; Polycizer DIDP); plastics/plastisols (generic, generic—C.P. Hall; Hatcol DIDP; Hexaflex DIDP; Jayflex DIDP, DIDP-E; Palatinol DIDP, DIDP-E; Polycizer DIDP; PX-120; Staflex DIDP); profile extrusions (Jayflex DIDP, DIDP-E); rubber (generic—C.P. Hall; Polycizer DIDP; Staflex DIDP); sheeting (PX-120)

PROPERTIES:
Form:
Liquid (generic—C.P. Hall)
Clear liquid (generic)
Color:
Water-white (generic—C.P. Hall)
APHA 25 (Staflex DIDP)
APHA 35 (Polycizer DIDP)
APHA 35 max. (Palatinol DIDP)

Diisodecyl phthalate (cont'd.)

Odor:
  Mild (generic)
  Mild, characteristic (generic—C.P. Hall; Palatinol DIDP)
Composition:
  99.0% min. ester content (Palatinol DIDP)
Solubility:
  < 0.01% in water @ 20 C (Palatinol DIDP)
M.W.:
  446 (generic—C.P. Hall; Hatcol DIDP; Staflex DIDP)
  446.7 (Palatinol DIDP)
  447 (generic—Monsanto; PX-120)
Sp.gr.:
  0.963–0.969 (20/20 C) (Palatinol DIDP)
  0.964 (23 C) (Staflex DIDP)
  0.966 (20/20 C) (generic, generic—Ashland, C.P. Hall, Unocal; Polycizer DIDP)
Density:
  8.0 lb/gal (generic; Staflex DIDP)
Visc.:
  86 cps (generic—C.P. Hall)
  108 cp (20 C) (generic)
  114.4 cps (20 C) (Palatinol DIDP)
  700 cSt (kinematic, 0 C) (Staflex DIDP)
F.P.:
  –50 C (generic)
B.P.:
  248 C (5 mm) (generic—C.P. Hall)
  250–257 C (4 mm) (generic)
  260 C (5 mm) (Palatinol DIDP)
M.P.:
  –53 C (generic—C.P. Hall)
Pour Pt.:
  –40 C (Palatinol DIDP; Polycizer DIDP)
  –40 F (Staflex DIDP)
Flash Pt.:
  232 C (generic; Polycizer DIDP); (COC) (Palatinol DIDP)
  465 F (Staflex DIDP)
Cloud Pt.:
  –8 F (Staflex DIDP)
Acid No.:
  0.07 max. (Palatinol DIDP)
Storage Stability:
  Unlimited storage life (Palatinol DIDP)

Diisodecyl phthalate *(cont'd.)*

Ref. Index:
  1.483 (25 C) (generic)
  1.4835 (20 C) (Palatinol DIDP); (25 C) (Polycizer DIDP)
  1.4836 (23 C) (Staflex DIDP)
STD. PKGS.:
  Avail. bulk (Palatinol DIDP)

Diisononyl adipate

SYNONYMS:
  DINA
STRUCTURE:
  $C_9H_{19}OOC(CH_2)_4COOC_9H_{19}$
TRADENAME EQUIVALENTS:
  Adimoll DN [Bayer AG]
  Jayflex DINA [Exxon]
  PX-209 [USS Chem.]
  Staflex DINA [Reichhold]
CATEGORY:
  Plasticizer, heat and light stabilizer
APPLICATIONS:
  Food applications: food wrap film (Jayflex DINA; PX-209; Staflex DINA)
  Industrial applications: coating (Adimoll DN); consumer goods (Jayflex DINA); electrical insulation (Adimoll DN; Jayflex DINA; PX-209; Staflex DINA); film and sheet (Adimoll DN; Jayflex DINA); gasketing (PX-209; Staflex DINA); hose and tubing (PX-209); plastics/plastisols (Adimoll DN); rubber (Adimoll DN); textile/ leather processing (Adimoll DN); upholstery (Jayflex DINA; PX-209; Staflex DINA)
PROPERTIES:
Color:
  APHA 10 (Staflex DINA)
  Hazen ≤ 50 (Adimoll DN)
M.W.:
  398 (PX-209; Staflex DINA)
  399 (Jayflex DINA)
Sp.gr.:
  0.919 (23 C) (Staflex DINA)
Density:
  0.910–0.920 g/cc (Adimoll DN)
  7.6 lb/gal (Staflex DINA)

Diisononyl adipate *(cont'd.)*

Visc.:
  22–25 mPa•s (20 C) (Adimoll DN)
  66 cSt (kinematic, 0 C) (Staflex DINA)
Pour Pt.:
  –59 F (Staflex DINA)
Flash Pt.:
  200–210 C (OC) (Adimoll DN)
  444 F (Staflex DINA)
Cloud Pt.:
  < –80 F (Staflex DINA)
Acid No.:
  ≤ 0.1 (Adimoll DN)
Saponification No.:
  270–290 (Adimoll DN)
Ref. Index:
  1.4465–1.4490 (20 C) (Adimoll DN)
  1.4502 (23 C) (Staflex DINA)

Diisononyl phthalate

SYNONYMS:
  DINP
TRADENAME EQUIVALENTS:
  Jayflex DINP [Exxon]
  Jayflex DINP-E [Exxon] (electrical grade, with Bisphenol-A)
  Palatinol N [Badische]
CATEGORY:
  Plasticizer
APPLICATIONS:
  Industrial applications: construction (Palatinol N); electrical insulation (Jayflex DINP, DINP-E); film and sheeting (Jayflex DINP); flooring (Jayflex DINP); hose and profiles (Jayflex DINP); lubricating/cutting oils; plastics/plastisols (Jayflex DINP, DINP-E); spray coating (Palatinol N)
PROPERTIES:
Form:
  Clear liquid (Palatinol N)
Color:
  APHA 30 max. (Palatinol N)

Odor:
Mild, characteristic (Palatinol N)
Composition:
99.0% min. ester content (Palatinol N)
Solubility:
Sol. < 0.01% in water (20 C) (Palatinol N)
M.W.:
418 (Palatinol N)
419 (Jayflex DINP, DINP-E)
Sp.gr.:
0.973–0.978 (20/20 C) (Palatinol N)
Visc.:
80 cps (20 C) (Palatinol N)
B.P.:
237 C (5 mm Hg) (Palatinol N)
Pour Pt.:
–49 C (Palatinol N)
Flash Pt.:
> 200 C (COC) (Palatinol N)
Acid No.:
0.07 max. (Palatinol N)
Stability:
Good (Palatinol N)
Storage Stability:
Can be kept indefinitely (Palatinol N)
Ref. Index:
1.484–1.488 (Palatinol N)
STORAGE/HANDLING:
Incompatible with oxidizing materials (Palatinol N)

Diisooctyl adipate

SYNONYMS:
DIOA
STRUCTURE:
$C_8H_{17}OOC(CH_2)_4COOC_8H_{17}$
TRADENAME EQUIVALENTS:
Monoplex DIOA [Rohm & Haas]
Staflex DIOA [Reichhold]

Diisooctyl adipate *(cont'd.)*

CATEGORY:
Plasticizer, heat and light stabilizer
APPLICATIONS:
Industrial applications: elastomers (Staflex DIOA); plastics/plastisols (Monoplex DIOA; Staflex DIOA)
PROPERTIES:
Color:
APHA 10 (Monoplex DIOA)
M.W.:
370 (Monoplex DIOA)
Sp.gr.:
0.924 (25/15 C) (Monoplex DIOA)
Density:
7.7 lb/gal (Monoplex DIOA)
Visc.:
0.17 poise (Monoplex DIOA)
F.P.:
< −60 F (Monoplex DIOA)
Acid No.:
0.08 (Monoplex DIOA)
Saponification No.:
300 (Monoplex DIOA)
Ref. Index:
1.446 (Monoplex DIOA)

Diisooctyl phthalate

SYNONYMS:
DIOP
STRUCTURE:
$(C_8H_{17}COO)_2C_6H_4$
CAS No.:
27554-26-3
TRADENAME EQUIVALENTS:
Hexaflex DIOP [Hexagon]
Staflex DIOP [Reichhold]
Generically sold by: [C.P. Hall]
CATEGORY:
Plasticizer

APPLICATIONS:
Industrial applications: calendered goods (Hexaflex DIOP); chlorinated rubbers (generic); coatings (Hexaflex DIOP); extruded goods (Hexaflex DIOP); film and sheeting (Hexaflex DIOP); molded goods (Hexaflex DIOP); plastics (generic; Hexaflex DIOP; Staflex DIOP); rubber (Staflex DIOP)

PROPERTIES:
Form:
Viscous liquid (generic)
Color:
Water-white (generic)
APHA 25 (Staflex DIOP)
Odor:
Mild, characteristic (generic)
Solubility:
Insol. in water (generic)
M.W.:
390 (generic; Hexaflex DIOP; Staflex DIOP)
Sp.gr.:
0.980–0983 (20/20 C) (generic)
0.985 (23 C) (Staflex DIOP)
Density:
8.2 lb/gal (Staflex DIOP)
Visc.:
≈ 83 cps (generic)
280 cSt (kinematic, 0 C) (Staflex DIOP)
B.P.:
≈ 228 C (generic)
M.P.:
≈ –50 C (generic)
Pour Pt.:
–50 F (Staflex DIOP)
Flash Pt.:
425 F (Staflex DIOP)
450 F (generic)
Cloud Pt.:
–36 F (Staflex DIOP)
Ref. Index:
1.4850 (23 C) (Staflex DIOP)

Diisopropyl adipate (CTFA)

SYNONYMS:
Bis (1-methylethyl) hexanedioate
Diester of isopropyl alcohol and adipic acid
Hexanedioic acid, bis (1-methylethyl) ester

EMPIRICAL FORMULA:
$C_{12}H_{22}O_4$

STRUCTURE:

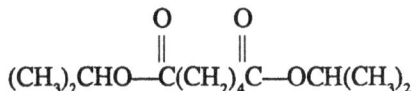

CAS No.:
6938-94-9

TRADENAME EQUIVALENTS:
Ceraphyl 230 [VanDyk]
Crodamol DA [Croda]
Lexol DIA [Inolex]
Radia 7194 [Oleofina S.A.]
Schercemol DIA [Scher]

CATEGORY:
Plasticizer, emollient, solvent, solubilizer, coupling agent, penetrant, lubricant, conditioner, superfatting agent, chemical intermediate

APPLICATIONS:
Bath products: bath oils (Ceraphyl 230; Schercemol DIA)
Cosmetic industry preparations: (Crodamol DA; Radia 7194; Schercemol DIA); creams and lotions (Crodamol DA; Schercemol DIA); hair sprays (Crodamol DA); makeup (Crodamol DA; Schercemol DIA); perfumery (Crodamol DA; Lexol DIA); shampoos (Crodamol DA); shaving preparations (Crodamol DA; Schercemol DIA); skin preparations (Crodamol DA; Lexol DIA)
Industrial applications: industrial processing (Radia 7194); lamination (Radia 7194); lubricating/cutting oils (Radia 7194); textile/leather processing (Radia 7194)
Pharmaceutical applications: (Radia 7194)

PROPERTIES:
Form:
Clear liquid (Schercemol DIA)
Very low viscosity liquid (Crodamol DA)
Low viscosity oil (Lexol DIA)
Color:
Water-white (Crodamol DA; Schercemol DIA)
Odor:
Neutral (Crodamol DA)
Faint ester (Schercemol DIA)

Solubility:
Miscible with acetoglycerides (Crodamol DA)
Sol. in alcohols (Lexol DIA; Schercemol DIA); sol. in aq. alcohol sol'ns. (Crodamol DA; Schercemol DIA)
Sol. in aliphatic hydrocarbons (Schercemol DIA)
Sol. in aromatic hydrocarbons (Schercemol DIA)
Sol. in chlorinated hydrocarbons (Schercemol DIA)
Sol. in esters (Schercemol DIA)
Sol. in 50% ethanol (Ceraphyl 230)
Sol. in higher glycols (Schercemol DIA)
Sol. in isopropyl myristate (Ceraphyl 230)
Sol. in min. oils (Ceraphyl 230; Lexol DIA; Radia 7194; Schercemol DIA); miscible (Crodamol DA)
Sol. in natural oils (Schercemol DIA)
Sol. in most organic solvents (Schercemol DIA)
Sol. in propylene glycol (Ceraphyl 230); partly sol. (Schercemol DIA)
Miscible with low-viscosity silicone oils (Crodamol DA)
Sol. in most solvents (Radia 7194)
Miscible with synthetic esters (Crodamol DA)
Miscible with triacetin (Crodamol DA)
Sol. in veg. oils (Lexol DIA; Radia 7194); miscible (Crodamol DA)
Insol. in water (Ceraphyl 230; Schercemol DIA)
M.W.:
230 avg. (Radia 7194)
Sp.gr.:
0.938 (37.8 C) (Radia 7194)
0.960 ± 0.01 (Schercemol DIA)
Density:
8.0 lb/gal (Schercemol DIA)
Visc.:
2.80 cps (37.8 C) (Radia 7194)
Flash Pt.:
131 C (COC) (Radia 7194)
> 170 C (Schercemol DIA)
Cloud Pt.:
−14.5 C (Radia 7194)
Acid No.:
0.5 max. (Schercemol DIA)
Iodine No.:
Nil (Schercemol DIA)
Saponification No.:
487 ± 10 (Schercemol DIA)

Diisopropyl adipate *(cont'd.)*

Ref. Index:
  1.423 ± 0.001 (Schercemol DIA)
  1.4237 (Radia 7194)
Surface Tension:
  30.00 dynes/cm (Radia 7194)
STD. PKGS.:
  55-gal (450 lb net) epoxy-lined, bung-head steel drums (Schercemol DIA)

Dilauryl thiodipropionate *(CTFA)*

SYNONYMS:
  Didodecyl 3,3′-thiodipropionate
  DLTDP
  Propanoic acid, 3,3′-thiobis-, didodecyl ester
EMPIRICAL FORMULA:
  $C_{30}H_{58}O_4S$
STRUCTURE:

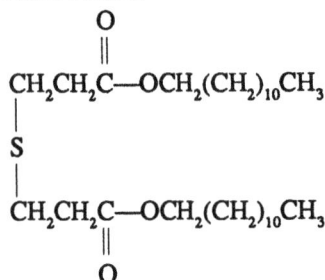

CAS No.:
  123-28-4
TRADENAME EQUIVALENTS:
  Argus DLTDP [Witco/Argus]
  Carstab DLTDP [Cincinnati Milacron; Morton Thiokol/Carstab]
  Cyanox LTDP [Amer. Cyanamid]
  Evanstab 12 [Evans Chemetics]
CATEGORY:
  Antioxidant, stabilizer, plasticizer, lubricant, preservative, thioester synergist
APPLICATIONS:
  Automotive casing (Cyanox LTDP)
  Food applications: food additives (Cyanox LTDP; Evanstab 12); food packaging
    (Cyanox LTDP; Evanstab 12)

112

Dilauryl thiodipropionate (cont'd.)

Industrial applications: adhesives/sealants (Cyanox LTDP); appliance casing (Cyanox LTDP); lubricating/cutting oils (Cyanox LTDP; Evanstab 12); plastics (Argus DLTDP; Carstab DLTDP; Cyanox LTDP; Evanstab 12); rubber (Argus DLTDP; Carstab DLTDP; Cyanox LTDP; Evanstab 12)

PROPERTIES:
Form:
Free-flowing crystalline flakes or powder (Evanstab 12)
Flakes (Carstab DLTDP)
Waxy crystalline flakes (Cyanox LTDP)
Free-flowing powder (Argus DLTDP)
Color:
White (Argus DLTDP; Carstab DLTDP; Cyanox LTDP; Evanstab 12)
Composition:
99% assay min. (Evanstab 12)
99% purity (Carstab DLTDP)
Solubility:
Sol. 51.2 g/100 g acetone (Cyanox LTDP); 51 g (Evanstab 12)
Slightly sol. in alcohols (Argus DLTDP)
Sol. in benzene (Argus DLTDP); sol. 30 g/100 g (Carstab DLTDP)
Sol. in chloroform (Argus DLTDP)
Sol. 7 g/100 g ethanol (Carstab DLTDP); sol. < 1 g/100 g 95% ethanol (Evanstab 12); < 0.5 g in 95% ethanol (Cyanox LTDP)
Sol. 39.4 g/100 g ethyl acetate (Cyanox LTDP); 39 g (Evanstab 12); slightly sol. (Argus DLTDP)
Sol. 41 g/100 g *n*-heptane (Evanstab 12); 40.5 g (Cyanox LTDP)
Sol. 51.2 g/100 g MEK (Cyanox LTDP); 51 g (Evanstab 12)
Sol. in toluene (Argus DLTDP); sol. 40 g/100 g toluene (90 C) (Evanstab 12); 39.4 g/100 g toluene (9 C) (Cyanox LTDP)
Slightly sol. in water (Argus DLTDP); insol. (Carstab DLTDP)
M.W.:
514 (Argus DLTDP; Carstab DLTDP; Cyanox LTDP; Evanstab 12)
Sp.gr.:
0.896 (80 C) (Carstab DLTDP)
0.915 (Cyanox LTDP)
1.01 (Argus DLTDP)
F.P.:
40 C min. (Evanstab 12)
41 C (Carstab DLTDP)
Set Pt.:
39 C (Argus DLTDP)
40 C (Cyanox LTDP)

Dilauryl thiodipropionate (cont'd.)

Acid No.:
   0.5 (Carstab DLTDP)
   1.0 (Cyanox LTDP)
   1.0 max. (Evanstab 12)
Stability:
   Good color stability during aging; improved melt stability during polymer fabrication; improved retention of physical and electrical properties during polymer fabrication and during environmental aging of the polymer (Evanstab 12)
   Thermally stable; resistant to discoloration and volatilization; relatively inert (Carstab DLTDP)
TOXICITY/HANDLING:
   Low oral and dermal toxicity (Carstab DLTDP)
STD. PKGS.:
   55-gal (200 lb net) polyethylene-lined fiber drums (Carstab DLTDP)

Dimethyl phthalate (CTFA)

SYNONYMS:
   1,2-Benzenedicarboxylic acid, dimethyl ester
   Dimethyl 1,2-benzenedicarboxylate
EMPIRICAL FORMULA:
   $C_{10}H_{10}O_4$
STRUCTURE:

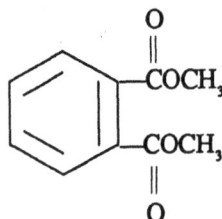

CAS No.:
   131-11-3
TRADENAME EQUIVALENTS:
   DMP [Dynamit-Nobel]
   Kodaflex DMP [Eastman]
   Staflex DMP [Reichhold]
CATEGORY:
   Plasticizer, solvent

114

APPLICATIONS:
Industrial applications: cellulosics (DMP; Kodaflex DMP; Staflex DMP); industrial processing (DMP); plastics (Kodaflex DMP; Staflex DMP); printing inks (Kodaflex DMP)

PROPERTIES:
Form:
Liquid (Kodaflex DMP)
Clear liquid (DMP)
Color:
Colorless (DMP)
Pt-Co 5 ppm (Kodaflex DMP)
Composition:
99.0% ester content min. (Kodaflex DMP)
Solubility:
Sol. 0.45 g/l in water (20 C) (Kodaflex DMP)
M.W.:
194 (theoret.) (Kodaflex DMP)
Sp.gr.:
1.192 (20/20 C) (Kodaflex DMP)
Density:
1.19 kg/l (20 C0 (Kodaflex DMP)
Visc.:
11.0 cP (D445) (Kodaflex DMP)
F.P.:
−1 C (Kodaflex DMP)
B.P.:
284 C (760 mm) (Kodaflex DMP)
Flash Pt.:
157 C (COC) (Kodaflex DMP)
Fire Pt.:
168 C (COC) (Kodaflex DMP)
Stability:
Good stability to heat and uv light (Kodaflex DMP)
Ref. Index:
1.513 (Kodaflex DMP)
Dissip. Factor:
0.0013 (1 mc, D150) (Kodaflex DMP)
Dielec. Const.:
7.5 (1 mc, D150) (Kodaflex DMP)
Vol. Resist.:
1.07×10^9 ohm-cm (D257) (Kodaflex DMP)

Dimethyl phthalate (cont'd.)

TOXICITY/HANDLING:
Eye contact with liquid may cause irritation; use with adequate ventilation (Kodaflex DMP)

Dimyristyl thiodipropionate (CTFA)

SYNONYMS:
Ditetradecyl 3,3′-thiobispropanoate
Propanoic acid, 3,3′-thiobis-, ditetradecyl ester
EMPIRICAL FORMULA:
$C_{34}H_{66}O_4S$
STRUCTURE:

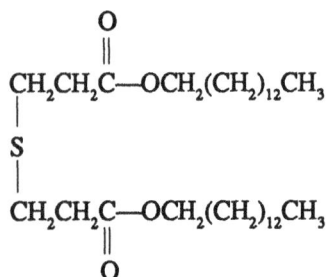

CAS No.:
16545-54-3
TRADENAME EQUIVALENTS:
Argus DMTDP [Witco/Argus]
Carstab DMTDP [Cincinnati Milacron]
Cyanox MTDP [Amer. Cyanamid]
Evanstab 14 [Evans Chemetics]
CATEGORY:
Antioxidant, stabilizer
APPLICATIONS:
Food applications: food additives (Cyanox MTDP); food packaging (Carstab DMTDP; Cyanox MTDP)
Industrial applications: plastics (Argus DMTDP; Carstab DMTDP; Cyanox MTDP; Evanstab 14); rubber (Argus DMTDP)
PROPERTIES:
Form:
Small flake (Carstab DMTDP)
Waxy crystalline flakes (Cyanox MTDP)
Free-flowing crystalline flakes or powder (Evanstab 14)

Color:
White (Carstab DMTDP; Cyanox MTDP; Evanstab 14)
Odor:
Excellent (Carstab DMTDP)
Composition:
97% purity (Cyanox MTDP)
98.5% purity (Evanstab 14)
99% purity (Carstab DMTDP)
Solubility:
Insol. in alcohols (Carstab DMTDP)
Sol. in benzene (Carstab DMTDP)
Sol. in chloroform (Carstab DMTDP)
Insol. in ethyl acetate (Carstab DMTDP)
Sol. in toluene (Carstab DMTDP)
Insol. in water (Carstab DMTDP)
M.W.:
570 (Carstab DMTDP; Evanstab 14)
Sp.gr.:
0.885 (80 C) (Carstab DMTDP)
Bulk Density:
3.99 lb/gal (Carstab DMTDP)
M.P.:
48–50 C (Evanstab 14)
51 C (Carstab DMTDP)
Set Pt.:
47 C (Cyanox MTDP)
Acid No.:
0.4 (Carstab DMTDP)
0.5 (Cyanox MTDP)
1.0 max. (Evanstab 14)
Saponification No.:
280–290 (Evanstab 14)
Stability:
Good color stability during aging; improved melt stability during polymer fabrication; improved retention of physical and electrical properties during polymer fabrication and during environmental aging of the polymer (Evanstab 14)
STD. PKGS.:
200-lb net polyethylene-lined fiber drums (Carstab DMTDP)

Dioctyl adipate (CTFA)

SYNONYMS:
Bis (2-ethylhexyl) adipate
Bis (2-ethylhexyl) hexanedioate
Di (2-ethylhexyl) adipate
DOA
Hexanedioic acid, bis (2-ethylhexyl) ester

EMPIRICAL FORMULA:
$C_{22}H_{42}O_4$

STRUCTURE:

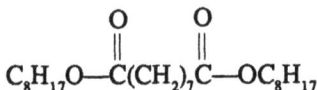

CAS No.:
103-23-1

TRADENAME EQUIVALENTS:
Adimoll DO [Bayer AG]
Hatcol DOA [Hatco]
Kodaflex DOA [Eastman]
Monoplex DOA [C.P. Hall]
Palatinol DOA [Badische]
Plasthall DOA [C.P. Hall]
Polycizer DOA [Harwick]
PX-238 [USS Chemical]
Staflex DOA [Reichhold]
Uniflex DOA [Union Camp]
Wickenol 158 [Wickhen Products]
 Generically sold by:
  [Ashland; Monsanto; Unocal]

CATEGORY:
Plasticizer, softener, emollient

APPLICATIONS:
Cosmetic industry preparations: (Wickenol 158)

Food applications: food packaging (generic—Monsanto; Hatcol DOA; Monoplex DOA; Palatinol DOA; Plasthall DOA; Polycizer DOA; Staflex DOA); indirect food additive (Uniflex DOA)

Industrial applications: adhesives (Hatcol DOA; Monoplex DOA; Plasthall DOA; Polycizer DOA); cellulosics (generic—Monsanto; Kodaflex DOA; Palatinol DOA); coatings (generic—Monsanto; Palatinol DOA); electrical insulation (Kodaflex DOA); plastics (generic—Monsanto; Adimoll DO; Hatcol DOA; Kodaflex DOA; Monoplex DOA; Palatinol DOA; PX-238; Staflex DOA; Uniflex DOA); plastisols (Palatinol DOA; Plasthall DOA; Polycizer DOA; Staflex DOA); rubber (generic—Monsanto; Adimoll DO; Hatcol DOA; Kodaflex DOA; Monoplex

DOA; Palatinol DOA; Plasthall DOA; Polycizer DOA; Uniflex DOA); synthetic lubricants (Uniflex DOA)

Pharmaceutical applications: topical preparations (Wickenol 158)

PROPERTIES:
Form:
Liquid (Hatcol DOA; Kodaflex DOA; Polycizer DOA)
Clear liquid (Monoplex DOA; Plasthall DOA)
Color:
APHA 10 (Staflex DOA; Uniflex DOA)
APHA 15 (Monoplex DOA)
APHA 25 max. (Palatinol DOA)
APHA 50 max. (Hatcol DOA; Plasthall DOA; Polycizer DOA)
Hazen ≤ 20 (Adimoll DO)
Pt-Co 20 ppm (Kodaflex DOA)
Odor:
Mild (Hatcol DOA; Palatinol DOA; Plasthall DOA; Polycizer DOA)
Composition:
99% min. ester content (Kodaflex DOA; Palatinol DOA)
Solubility:
Sol. in alcohols (Wickenol 158)
Sol. in animal oils (Wickenol 158)
Sol. in min. oil (Wickenol 158)
Sol. in veg. oil (Wickenol 158)
Insol. in water (Wickenol 158); sol. < 0.1 g/l @ 20 C (Kodaflex DOA); < 0.1% @ 20 C (Palatinol DOA)
M.W.:
370 (Kodaflex DOA; Monoplex DOA)
370.58 (Palatinol DOA)
370.6 (Hatcol DOA; Polycizer DOA)
371 (generic—Monsanto; PX-238; Staflex DOA)
373 (Plasthall DOA)
Sp.gr.:
0.924 (generic—Unocal); (25/15 C) (Monoplex DOA)
0.924–0.930 (20/20 C) (Palatinol DOA)
0.926 (23 C) (Staflex DOA)
0.927 (Hatcol DOA; Polycizer DOA); (20/20 C) (generic—Ashland; Kodaflex DOA; Uniflex DOA)
Density:
0.923–0.926 g/cc (Adimoll DO)
0.924 kg/l (20 C) (Kodaflex DOA)
7.7 lb/gal (Monoplex DOA; Staflex DOA)

Dioctyl adipate (cont'd.)

Visc.:
12–14 mPa•s (20 C) (Adimoll DO)
12 cps (Kodaflex DOA)
13.5 cps (Hatcol DOA; Polycizer DOA)
13.7 cps (20 C) (Palatinol DOA)
15–18 cps (Monoplex DOA)
18 cps (Plasthall DOA)
12 cSt (Uniflex DOA)
30 cSt (kinematic, 0 C) (Staflex DOA)
F.P.:
< –70 C (Kodaflex DOA)
–70 C (Uniflex DOA); (gel) (Hatcol DOA; Polycizer DOA)
–65 C (gel) (Plasthall DOA)
< –60 F (Monoplex DOA)
B.P.:
214 C (5 mm) (Hatcol DOA; Palatinol DOA; Plasthall DOA; Polycizer DOA)
417 C (760 mm) (Kodaflex DOA)
M.P.:
–50 C (Monoplex DOA)
Pour Pt.:
–75 C (Palatinol DOA)
< –80 F (Staflex DOA)
Flash Pt.:
196 C (COC) (Palatinol DOA)
200–210 C (OC) (Adimoll DO)
206 C (COC) (Kodaflex DOA)
207 C (COC) (Uniflex DOA)
410 F (Staflex DOA)
415 F (Monoplex DOA)
Fire Pt.:
229 C (COC) (Kodaflex DOA)
230 C (COC) (Uniflex DOA)
Cloud Pt.:
< –80 F (Staflex DOA)
Acid No.:
0.07 max. (Palatinol DOA)
0.08 (Monoplex DOA)
< 0.1 (Uniflex DOA)
≤ 0.1 (Adimoll DO)
Saponification No.:
270–290 (Adimoll DO)
300 (Monoplex DOA)

Stability:
Good resistance to weathering; good stability to heat and uv light (Kodaflex DOA)
Chemically stable; resistant to discoloration on extended exposure to heat and uv light (Palatinol DOA)
Storage Stability:
Indefinite storage life (Palatinol DOA)
Ref. Index:
1.445 (Monoplex DOA)
1.4462 (23 C) (Staflex DOA)
1.4465–1.4480 (20 C) (Adimoll DO)
1.4466 (20 C) (Palatinol DOA)
1.447 (Uniflex DOA)
1.4472 (Kodaflex DOA)
Dissipation Factor:
0.04×10^{-2} (1 mc) (Kodaflex DOA)
Dielectric Constant:
3.9 (1 mc) (Kodaflex DOA)
Volume Resistivity:
9.3×10^{11} ohm-cm (Kodaflex DOA)
TOXICITY/HANDLING:
May cause slight eye and skin irritation (Kodaflex DOA)
Avoid repeated/prolonged skin and eye contact; avoid breathing vapors; use with adequate ventilation (Palatinol DOA)
STD. PKGS.:
Drum, tank truck, tank car (Kodaflex DOA)
Bulk (Palatinol DOA)

Dioctyl azelate

SYNONYMS:
Di (2-ethylhexyl) azelate
DOZ
STRUCTURE:
$(CH_2)_7[COOCH_2(C_2H_5)C_4H_9]_2$
TRADENAME EQUIVALENTS:
Plasthall DOZ [C.P. Hall]
Staflex DOZ [Reichhold]
CATEGORY:
Plasticizer, softener

Dioctyl azelate *(cont'd.)*

APPLICATIONS:
Industrial applications: consumer goods (Staflex DOZ); plastics (Plasthall DOZ); rubber (Plasthall DOZ; Staflex DOZ); upholstery (Staflex DOZ)

PROPERTIES:

Form:
Clear liquid (Plasthall DOZ)

Color:
APHA 45 (Staflex DOZ)
APHA 75 max. (Plasthall DOZ)

Odor:
Essentially odorless (Plasthall DOZ)

M.W.:
412 (Plasthall DOZ; Staflex DOZ)

Sp.gr.:
0.915 (Plasthall DOZ)
0.920 (23 C) (Staflex DOZ)

Density:
7.6 lb/gal (Staflex DOZ)

Visc.:
25 cps (Plasthall DOZ)
47 cSt (kinematic, 0 C) (Staflex DOZ)

B.P.:
376 C (Plasthall DOZ)

Pour Pt.:
< -80 F (Staflex DOZ)

Flash Pt.:
390 F (COC) (Plasthall DOZ)
455 F (Staflex DOZ)

Cloud Pt.:
-60 F (Staflex DOZ)

Ref. Index:
1.448 (Plasthall DOZ)
1.4500 (23 C) (Staflex DOZ

Dioctyl maleate *(CTFA)*

SYNONYMS:
Bis (2-ethylhexyl) 2-butenedioate
2-Butenedioic acid, bis (2-ethylhexyl) ester
Di-2-ethylhexyl maleate
DOM

EMPIRICAL FORMULA:
$C_{20}H_{36}O_4$

STRUCTURE:

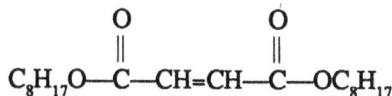

CAS No.:
142-16-5

TRADENAME EQUIVALENTS:
PX-538 [USS Chem.]
Staflex DOM [Reichhold]

CATEGORY:
Internal plasticizer, intermediate

APPLICATIONS:
Household detergents: surfactant intermediate (PX-538; Staflex DOM)
Industrial applications: adhesives (PX-538); coatings (PX-538); oil additives (PX-538); paint mfg. (PX-538; Staflex DOM); paper coatings (PX-538); plastics (PX-538; Staflex DOM); textile/leather processing (PX-538)

PROPERTIES:

Form:
Liquid (PX-538)

Color:
Practically colorless (PX-538)
APHA 50 (Staflex DOM)

M.W.:
340 (PX-538; Staflex DOM)

Sp.gr.:
0.943 (23 C) (Staflex DOM)

Density:
7.8 lb/gal (Staflex DOM)

Visc.:
15 cSt (100 F, kinematic) (Staflex DOM)

Pour Pt.:
< –70 F (Staflex DOM)

Flash Pt.:
370 F (Staflex DOM)

Cloud Pt.:
–50 F (Staflex DOM)

Ref. Index:
1.4540 (23 C) (Staflex DOM)

Dioctyl phthalate (CTFA)

SYNONYMS:
Di (2-ethylhexyl) phthalate
DOP

STRUCTURE:

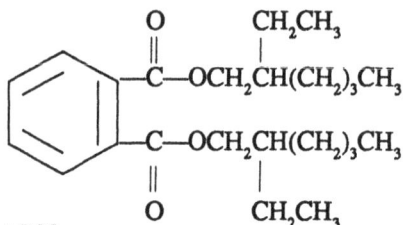

CAS No.:
117-81-7

TRADENAME EQUIVALENTS:
DOP Dry Liquid Conc. [Polymerics]
Hatcol DOP [Hatco]
Hexaflex DOP [Hexagon]
Kodaflex DOP [Eastman]
Palatinol DOP [Badische]
Palatinol DOP-SG [Badische] (special grade)
Polycizer 162, DOP [Harwick]
PX-138 [USS Chem.]
Rilanit DNOP [Henkel KGaA]
Staflex DOP [Reichhold]
    Generically sold by:
        [Ashland; C.P. Hall; Monsanto; Union Oil; Unocal]

CATEGORY:
Plasticizer, softener, lubricant

APPLICATIONS:
Food applications: food packaging (generic—C.P. Hall, Union Oil; Hatcol DOP; Hexaflex DOP; Kodaflex DOP; Palatinol DOP; Polycizer DOP)

Household products: film/sheeting for clothing, shower curtains, molded products (Hexaflex DOP; Kodaflex DOP; Staflex DOP); packaging (Staflex DOP)

Industrial applications: adhesives (Staflex DOP); capacitor applications (Palatinol DOP-SG); cellulosics (generic—C.P. Hall, Union Oil; Kodaflex DOP; Polycizer DOP; PX-138); ceramics (Rilanit DNOP); electrical insulation (Hexaflex DOP; Kodaflex DOP); lubricating/cutting oils (Rilanit DNOP); medical tubing (Hatcol DOP); metalworking fluids (Rilanit DNOP); molded goods (Hexaflex DOP); paper coatings (Hexaflex DOP; Kodaflex DOP); petroleum products (Rilanit DNOP); plastics (generic—Monsanto; Hatcol DOP; Polycizer 162; PX-138); rubber (generic—C.P. Hall, Monsanto, Union Oil; DOP Dry Liquid Concentrate; Hatcol DOP; Kodaflex DOP; Polycizer 162, DOP; PX-138); textile/leather processing (Rilanit DNOP; Staflex DOP)

PROPERTIES:

Form:
  Liquid (generic—C.P. Hall, Union Oil; Hatcol DOP; Kodaflex DOP; Polycizer DOP;
    Rilanit DNOP)
  Dustless powder (DOP Dry Liquid Concentrate)
Color:
  Off-white (DOP Dry Liquid Concentrate)
  APHA 25 max. (Hatcol DOP)
  APHA 30 max. (Palatinol DOP)
  APHA 35 (Staflex DOP)
  Pt-Co 15 ppm (Kodaflex DOP)
Odor:
  Essentially odorless (Hatcol DOP; Kodaflex DOP)
  Mild, characteristic (generic—C.P. Hall, Union Oil; Palatinol DOP; Polycizer DOP)
Composition:
  72% DOP in a microcel binder (DOP Dry Liquid Concentrate)
  99% assay (Kodaflex DOP)
  99% min. ester content (Palatinol DOP)
Solubility:
  Sol. < 0.1 g/l in water @ 20 C (Kodaflex DOP); < 0.01% @ 20 C (Palatinol DOP)
M.W.:
  390 (Hatcol DOP; PX-138; Staflex DOP)
  390.57 (calc.) (Kodaflex DOP; Palatinol DOP)
Sp.gr.:
  0.9800–0.9865 (Hatcol DOP)
  0.980–0.989 (Polycizer DOP)
  0.982 (generic—C.P. Hall, Union Oil, Unocal)
  0.982–0.986 (20/20 C) (Palatinol DOP)
  0.985 (23 C) (Staflex DOP)
  0.9852 (20/20 C) (Kodaflex DOP)
  0.986 (20/20 C) (generic—Ashland)
  1.182 (DOP Dry Liquid Concentrate)
Density:
  0.982 kg/l (20 C) (Kodaflex DOP)
  8.2 lb/gal (Staflex DOP)
Visc.:
  56.5 cps (Hatcol DOP; Kodaflex DOP)
  57 cps (generic—C.P. Hall, Union Oil)
  81–88 cps (Polycizer DOP)
  81.4 cps (20 C) (Palatinol DOP)
  375 cSt (0 C, kinematic) (Staflex DOP)

Dioctyl phthalate (cont'd.)

F.P.:
    −55 C (to a stiff gel) (generic—C.P. Hall, Union Oil; Polycizer DOP)
    −50 C (Kodaflex DOP)
B.P.:
    230 C (generic—C.P. Hall, Union Oil)
    231 C (5 mm Hg) (Palatinol DOP)
    350 C (Polycizer DOP)
    384 C (760 mm) (Kodaflex DOP)
Pour Pt.:
    −46 C (Palatinol DOP)
    −60 F (Staflex DOP)
Flash Pt.:
    216 C (COC) (Kodaflex DOP)
    218 C (COC) (Palatinol DOP)
    410 F (Hatcol DOP)
    410–415 F (Polycizer DOP)
    420 F (generic—C.P. Hall, Union Oil; Staflex DOP)
Fire Pt.:
    240 C (COC) (Kodaflex DOP)
Cloud Pt.:
    < −60 F (Staflex DOP)
Acid No.:
    0.07 max. (Palatinol DOP)
Stability:
    Excellent heat and light stability (Palatinol DOP)
    Good stability to heat and uv light; excellent resistance to hydrolysis (Kodaflex DOP)
Storage Stability:
    Indefinite storage life (Palatinol DOP)
Ref. Index:
    1.4836 (Kodaflex DOP)
    1.4850 (Hatcol DOP)
    1.4851 (23 C) (Staflex DOP)
    1.4859 (20 C) (Palatinol DOP)
Dissipation Factor:
    0.64×10^{-2} (Kodaflex DOP)
Dielectric Constant:
    4.8 (1 mc) (Kodaflex DOP)
Volume Resistivity:
    2.2×10^{11} ohm-cm (Kodaflex DOP)
TOXICITY/HANDLING:
    Nontoxic (generic—C.P. Hall, Union Oil; Polycizer DOP)
    Avoid repeated/prolonged skin and eye contact; avoid breathing vapors; use with
        adequate ventilation (Palatinol DOP)

126

Dioctyl sebacate (CTFA)

SYNONYMS:
Bis (2-ethylhexyl) decanedioate
Decanedioic acid, bis (2-ethylhexyl) ester
Di-(2-ethylhexyl) sebacate

EMPIRICAL FORMULA:
$C_{26}H_{50}O_4$

STRUCTURE:

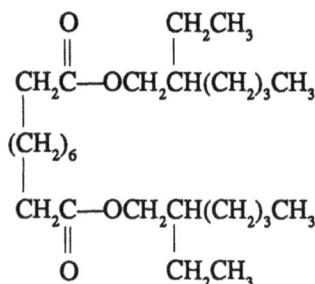

CAS No.:
122-62-3

TRADENAME EQUIVALENTS:
Hatcol DOS [Hatco]
Monoplex DOS [C.P. Hall]
Plasthall DOS [C.P. Hall]
Polycizer DOS [Harwick]
Rilanit DEHS [Henkel KGaA]
Uniflex DOS [Union Camp]
  Generically sold by: [Hexagon]

CATEGORY:
Plasticizer, lubricant, softener, processing aid, stabilizer, emulsifier, dispersant

APPLICATIONS:
Industrial applications: ceramics (Rilanit DEHS); electrical compounds (Monoplex DOS) ; sheet and film (Monoplex DOS; Polycizer DOS); lubricating/cutting oils and greases (generic; Rilanit DEHS; Uniflex DOS); metalworking (Rilanit DEHS; Uniflex DOS); motor oils (Rilanit DEHS); plastics and resins (generic; Hatcol DOS; Monoplex DOS; Plasthall DOS; Polycizer DOS; Uniflex DOS); rubber (Plasthall DOS; Polycizer DOS; Uniflex DOS); textile/leather processing (Rilanit DEHS)

PROPERTIES:
Form:
Liquid (Hatcol DOS; Polycizer DOS; Rilanit DEHS)
Clear liquid (Monoplex DOS; Plasthall DOS)
Color:
Water-white to pale straw, APHA 50 max. (Hatcol DOS; Plasthall DOS)

Dioctyl sebacate *(cont'd.)*

APHA 35 (Uniflex DOS)
Gardner 4 (Monoplex DOS)
Hazen 50 max. (Polycizer DOS)
Odor:
Mild (Hatcol DOS; Plasthall DOS)
M.W.:
426 (Hatcol DOS)
430 (Monoplex DOS; Plasthall DOS)
Sp.gr.:
0.911 (25/15 C) (Monoplex DOS)
0.913 ± 0.003 (Hatcol DOS; Plasthall DOS; Polycizer DOS)
0.915 (20/20 C) (Uniflex DOS)
Density:
7.6 lb/gal (Hatcol DOS; Monoplex DOS; Plasthall DOS)
Visc.:
0.21 poises (Monoplex DOS)
19 cSt (Uniflex DOS)
25 cps (Plasthall DOS)
27 cps (Hatcol DOS)
F.P.:
–55 C (Uniflex DOS)
–54 F (Monoplex DOS)
B.P.:
249 C (4 mm) (Hatcol DOS; Plasthall DOS)
M.P.:
–48 C (Hatcol DOS; Plasthall DOS)
Flash Pt.:
227 C (COC) (Uniflex DOS)
425 F min. (COC) (Polycizer DOS)
432 F (Monoplex DOS)
Fire Pt.:
260 C (COC) (Uniflex DOS)
Acid No.:
0.05 (Monoplex DOS)
< 0.1 (Uniflex DOS)
Saponification No.:
260 (Hatcol DOS; Monoplex DOS)
Stability:
Excellent resistance to water, soaps, and detergents (Monoplex DOS)
Excellent resistance to extraction by water and various soaps and detergents; excellent heat, light, and weathering stability (generic; Uniflex DOS)

Ref. Index:
1.448 (Uniflex DOS)
1.451 (Monoplex DOS)
STD. PKGS.:
Drum, tankcar, tankwagon (Polycizer DOS)

Dipropylene glycol dibenzoate

SYNONYMS:
3,3′-Oxydyl-1-propanol dibenzoate
POP (2) dibenzoate
PPG-2 dibenzoate (CTFA)
PPG (2) dibenzoate
EMPIRICAL FORMULA:
$C_{20}H_{22}O_5$
STRUCTURE:

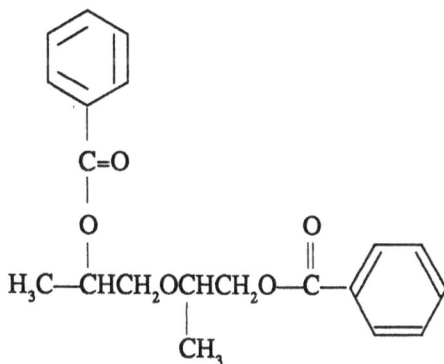

CAS No.:
94-51-9
TRADENAME EQUIVALENTS:
Benzoflex 9-88, 9-88 SG [Velsicol]
Generically sold by:
[Ashland; Nuodex; Unitex]
CATEGORY:
Plasticizer
APPLICATIONS:
Industrial applications: adhesives (Benzoflex 9-88); mechanical goods, printing rolls, gaskets and seals (Benzoflex 9-88 SG); plastics (Benzoflex 9-88, 9-88 SG); rubber (Benzoflex 9-88)

Dipropylene glycol dibenzoate *(cont'd.)*

PROPERTIES:
Form:
  Clear oily liquid (Benzoflex 9-88, 9-88 SG)
Color:
  APHA 100 max. (Benzoflex 9-88)
  APHA 150 max. (Benzoflex 9-88 SG)
Odor:
  Mild ester (Benzoflex 9-88, 9-88 SG)
Composition:
  98% purity (Benzoflex 9-88)
  99% purity (Benzoflex 9-88 SG)
Solubility:
  Sol. in aliphatic hydrocarbons (Benzoflex 9-88, 9-88 SG)
  Sol. in aromatic hydrocarbons (Benzoflex 9-88, 9-88 SG)
  Sol. < 0.01% in water (Benzoflex 9-88, 9-88 SG)
M.W.:
  342.3 (theoret.) (Benzoflex 9-88, 9-88 SG)
Sp.gr.:
  1.120 (Benzoflex 9-88, 9-88 SG)
Density:
  9.346 lb/gal (Benzoflex 9-88, 9-88 SG)
Visc.:
  215 cps (Benzoflex 9-88, 9-88 SG)
F.P.:
  —40 C (becomes a glass) (Benzoflex 9-88, 9-88 SG)
B.P.:
  232 C (5 mm Hg) (Benzoflex 9-88, 9-88 SG)
Pour Pt.:
  −19 C (Benzoflex 9-88, 9-88 SG)
Flash Pt.:
  199 C (COC) (Benzoflex 9-88, 9-88 SG)
Hydroxyl No.:
  15 max. (Benzoflex 9-88, 9-88 SG)
Stability:
  Hydrolytically stable under normal conditions (Benzoflex 9-88, 9-88 SG)
Ref. Index:
  1.5282 (Benzoflex 9-88, 9-88 SG)
Dielectric Constant:
  7.5 (1 kHz) (Benzoflex 9-88, 9-88 SG)
Volume Resistivity:
  10^{10} ohm cm (Benzoflex 9-88, 9-88 SG)

Dipropylene glycol dibenzoate (cont'd.)

TOXICITY/HANDLING:
Repeated exposure may cause defatting of skin and mild irritation to eyes and nose; avoid skin and eye contact and inhalation of vapor (Benzoflex 9-88, 9-88 SG)

STD. PKGS.:
55-gal (500 lb net) epoxy-phenolic-lined steel drums; 5-gal (45 lb net) lacquer-lined steel pails; bulk (Benzoflex 9-88, 9-88 SG)

Distearyl thiodipropionate (CTFA)

SYNONYMS:
Propanoic acid, 3,3´-thiobis-, dioctadecyl ester
3,3´-Thiobispropanoic acid, dioctadecyl ester

EMPIRICAL FORMULA:
$C_{42}H_{82}O_4S$

STRUCTURE:

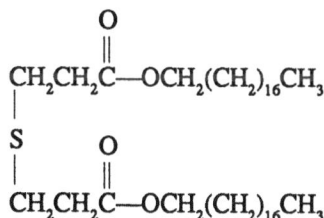

CAS No.:
693-36-7

TRADENAME EQUIVALENTS:
Argus DSTDP [Argus]
Argus DXTDP [Witco/Argus]
Carstab DSTDP [Cincinnati Milacron; Morton Thiokol]
Cyanox STDP [Amer. Cyanamid]
Evanstab 18 [Evans Chemetics]

CATEGORY:
Antioxidant, stabilizer, plasticizer, softener, synergist

APPLICATIONS:
Automotive applications (Cyanox STDP)
Food applications: food additives (Cyanox STDP; Evanstab 18); food packaging (Cyanox STDP; Evanstab 18)
Industrial applications: adhesives/sealants (Cyanox STDP); appliances (Cyanox STDP); plastics (Argus DSTDP, DXTDP; Carstab DSTDP; Cyanox STDP; Evanstab 18); rubber (Argus DSTDP, DXTDP; Carstab DSTDP)

Distearyl thiodipropionate *(cont'd.)*

PROPERTIES:
Form:
   Flakes (Argus DSTDP)
   Waxy crystalline flakes (Cyanox STDP)
   Free-flowing crystalline flakes or powder (Evanstab 18)
   Crystalline flake or powder (Carstab DSTDP)
Color:
   White (Argus DSTDP; Carstab DSTDP; Cyanox STDP; Evanstab 18)
Odor:
   Mild fatty (Argus DSTDP)
Composition:
   98% purity (Carstab DSTDP; Evanstab 18)
Solubility:
   Sol. 2 g/100 g acetone (Carstab DSTDP)
   Slightly sol. in alcohol (Argus DSTDP)
   Sol. in benzene (Argus DSTDP); sol. 29 g/100 g (Carstab DSTDP)
   Sol. in chloroform (Argus DSTDP)
   Sol. > 1 g/100 g ethanol (Carstab DSTDP); < 1 g/100 g 95% ethanol (Evanstab 18); <
      0.5 g/100 g (Cyanox STDP)
   Slightly sol. in ethyl acetate (Argus DSTDP); sol. > 1 g/100 g (Carstab DSTDP); sol.
      < 1 g/100 g (Evanstab 18); < 0.5 g/100 g (Cyanox STDP)
   Sol. 2 g/100 g *n*-heptane (Evanstab 18); 1.5 g/100 g (Cyanox STDP)
   Sol. 2 g/100 g hexane (Carstab DSTDP)
   Sol. < 1 g/100 g MEK (Evanstab 18); < 0.5 g/100 g (Cyanox STDP)
   Sol. in toluene (Argus DSTDP); sol. 11 g/100 g (Evanstab 18); sol. 10.7 g/100 g
      (Cyanox STDP)
   Slightly sol. in water (Argus DSTDP); sol. < 1 g/100 g (Evanstab 18); insol. (Carstab
      DSTDP)
M.W.:
   680 (Argus DSTDP)
   683 (Cyanox STDP; Evanstab 18)
Sp.gr.:
   0.858 (80 C) (Carstab DSTDP)
   0.985 (Argus DSTDP)
   1.027 (Cyanox STDP)
F.P.:
   64 C min. (Evanstab 18)
Set Pt.:
   64 C (Cyanox STDP)
   64 C min. (Argus DSTDP)
Acid No.:
   1.0 (Carstab DSTDP)

1.0 max. (Argus DSTDP)
1.5 max. (Cyanox STDP)
Stability:
>Good color stability during aging; improved melt stability during polymer fabrication; improved retention of physical and electrical properties during polymer fabrication and during environmental aging of the polymer (Evanstab 18)
>Thermally stable; resistant to discoloration and volatilization; relatively inert (Carstab DSTDP)

TOXICITY/HANDLING:
>Low oral and dermal toxicity (Carstab DSTDP)

STD. PKGS.:
>55-gal (200 lb net) fiber drums (Carstab DSTDP)

Ditridecyl phthalate

SYNONYMS:
>DTDP

STRUCTURE:
>$C_6H_4(COOC_{13}H_{27})_2$

CAS No.:
>119-06-2

TRADENAME EQUIVALENTS:
>Jayflex DTDP [Exxon] (with Bisphenol-A)
>PX-126 [USS Chem.]
>Staflex DTDP [Reichhold]

CATEGORY:
>Plasticizer

APPLICATIONS:
>Automotive upholstery: (PX-126)
>Industrial applications: electrical insulation (Jayflex DTDP; PX-126; Staflex DTDP); film and sheeting (Jayflex DTDP); gasketing (PX-126); plastics (Jayflex DTDP; PX-126; Staflex DTDP)

PROPERTIES:
Color:
>APHA 40 (Staflex DTDP)

M.W.:
>530 (PX-126)
>531 (Jayflex DTDP; Staflex DTDP)

Ditridecyl phthalate (cont'd.)

Sp.gr.:
   0.952 (23 C) (Staflex DTDP)
Density:
   7.9 lb/gal (Staflex DTDP)
Visc.:
   1900 cSt (kinematic, 0 C) (Staflex DTDP)
Pour Pt.:
   –48 F (Staflex DTDP)
Flash Pt.:
   500 F (Staflex DTDP)
Cloud Pt.:
   < –48 F (Staflex DTDP)
Ref. Index:
   1.4833 (23 C) (Staflex DTDP)

Ditridecyl thiodipropionate (CTFA)

SYNONYMS:
   Di (tridecyl) thiodipropionate
   DTDTDP
   Propanoic acid, 3,3´-thiobis-, ditridecyl ester
   3,3´-Tetramethylnonyl thiodipropionate
   3,3´-Thiobispropanoic acid, ditridecyl ester
EMPIRICAL FORMULA:
   $C_{32}H_{62}O_4S$
STRUCTURE:

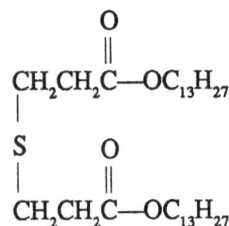

CAS No.:
   10595-72-9
TRADENAME EQUIVALENTS:
   Argus DTDTDP [Witco/Argus]
   Cyanox 711 [Amer. Cyanamid]
   Evanstab 13 [Evans Chemetics]

Ditridecyl thiodipropionate (cont'd.)

CATEGORY:
Antioxidant, stabilizer

APPLICATIONS:
Industrial applications: latexes (Argus DTDTDP; Cyanox 711; Evanstab 13); petroleum lubricants (Cyanox 711); plastics (Argus DTDTDP; Cyanox 711; Evanstab 13); rubber (Argus DTDTDP; Cyanox 711; Evanstab 13)

PROPERTIES:

Form:
Liquid (Argus DTDTDP)
Clear liquid (Cyanox 711; Evanstab 13)

Color:
Colorless to slightly yellow (Cyanox 711; Evanstab 13)

Composition:
99% purity (Evanstab 13)

Solubility:
Sol. 14 g/100 g 95% ethanol (Evanstab 13); 13.8 g/100 g 95% ethanol (Cyanox 711)
Completely miscible with ethyl acetate (Cyanox 711; Evanstab 13)
Completely miscible with n-heptane (Cyanox 711; Evanstab 13)
Completely miscible with MEK (Cyanox 711; Evanstab 13)
Completely miscible with toluene (Cyanox 711; Evanstab 13)

M.W.:
543 (Cyanox 711; Evanstab 13)

Sp.gr.:
0.936 (Cyanox 711)

B.P.:
265 C (0.25 mm) (Evanstab 13)

Acid No.:
3.0 (Cyanox 711)
3.0 max. (Evanstab 13)

Saponification No.:
200–210 (Cyanox 711; Evanstab 13)

Stability:
Good color stability during aging; improved melt stability during polymer fabrication; improved retention of physical and electrical properties during polymer fabrication and during environmental aging of the polymer (Evanstab 13)

TOXICITY/HANDLING:
No eye or skin irritation was produced during primary irritation studies (Cyanox 711)
Considered relatively nontoxic by ingestion in single doses and by single skin applications; small quantities are not irritating to the skin or eyes of rabbits (Evanstab 13)

STD. PKGS.:
Drum or bulk (Evanstab 13)

135

Epoxidized linseed oil

TRADENAME EQUIVALENTS:
  Admex ELO [Sherex]
  Drapex 10.4 [Argus]
  Epoxol 9-5 [Swift]
  Flexol Plasticizer LOE [Union Carbide]

CATEGORY:
  Plasticizer, stabilizer

APPLICATIONS:
  Food applications: food packaging (Admex ELO; Drapex 10.4; Epoxol 9-5); food
    processing materials (Admex ELO)
  Industrial applications: plastics (Admex ELO; Drapex 10.4; Epoxol 9-5)

PROPERTIES:
Form:
  Liquid (Flexol Plasticizer LOE)
Color:
  APHA 150 (Admex ELO)
  Co-Pt 70 (Epoxol 9-5)
  Co-Pt 150 max. (Flexol Plasticizer LOE)
  Gardner 1 (Drapex 10.4)
Odor:
  Faintly fatty (Drapex 10.4)
  Low (Epoxol 9-5)
Taste:
  Low (Epoxol 9-5)
Solubility:
  Miscible with aromatic hydrocarbons (Epoxol 9-5)
  Miscible with chlorinated hydrocarbons (Epoxol 9-5)
  Miscible with esters (Epoxol 9-5)
  Miscible with ethyl ether (Epoxol 9-5)
  Miscible with heptane (Epoxol 9-5)
  Miscible with ketones (Epoxol 9-5)
  Miscible with most organic solvents (Epoxol 9-5)
  Miscible with vinyl plasticizers (Epoxol 9-5)
  Sol. 0.01% in water (20 C) (Drapex 10.4)
M.W.:
  1000 (Drapex 10.4)

136

Sp.gr.:
  1.03 (Epoxol 9-5; Flexol Plasticizer LOE)
  1.032 (Admex ELO)
  1.032 ± 0.010 (Drapex 10.4)
Density:
  8.58 lb/gal (Admex ELO; Epoxol 9-5)
Visc.:
  619 cps (Epoxol 9-5)
  750 cps (Drapex 10.4)
  815 cps (Admex ELO)
Pour Pt.:
  290 C (COC) (Drapex 10.4)
  30 F (Admex ELO; Epoxol 9-5)
Flash Pt.:
  320 C (COC) (Drapex 10.4)
  590 F (COC) (Admex ELO)
  595 F (COC) (Epoxol 9-5)
Fire Pt.:
  635 F (COC) (Admex ELO)
Acid No.:
  0.12 (Epoxol 9-5)
  0.5 (Drapex 10.4)
  1.0 (Admex ELO)
Iodine No.:
  2.5 (Drapex 10.4)
Saponification No.:
  172 (Epoxol 9-5)
Hydroxyl No.:
  12 (Epoxol 9-5)
Ref. Index:
  1.4715 (Epoxol 9-5)
  1.477 (Admex ELO)
  1.4788 (Drapex 10.4)

Epoxidized octyl tallate

SYNONYMS:
  Octyl epoxy tallate
TRADENAME EQUIVALENTS:
  Drapex 4.4 [Argus]
  Epoxol 5-2E [Swift]

Epoxidized octyl tallate (cont'd.)

TRADENAME EQUIVALENTS *(cont'd.):*
Flexol Plasticizer EP-8 [Union Carbide]
Monoplex S-73 [Rohm & Haas]

CATEGORY:
Plasticizer, softener, costabilizer, extrusion aid

APPLICATIONS:
Industrial applications: plastics/plastisols (Drapex 4.4; Epoxol 5-2E; Flexol Plasticizer EP-8; Monoplex S-73); rubber (Flexol Plasticizer EP-8)

PROPERTIES:

Form:
Liquid (Drapex 4.4; Flexol Plasticizer EP-8)
Clear liquid (Monoplex S-73)

Color:
Pale yellow (Drapex 4.4)
Gardner 1 (Epoxol 5-2E; Monoplex S-73)
Pt-Co 75 max. (Flexol Plasticizer EP-8)

Odor:
Slight fatty (Drapex 4.4)

M.W.:
420 (Monoplex S-73)

Sp.gr.:
0.922 (Drapex 4.4)
0.927 (25/15 C) (Monoplex S-73)
0.9232 (Flexol Plasticizer EP-8)
0.9244 (Epoxol 5-2E)

Density:
7.6 lb/gal (Monoplex S-73)

Visc.:
A (Gardner) (Epoxol 5-2E)
0.4 poise (Monoplex S-73)

F.P.:
–22 C (Drapex 4.4)
–8.5 C (Flexol Plasticizer EP-8)
19 F (Monoplex S-73)

Flash Pt.:
220 C (Drapex 4.4)
435 F (Monoplex S-73)

Acid No.:
0.5 (Drapex 4.4; Epoxol 5-2E)
0.7 (Monoplex S-73)

Epoxidized octyl tallate *(cont'd.)*

Saponification No.:
139.8 (Epoxol 5-2E)
140 (Monoplex S-73)
Ref. Index:
1.458 (Monoplex S-73)
STD. PKGS.:
Drum, TT, TC (Flexol Plasticizer EP-8)

Epoxidized soybean oil

SYNONYMS:
Chinese bean oil, epoxidized
ESO
Soy oil, epoxidized
Soya oil, epoxidized
Soyabean oil, epoxidized
Soybean oil, epoxidized
TRADENAME EQUIVALENTS:
Admex 710, 711 [Sherex]
Drapex 6.8 [Argus]
Epoxol 7-4 [Swift]
Estabex 138-A [Interstab]
Flexol Plasticizer EPO [Union Carbide]
Paraplex G-60, G-62 [C.P. Hall; Rohm & Haas]
Plas-Chek 775 [Ferro]
Plastoflex 2307 [Interstab]
    Generically sold by:
        [Ashland; FMC Corp.]
CATEGORY:
Plasticizer, heat and light stabilizer, acid scavenger
APPLICATIONS:
Automotive applications: (generic; Flexol Plasticizer EPO)
Farm applications: insecticides (Flexol Plasticizer EPO)
Food applications: beverage tubing (Admex 710, 711; Flexol Plasticizer EPO; Plas-Chek 775); food packaging (Admex 710, 711; Drapex 6.8; Epoxol 7-4; Estabex 138-A; Flexol Plasticizer EPO; Paraplex G-62; Plas-Chek 775)
Industrial applications: chlorinated paraffin (Plas-Chek 775; Plastoflex 2307); coatings (generic; Admex 710, 711; Flexol Plasticizer EPO); consumer goods (generic; Plas-Chek 775); electrical insulation (generic; Plastoflex 2307); film and sheeting

Epoxidized soybean oil (cont'd.)

(generic; Admex 710, 711; Flexol Plasticizer EPO; Plastoflex 2307); flooring (generic; Admex 710, 711; Flexol Plasticizer EPO; Plas-Chek 775; Plastoflex 2307); foam applications (Plas-Chek 775); gasketing (Admex 710, 711; Flexol Plasticizer EPO); hose (Flexol Plasticizer EPO; Plastoflex 2307) ; pigment dispersions (Flexol Plasticizer EPO; Plas-Chek 775; Plastoflex 2307); plastics/plastisols (generic; Admex 710, 711; Drapex 6.8; Epoxol 7-4; Estabex 138-A; Flexol Plasticizer EPO; Paraplex G-60, G-62; Plas-Chek 775; Plastoflex 2307); rubber (Flexol Plasticizer EPO; Paraplex G-62; Plastoflex 2307); upholstery (Plas-Chek 775)

Pharmaceutical applications: medical tubing and bags (Plas-Chek 775)

PROPERTIES:
Form:
Liquid (Flexol Plasticizer EPO)

Clear liquid (Paraplex G-60; Plastoflex 2307)

Clear viscous oily liquid (Paraplex G-62)

Clear liquid; may become cloudy during cold weather—clears on warming to R.T. (Estabex 138-A)

Polymeric liquid (Drapex 6.8)

Color:
Pale yellow (Drapex 6.8)

APHA 100 (Paraplex G-62)

APHA 150 (Admex 710, 711)

Gardner 1+ (Paraplex G-60)

Gardner 1 max. (Plas-Chek 775)

Gardner 2 (Estabex 138-A)

Gardner 3 max. (Plastoflex 2307)

Pt-Co 70 (Epoxol 7-4)

Pt-Co 75 max. (Flexol Plasticizer EPO)

Odor:
Negligible (Plas-Chek 775)

Slight fatty (Drapex 6.8)

Low (Epoxol 7-4; Paraplex G-62)

Taste:
Low (Epoxol 7-4; Paraplex G-62)

Solubility:
Sol. in higher alcohols (Estabex 138-A)

Sol. in aliphatic hydrocarbons (Estabex 138-A)

Sol. in aromatic hydrocarbons (Estabex 138-A); miscible with aromatic hydrocarbons (Epoxol 7-4)

Miscible with chlorinated hydrocarbons (Epoxol 7-4)

Sol. in esters (Estabex 138-A); miscible with esters (Epoxol 7-4)

Sol. in ethers (Estabex 138-A)

Miscible with ethyl ether (Epoxol 7-4)
Miscible with heptane (Epoxol 7-4)
Miscible with ketones (Epoxol 7-4)
Miscible with most organic solvents (Epoxol 7-4)
Miscible with vinyl plasticizers (Epoxol 7-4)

M.W.:

1000 (Paraplex G-60, G-62; Plas-Chek 775)

Sp.gr.:

0.980 (25/15 C) (Paraplex G-60)
0.982 (Estabex 138-A)
0.99 (Drapex 6.8)
0.992 min. (Plas-Chek 775)
0.993 (Paraplex G-62)
0.994 (generic—Ashland; Admex 710, 711; Epoxol 7-4)
0.995 (Flexol Plasticizer EPO)

Density:

0.99 g/ml (20 C) (Plastoflex 2307)
8.18 lb/gal (Estabex 138-A)
8.2 lb/gal (Paraplex G-60)
8.26 lb/gal (Admex 710, 711)
8.28 lb/gal (Epoxol 7-4)
8.3 lb/gal (Paraplex G-62)

Visc.:

G (Gardner-Holdt) (Estabex 138-A)
Q (Gardner-Holdt) (Paraplex G-62)
3–7 poise (20 C) (Plastoflex 2307)
3.5 poise (Paraplex G-60)
314 cps (Epoxol 7-4)
368 cps (Admex 710, 711)
5 stokes max. (Plas-Chek 775)

F.P.:

0 C (Drapex 6.8)
5 C (Paraplex G-62)
41 F (Paraplex G-60)

Pour Pt.:

–5 C (Paraplex G-62)
–4 C (Flexol Plasticizer EPO)
25 F (Epoxol 7-4; Plas-Chek 775)
25–35 F (Admex 710, 711)

Flash Pt.:

290 C (Drapex 6.8)
590 F (Paraplex G-60); (COC) (Admex 710, 711; Epoxol 7-4; Paraplex G-62)

Epoxidized soybean oil (cont'd.)

    600 F (Plas-Chek 775)
Fire Pt.:
    600 F (COC) (Admex 710, 711)
    610 F (Plas-Chek 775)
Acid No.:
    0.1 (Epoxol 7-4)
    0.2 (Paraplex G-62)
    0.3 (Admex 710, 711)
    0.5 (Drapex 6.8); 0.5 max. (Plas-Chek 775)
    0.6 (Paraplex G-60)
    1.0 max. (Plastoflex 2307)
Iodine No.:
    0.9 (Paraplex G-62)
    2.5 max. (Plas-Chek 775)
Saponification No.:
    178.1 (Epoxol 7-4)
    182 (Paraplex G-60)
    183 (Paraplex G-62)
Hydroxyl No.:
    6.5 (Epoxol 7-4)
Ref. Index:
    1.4705 (Epoxol 7-4)
    1.471 (Admex 710, 711; Paraplex G-62)
    1.472 (Paraplex G-60; Plas-Chek 775)
    1.473 (Plastoflex 2307)
STD. PKGS.:
    Drums, tank wagons (Epoxol 7-4)
    Drum, TT, TC (Flexol Plasticizer EPO)

Ethyl dihydroxypropyl PABA (CTFA)

SYNONYMS:
    Benzoic acid, 4-[bis (2-hydroxypropyl) amino]-, ethyl ester
    4-[Bis (2-hydroxypropyl) amino] benzoic acid, ethyl ester
    Ethyl dihydroxypropyl p-aminobenzoate
    Propoxylated (2 moles) ethyl para-aminobenzoate
EMPIRICAL FORMULA:
    $C_{15}H_{23}NO_4$

STRUCTURE:

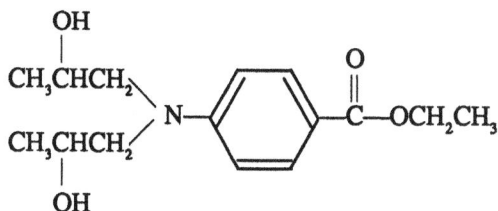

CAS No.:
58882-17-0

TRADENAME EQUIVALENTS:
Amerscreen P, P80/20 [Amerchol]

CATEGORY:
UV absorber

APPLICATIONS:
Cosmetic industry preparations: (Amerscreen P, P 80/20)
Pharmaceutical applications: (Amerscreen P, P 80/20); dermatological products (Amerscreen P, P 80/20); sunscreens (Amerscreen P, P 80/20)

PROPERTIES:
Form:
Liquid (crystallizes on aging) (Amerscreen P 80/20)
Gel (crystallizes on aging) (Amerscreen P)
Color:
Faint yellow (Amerscreen P, P 80/20)
Odor:
Practically odorless (Amerscreen P, P 80/20)
Composition:
80% active in propylene glycol (Amerscreen P 80/20)
100% active (Amerscreen P)
Solubility:
Sol. in castor oil (Amerscreen P, P 80/20)
Sol. in 2.5% corn oil (Amerscreen P, P 80/20)
Sol. in ester (Amerscreen P, P 80/20)
Sol. in ethanol (Amerscreen P, P 80/20); sol. in 50% aq. ethanol (Amerscreen P, P 80/20)
Insol. in glycerin (Amerscreen P, P 80/20)
Sol. in isopropanol (Amerscreen P, P 80/20)
Sol. in 2.5% isopropyl myristate (Amerscreen P, P 80/20)
Insol. in min. oil (70 visc.) (Amerscreen P, P 80/20)
Sol. in propylene glycol (Amerscreen P, P 80/20)
Sol. in veg. oils (Amerscreen P, P 80/20)
Insol. in water (Amerscreen P, P 80/20)

Ethyl dihydroxypropyl PABA *(cont'd.)*

Acid No.:
1 max. (Amerscreen P, P 80/20)
Saponification No.:
160–180 (Amerscreen P 80/20)
195–215 (Amerscreen P)
Hydroxyl No.:
380–400 (Amerscreen P)

Ethylene glycol monostearate

SYNONYMS:
EGMS
Glycol monostearate
Glycol stearate (CTFA)
2-Hydroxy-ethyl octadecanoate
Octadecanoic acid, 2-hydroxyethyl ester
EMPIRICAL FORMULA:
$C_{20}H_{40}O_3$
STRUCTURE:

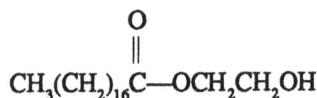

CAS No.:
111-60-4
TRADENAME EQUIVALENTS:
AMS-33 [Hefti]
Cerasynt M [Van Dyk]
Cerasynt MN [Van Dyk] (self-emulsifying)
Cithrol EGMS N/E [Croda]
Cithrol EGMS S/E [Croda] (self-emulsifying)
Clindrol SEG [Clintwood]
Cyclochem EGMS [Cyclo]
Drewmulse EGMS [Drew Produtos]
Emerest 2350 [Emery]
Empilan EGMS [Albright&Wilson/Marchon]
Hodag EGMS [Hodag]
Kessco Ethylene Glycol Monostearate [Armak]
Kessco Ethylene Glycol Monostearate 70 [Armak]
Lexemul EGMS [Inolex]

Ethylene glycol monostearate (cont'd.)

CATEGORY:
Plasticizer, viscosity builder, emulsifier, dispersant, opacifier, pearling agent, antistat, chemical intermediate, defoamer, emollient, lubricant, rust inhibitor, scouring and detergent aid, solubilizer, w/o emulgent, wetting agent

APPLICATIONS:
Cleansers: hand cleanser (Clindrol SEG)
Cosmetic industry preparations: (Emerest 2350; Mapeg EGMS; Lipo EGMS; Radiasurf 7270); conditioners (AMS-33; Cerasynt M, MN; Empilan EGMS; Kessco Ethylene Glycol Monostearate; Lexemul EGMS; Schercemol EGMS); lotions (Cyclochem EGMS); personal care products (AMS-33; Cerasynt M, MN; Clindrol SEG; Empilan EGMS; Kessco Ethylene Glycol Monostearate; Lexemul EGMS; Schercemol EGMS); shampoos (AMS-33; Cerasynt M, MN; Clindrol SEG; Cyclochem EGMS; Empilan EGMS; Kessco Ethylene Glycol Monostearate, 70; Lexemul EGMS; Schercemol EGMS)
Farm products: insecticides/pesticides (Radiasurf 7270)
Household detergents: (Emerest 2350); dishwashing (Clindrol SEG)
Industrial applications: lubricating/cutting oils (Cithrol EGMS N/E, EGMS S/E; Radiasurf 7270); metalworking (Mapeg EGMS); paint mfg. (Radiasurf 7270); paper mfg. (Cithrol EGMS N/E, EGMS S/E); plastics (Radiasurf 7270); polishes and waxes (Cithrol EGMS N/E, EGMS S/E; Radiasurf 7270); printing inks (Radiasurf 7270); textile/leather processing (Cithrol EGMS N/E, EGMS S/E; Mapeg EGMS; Radiasurf 7270)
Industrial cleaners: metal processing surfactants (Cithrol EGMS N/E, EGMS S/E)
Pharmaceutical applications: (Mapeg EGMS; Radiasurf 7270)

PROPERTIES:
Form:
Solid (Cithrol EGMS N/E, EGMS S/E; Emerest 2350; Hodag EGMS; Mapeg EGMS; Radiasurf 7270; Schercemol EGMS)
Beads (Lipo EGMS)
Flake (AMS-33; Cerasynt M, MN; Clindrol SEG; Cyclochem EGMS; Drewmulse EGMS; Empilan EGMS; Kessco Ethylene Glycol Monostearate, 70; Lexemul EGMS; Lipo EGMS; Mapeg EGMS; Pegosperse 50MS; Schercemol EGMS)
Powder (Tegin G6100)
Color:
White (Clindrol SEG; Cyclochem EGMS; Empilan EGMS; Pegosperse 50MS; Radiasurf 7270; Schercemol EGMS)

Ethylene glycol monostearate *(cont'd.)*

White to off-white (Lipo EGMS)
White to cream (Mapeg EGMS)
Gardner 1 (Emerest 2350)
Odor:
Slightly fatty (Clindrol SEG)
Composition:
48% active (Drewmulse EGMS)
65% diester, 30% monoester (Empilan EGMS)
100% active (Clindrol SEG; Emerest 2350)
100% conc. (AMS-33; Cerasynt M, MN; Cithrol EGMS N/E, EGMS S/E; Lexemul
    EGMS; Mapeg EGMS; Pegosperse 50MS; Tegin G6100)
Solubility:
Sol. in acetone (Pegosperse 50MS)
Sol. in alcohols (Clindrol SEG; Schercemol EGMS)
Sol. in aliphatic hydrocarbons (Clindrol SEG)
Sol. in aromatic hydrocarbons (Clindrol SEG; Schercemol EGMS)
Sol. in benzene (Radiasurf 7270)
Sol. in chlorinated hydrocarbons (Schercemol EGMS)
Sol. in esters (Clindrol SEG; Schercemol EGMS)
Sol. in ethanol (Pegosperse 50MS)
Sol. in ethyl acetate (Pegosperse 50MS)
Disp. in glycerols (Schercemol EGMS)
Disp. in glycols (Clindrol SEG; Schercemol EGMS)
Sol. in isopropanol (Emerest 2350; Radiasurf 7270)
Sol. in ketones (Clindrol SEG; Schercemol EGMS)
Sol. in methanol (Pegosperse 50MS)
Sol. in min. oil (Emerest 2350; Empilan EGMS; Mapeg EGMS; Pegosperse 50MS;
    Schercemol EGMS)
Sol. in naphtha (Pegosperse 50MS)
Sol. in soybean oil (Mapeg EGMS)
Sol. in toluene (Emerest 2350)
Sol. in toluol (Mapeg EGMS; Pegosperse 50MS)
Sol. in trichlorethylene (Radiasurf 7270)
Disp. in triols (Schercemol EGMS)
Sol. in veg. oils (Radiasurf 7270)
Insol. in water (Emerest 2350; Schercemol EGMS)
Sol. in xylene (Emerest 2350)
Ionic Nature:
Nonionic (AMS-33; Cerasynt M, MN; Cithrol EGMS N/E; Clindrol SEG; Drewmulse
    EGMS; Emerest 2350; Hodag EGMS; Kessco Ethylene Glycol Monostearate, 70;
    Lexemul EGMS; Lipo EGMS; Mapeg EGMS; Pegosperse 50MS; Radiasurf 7270;
    Schercemol EGMS; Tegin G6100)

Anionic (Cithrol EGMS S/E)
M.W.:
  445 (Radiasurf 7270)
Sp.gr.:
  0.851 (98.9 C) (Radiasurf 7270)
  0.96 (Pegosperse 50MS)
  0.998 (Clindrol SEG)
Visc.:
  4.80 cps (98.9 C) (Radiasurf 7270)
M.P.:
  50 C (Emerest 2350)
  52–56 C (Kessco Ethylene Glycol Monostearate 70)
  52–60 C (Clindrol SEG)
  55–60 C (Pegosperse 50MS; Schercemol EGMS)
  56 C (Cyclochem EGMS; Empilan EGMS; Mapeg EGMS)
  56–60 C (Kessco Ethylene Glycol Monostearate)
  61 C (Radiasurf 7270)
Flash Pt.:
  214 C (COC) (Radiasurf 7270)
  370 F (COC) (Kessco Ethylene Glycol Monostearate 70)
  390 F (Emerest 2350; Kessco Ethylene Glycol Monostearate)
HLB:
  0.9 ± 0.5 (Pegosperse 50MS)
  1.0 (Lipo EGMS)
  1.8 (Radiasurf 7270)
  2.0 (AMS-33)
  2.0 ± 1 (Lipo EGMS)
  2.2 (Emerest 2350)
  2.3 (Lexemul EGMS)
  2.4 (Mapeg EGMS)
  2.9 (Kessco Ethylene Glycol Monostearate)
  3.2 (Tegin G6100)
  3.5 (Hodag EGMS)
Acid No.:
  2.0 max. (Kessco Ethylene Glycol Monostearate, 70; Radiasurf 7270; Schercemol EGMS)
  4.0 max. (Cyclochem EGMS; Mapeg EGMS)
  < 5 (Pegosperse 50MS)
  6 max. (Lipo EGMS)
Iodine No.:
  0.5 max. (Kessco Ethylene Glycol Monostearate, 70; Mapeg EGMS)
  1.0 max. (Radiasurf 7270; Schercemol EGMS)

Ethylene glycol monostearate *(cont'd.)*

< 2 (Pegosperse 50MS)

Saponification No.:
170–190 (Schercemol EGMS)
175–190 (Lipo EGMS)
180–187 (Pegosperse 50MS)
180–188 (Mapeg EGMS)
180–190 (Radiasurf 7270)
182 (Cyclochem EGMS)

Stability:
Stable over a wide pH range (Schercemol EGMS)

pH:
4.0–6.0 (3% aq. disp.) (Pegosperse 50MS)

STD. PKGS.:
25-kg net multiply paper bags or bulk (Radiasurf 7270)
55-gal (200 lb net) open-head fiber drums (Clindrol SEG)

Ethyl toluenesulfonamide (CTFA)

SYNONYMS:
N-Ethyl *o,p*-toluene sulfonamide

EMPIRICAL FORMULA:
$C_9H_{13}NO_2S$

STRUCTURE:

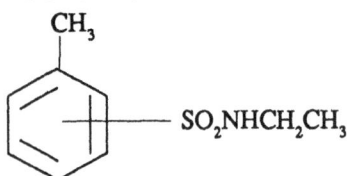

CAS No.:
80-39-7; 1077-56-1

TRADENAME EQUIVALENTS:
Santicizer 8 [Monsanto]

CATEGORY:
Plasticizer, processing aid

APPLICATIONS:
Food applications: food packaging
Industrial applications: adhesives; cellulosics; coatings and lacquers; plastics; resins

Ethyl toluenesulfonamide *(cont'd.)*

PROPERTIES:
Form:
Viscous liquid
Color:
Light yellow
Odor:
Slight, characteristic
Solubility:
Negligible sol. in petroleum hydrocarbons
Readily miscible with all other common organic solvents
Sol. 0.13% in water @ 23 C
M.W.:
199
Sp.gr.:
1.190
Density:
9.92 lb/gal
Visc.:
358 cSt
B.P.:
196 C (10 mm Hg)
Crystal. Pt.:
< 40 C
Solidification Pt.:
0 C
Flash Pt.:
345 F (COC)
Ref. Index:
1.540
Surface Tension:
44.5 dynes/cm

Glycerin (CTFA)

SYNONYMS:
Glycerine
Glycerol
Glycyl alcohol
1,2,3-Propanetriol

EMPIRICAL FORMULA:
$C_3H_8O_3$

STRUCTURE:
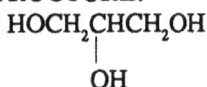

CAS No.:
56-81-5

TRADENAME EQUIVALENTS:
Croderol G7000 [Croda]
Generically sold by:
[Dow; Housmex; Procter & Gamble]

CATEGORY:
Plasticizer, humectant, moisturizer, diluent, solvent, binder, lubricant, softener, bacteriostat, penetrant, emulsifier

APPLICATIONS:
Cosmetic industry preparations: (generic; Croderol G7000); hair preparations (Croderol G7000); perfumery (generic); shaving preparations (Croderol G7000); skin preparations (Croderol G7000)
Food applications: (generic); liquors (generic)
Industrial applications: cellulosics (generic); cements (generic); dynamite (generic); hydraulic fluid (generic); latexes (generic); mold release (generic); paint mfg. (generic); polyurethane polyols (generic); printing inks (generic); resins (generic); rubber (generic)
Industrial cleaners: soaps (generic)
Pharmaceutical applications: (generic; Croderol G7000)

PROPERTIES:
Form:
Clear liquid (generic)
Color:
Colorless (generic)

Odor:
  Odorless (generic)
Taste:
  Sweet (generic)
Solubility:
  Sol. in alcohols (generic)
  Sol. in water (generic)
Sp.gr.:
  1.2620 (generic dynamite grade)
  > 1.249 (generic USP grade)
  1.2653 (anhyd. generic)
B.P.:
  290 C (generic)
M.P.:
  18 C (generic)
Flash Pt.:
  160 C (generic)
STORAGE/HANDLING:
  Combustible (generic)

Glyceryl dilaurate (CTFA)

SYNONYMS:
  Dilaurin
  Dodecanoic acid, diester with 1,2,3-propanetriol
EMPIRICAL FORMULA:
  $C_{27}H_{52}O_5$
STRUCTURE:

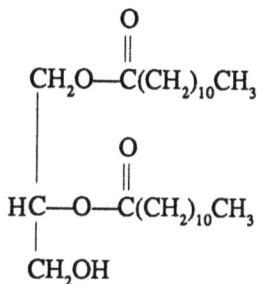

CAS No.:
  27638-00-2

Glyceryl dilaurate *(cont'd.)*

TRADENAME EQUIVALENTS:
Cithrol GDL S/E [Croda Chem. Ltd.] (self-emulsifying)
Emulsynt GDL [Van Dyk]
CATEGORY:
Thickener, emollient, antistat, emulsifier, dispersant, lubricant
APPLICATIONS:
Cosmetic industry preparations: creams and lotions (Emulsynt GDL)
Industrial applications: lubricating/cutting oils (Cithrol GDL S/E); paper mfg. (Cithrol GDL S/E); polishes (Cithrol GDL S/E); textile/leather processing (Cithrol GDL S/E)
Industrial cleaners: metal cleaners (Cithrol GDL S/E)
PROPERTIES:
Form:
Liquid (Cithrol GDL S/E)
Composition:
100% conc. (Cithrol GDL S/E)
Solubility:
Sol. in oil (Emulsynt GDL)
Disp. in hot water (Emulsynt GDL)
Ionic Nature:
Anionic (Cithrol GDL S/E)

Glyceryl hydroxystearate *(CTFA)*

SYNONYMS:
Hydroxystearic acid, monoester with glycerol
Stearic acid, hydroxy-, monoester with glycerol
EMPIRICAL FORMULA:
$C_{21}H_{42}O_5$
STRUCTURE:

$$CH_3(CH_2)_5\underset{\underset{OH}{|}}{CH}(CH_2)_{10}\overset{\overset{O}{\|}}{C}-OCH_2\underset{\underset{OH}{|}}{CH}CH_2OH$$

CAS No.:
1323-42-8

TRADENAME EQUIVALENTS:
Naturechem GMHS [CasChem]
Paricin 13 [CasChem]
Softigen 701 [Dynamit-Nobel]
CATEGORY:
Thickener, bodying agent, wax modifier, refatting agent, skin protective agent
APPLICATIONS:
Bath products: (Softigen 701)
Cosmetic industry preparations: (Naturechem GMHS; Softigen 701); creams and lotions (Naturechem GMHS; Softigen 701); shaving preparations (Softigen 701); soaps (Softigen 701)
PROPERTIES:
Form:
Flake (Naturechem GMHS)
Color:
White (Naturechem GMHS)
Composition:
100% active (Naturechem GMHS; Paricin 13)
Ionic Nature:
Nonionic (Naturechem GMHS)
M.P.:
69 C (Paricin 13)

Glyceryl monococoate

SYNONYMS:
Glycerides, coconut oil mono-
Glycerol mono coconut oil
Glyceryl cocoate (CTFA)
Glyceryl coconate
STRUCTURE:

$$RC\underset{\underset{O}{\|}}{-}OCH_2\underset{\underset{OH}{|}}{C}HCH_2OH$$

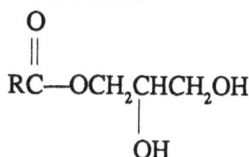

where RCO⁻ represents the coconut radical
CAS No.
61789-05-7

Glyceryl monococoate (cont'd.)

TRADENAME EQUIVALENTS:
Aldo MC [Glyco]
Drewmulse 75 [PVO]
Drewmulse CNO [PVO]
Radiasurf 7144 [Oleofina SA]

CATEGORY:
Emulsifier, stabilizer, opacifier, viscosity builder

APPLICATIONS:
Cosmetic industry preparations: (Drewmulse CNO); conditioners (Drewmulse CNO); cosmetic base (Drewmulse CNO); creams and lotions (Drewmulse CNO); hair rinses (Drewmulse CNO)
Food applications: food emulsifying (Drewmulse 75)
Pharmaceutical applications: (Drewmulse CNO)

PROPERTIES:

Form:
Paste (Drewmulse 75; Radiasurf 7144)
Soft solid (Aldo MC)
Solid (Drewmulse CNO)

Color:
Cream (Aldo MC)
Gardner 4 (Drewmulse CNO)

Composition:
50% mono content (Drewmulse CNO)
100% active (Drewmulse 75)
100% conc. (Aldo MC)

Solubility:
Sol. in ethanol (Aldo MC)
Sol. in ethyl acetate (Aldo MC)
Sol. in min. oils (Aldo MC)
Sol. in naphtha (Aldo MC)
Sol. in toluol (Aldo MC)
Sol. in veg. oil (Aldo MC)
Disp. in water (Aldo MC)

Ionic Nature: Nonionic

Sp.gr.:
0.97 (Aldo MC)

M.P.:
20–26 C (Aldo MC)

HLB:
3.0 (Radiasurf 7144)
3.8 (Drewmulse CNO)
6.8 (Aldo MC)

Acid No.:
  < 6 (Aldo MC)
Iodine No.:
  < 10 (Aldo MC)
Saponification No.:
  180–190 (Aldo MC)
pH:
  7.5–8.5 (5% aq.) (Aldo MC)

Glyceryl monoricinoleate

SYNONYMS:
  Glycerol monoricinoleate
  Glyceryl ricinoleate (CTFA)
  12-Hydroxy-9-octadecenoic acid, monoester with 1,2,3-propanetriol
  Monoricinolein
  9-Octadecenoic acid, 12-hydroxy-, monoester with 1,2,3-propanetriol
  Ricinolein, 1-mono-
EMPIRICAL FORMULA:
  $C_{21}H_{40}O_5$
STRUCTURE:

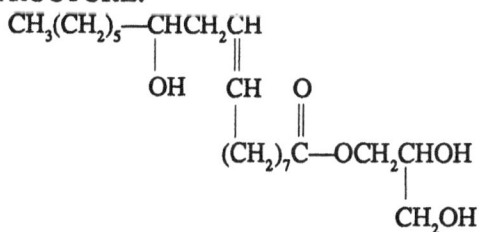

CAS No.:
  141-08-2; 1323-38-2
TRADENAME EQUIVALENTS:
  Aldo MR [Glyco]
  Alkamuls GMR [Alkaril]
  Drewmulse GMRO [Drew Produtos]
  Flexricin 13 [NL Industries]
  Hodag GMR, GMR-D [Hodag]
  Radiasurf 7153 [Oleofina]
  Softigen 701 [Dynamit Nobel]

Glyceryl monoricinoleate *(cont'd.)*

CATEGORY:
Emulsifier, solvent, plasticizer, lubricant, refatting agent, stabilizer, opacifier, pearlescent, wetting agent, intermediate, detergent, lubricant, release agent, antistat, antifogging agent, antifoam agent

APPLICATIONS:
Bath products: (Softigen 701)
Cleansers: (Softigen 701)
Cosmetic industry preparations: (Flexricin 13; Hodag GMR, GMR-D; Radiasurf 7153; Softigen 701); creams and lotions (Softigen 701); ointments (Softigen 701); shampoos (Drewmulse GMRO)
Food applications: (Alkamuls GMR; Hodag GMR, GMR-D)
Household applications (Flexricin 13)
Household detergents: (Radiasurf 7153)
Industrial applications: dyes and pigments (Drewmulse GMRO; Radiasurf 7153); industrial processing (Drewmulse GMRO); lubricating/cutting oils (Radiasurf 7153); paint mfg. (Radiasurf 7153); paper mfg. (Radiasurf 7153); plastics (Radiasurf 7153); polishes and waxes (Flexricin 13); polymers/polymerization; printing inks (Radiasurf 7153); rubber (Flexricin 13); textile/leather processing (Alkamuls GMR; Flexricin 13; Radiasurf 7153)
Pharmaceutical applications: (Hodag GMR, GMR-D; Radiasurf 7153; Softigen 701); ointments (Softigen 701); suppositories (Softigen 701)

PROPERTIES:
Form:
Liquid (Aldo MR; Alkamuls GMR; Drewmulse GMRO; Flexricin 13; Hodag GMR, GMR-D; Radiasurf 7153)
Clear oily liquid (Softigen 701)
Color:
Yellowish-white (Softigen 701)
Yellow (Aldo MR)
Gardner 2 min. (Flexricin 13)
Odor:
Characteristic (Softigen 701)
Composition:
40% monoester content (Radiasurf 7153)
100% active (Alkamuls GMR; Flexricin 13)
100% conc. (Aldo MR; Hodag GMR, GMR-D; Softigen 701)
Solubility:
Sol. in benzene (Softigen 701); (@ 10%) (Radiasurf 7153)
Sol. in butyl acetate (Flexricin 13)
Sol. in chloroform (Softigen 701)
Sol. in dichlorethylene (Softigen 701)
Sol. in ethanol (Aldo MR; Flexricin 13)
Sol. in ether (Softigen 701)

Sol. in ethyl acetate (Aldo MR)
Sol. in isopropanol (@ 10%) (Radiasurf 7153)
Sol. in MEK (Flexricin 13)
Sol. in methanol (Aldo MR)
Sol. cloudy in min. oils (@ 10%) (Radiasurf 7153)
Sol. in toluene (Flexricin 13; Softigen 701)
Sol. in toluol (Aldo MR)
Sol. cloudy in trichlorethylene (@ 10%) (Radiasurf 7153)
Sol. in veg. oil (Aldo MR); sol. cloudy (@ 10%) (Radiasurf 7153)
Disp. in water (Aldo MR)
Sol. in xylene (Softigen 701)

Ionic Nature:
Nonionic

M.W.:
554 avg. (Radiasurf 7153)

Sp.gr.:
0.979–0.981 (30 C) (Softigen 701)
0.982 (37.8 C) (Radiasurf 7153)
0.985 (Flexricin 13)
1.02 (Aldo MR)

Visc.:
8.8 stokes (Flexricin 13)
356.20 cps (37.8 C) (Radiasurf 7153)
640–660 cps (30 C) (Softigen 701)

M.P.:
< –8 C (Aldo MR)

Flash Pt.:
229 C (COC) (Radiasurf 7153)

Cloud Pt.:
–16 C (Radiasurf 7153)

HLB:
2.5 avg. (Radiasurf 7153)
5.6 (Aldo MR)

Acid No.:
3 max. (Softigen 701)
< 4 (Aldo MR)

Iodine No.:
66–69 (Aldo MR)
72–78 (Softigen 701)

Saponification No.:
122–132 (Aldo MR)
155–170 (Softigen 701)

Glyceryl monoricinoleate *(cont'd.)*

Ref. Index:
1.4774 (Radiasurf 7153)
pH:
8.3–9.3 (Aldo MR)
STORAGE/HANDLING:
Keep containers closed; avoid contact with oxidizing materials (Flexricin 13)
STD. PKGS.:
25-kg net tinned casks, 200-kg net barrels (Softigen 701)
55-gal drums (Flexricin 13)

Glyceryl triacetate

SYNONYMS:
Acetic, 1,2,3-propanetriyl ester
Glycerol triacetate
1,2,3-Propanetriol, triacetate
Triacetin (CTFA)
EMPIRICAL FORMULA:
$C_9H_{14}O_6$
STRUCTURE:

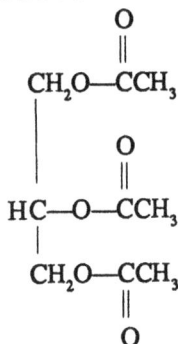

CAS No.:
102-76-1
TRADENAME EQUIVALENTS:
Kessco Triacetin [Armak]
Kodaflex Triacetin [Eastman]
CATEGORY:
Emulsifier, stabilizer, plasticizer, solvent, fixative, emollient, opacifier, bodying agent, topical antifungal

158

APPLICATIONS:

Cosmetic industry preparations: (Kessco Triacetin); perfumery (Kessco Triacetin)

Food applications: flavor/fragrances (Kessco Triacetin); food packaging (Kodaflex Triacetin)

Industrial applications: adhesives (Kodaflex Triacetin); paper/paperboard (Kodaflex Triacetin); plastics (Kodaflex Triacetin); resinous and polymeric coatings (Kodaflex Triacetin)

PROPERTIES:

Form:

Liquid (Kodaflex Triacetin)

Oily liquid (Kessco Triacetin)

Color:

APHA 50 max. (Kessco Triacetin)

Pt-Co 5 (Kodaflex Triacetin)

Odor:

Slight, fatty (Kodaflex Triacetin)

Solubility:

Sol. in benzol (Kessco Triacetin)

Partly sol. in castor oil (Kessco Triacetin)

Sol. in ethyl acetate (Kessco Triacetin)

Sol. in isopropanol (Kessco Triacetin)

Sol. in MEK (Kessco Triacetin)

Sol. in methanol (Kessco Triacetin)

Sol. in propylene glycol (Kessco Triacetin)

Moderate sol. in water (Kodaflex Triacetin); 6.1% sol. (Kessco Triacetin)

M.W.:

218 (theoret.) (Kodaflex Triacetin)

Sp.gr.:

1.152–1.158 (25/20 C) (Kessco Triacetin)

1.160 (20/20 C) (Kodaflex Triacetin)

Density:

1.16 kg/l (20 C) (Kodaflex Triacetin)

9.6 lb/gal (Kessco Triacetin)

Visc.:

16 cps (Kessco Triacetin)

17 cP (Kodaflex Triacetin)

F.P.:

−50 C (gel) (Kessco Triacetin)

B.P.:

258 C (760 mm) (Kodaflex Triacetin)

Flash Pt.:

290 F (COC) (Kessco Triacetin)

Glyceryl triacetate (cont'd.)

Acidity:
   0.005% (as acetic) (Kessco Triacetin)
Ref. Index:
   1.429 (Kessco Triacetin)
STD. PKGS.:
   55-gal bung-type steel drum (Kessco Triacetin)

Glyceryl triacetyl ricinoleate (CTFA)

SYNONYMS:
   9-Octadecenoic acid, 12-(acetyloxy)-, 1,2,3-propanetriol ester
   1,2,3-Propanetriyl 12-(acetoxy)-9-octadecenoate
EMPIRICAL FORMULA:
   $C_{63}H_{110}O_{12}$
STRUCTURE:

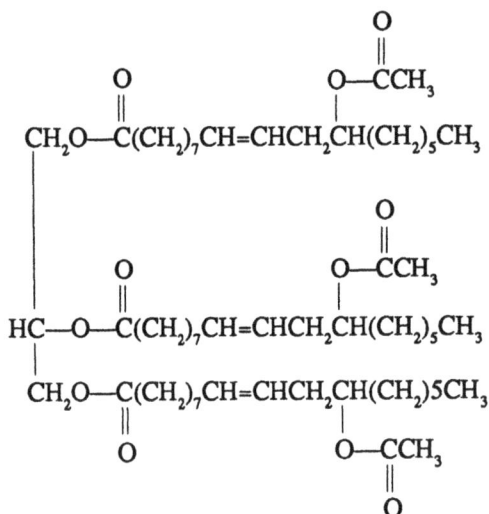

CAS No.:
   101-34-8
TRADENAME EQUIVALENTS:
   Flexricin P-8 [CasChem Inc.]
CATEGORY:
   Stabilizer

Glyceryl triacetyl ricinoleate (cont'd.)

APPLICATIONS:
  Industrial applications: anhydrous pigmented systems; plastics
PROPERTIES:
Form:
  Liquid
Composition:
  100% active
Solubility:
  Sol. in most organic solvents

Glyceryl tri (12-hydroxystearate)

SYNONYMS:
  12-Hydroxyoctadecanoic acid, 1,2,3-propanetriol ester
  Octadecanoic acid, 12-hydroxy-1,2,3,-propanetriol ester
  Trihydroxystearin (CTFA)
EMPIRICAL FORMULA:
  $C_{57}H_{110}O_9$
STRUCTURE:

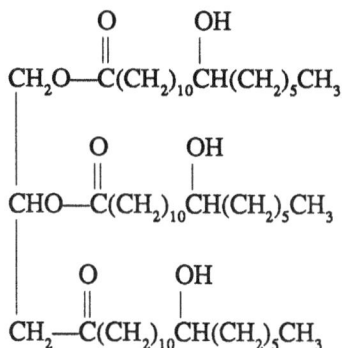

CAS No.:
  139-44-6
TRADENAME EQUIVALENTS:
  Thixcin E, R [NL Industries]
CATEGORY:
  Thickener, thixotrope, gellant
APPLICATIONS:
  Industrial applications: (Thixcin E, R); coatings/paints (Thixcin R); pigmented systems (Thixcin E, R)

Glyceryl tri (12-hydroxystearate) *(cont'd.)*

PROPERTIES:
Form:
  Fine powder (Thixcin R)
Color:
  White (Thixcin R)
Composition:
  100% conc. (Thixcin R)
Density:
  1.02 g/cm^3 (Thixcin R)

Guar gum (CTFA)

SYNONYMS:
  Guar flour
CAS No.:
  9000-30-0
TRADENAME EQUIVALENTS:
  Galactasol Series [Henkel]
  Indalca CD 30 [Hercules]
  Progacyl COS-1 [Lyndal]
  Supercol U [Henkel]
    Generically sold by:
      [Fallek; Hercules]
CATEGORY:
  Thickener, bodying agent, stabilizer, binder, lubricant, emulsifier, flocculant, suspending aid, hydration aid
APPLICATIONS:
  Cosmetic industry preparations: (generic—Fallek); personal care products (Progacyl COS-1)
  Food applications: (generic—Hercules, Fallek)
  Industrial applications: industrial processing (aq. systems) (Progacyl COS-1); paper coating (generic); petroluem industry (generic); polishing (generic); printing (generic); textiles (generic)
  Pharmaceutical applications: (generic—Fallek); tablet mfg. (generic)
PROPERTIES:
Form:
  Powder (coarse to fine grinds avail.) (generic—Hercules)
  Free-flowing powder (generic)

Guar gum *(cont'd.)*

Color:
  Yellowish-white (generic)
Solubility:
  Practically insol. in esters (generic)
  Practically insol. in hydrocarbons (generic)
  Practically insol. in ketones (generic)
  Practically insol. in oils, greases (generic)
  Sol. in water (generic—Fallek; Progacyl COS-1)
Visc.:
  1000–> 6000 cps (generic—Hercules)
Stability:
  Little affected by variations in pH between 2.0 and 9.0; relatively unaffected by common salts in sol'n. (Galactasol Series)

Guar hydroxypropyltrimonium chloride (CTFA)

SYNONYMS:
  Guar, gum, 2-hydroxy-3-(trimethylammonio) propyl ether, chloride
  Guar hydroxypropyl trimethyl ammonium chloride
CAS No.:
  65497-29-2
TRADENAME EQUIVALENTS:
  Cationic Guar C-261 [Henkel/Henkel Canada]
  Cosmedia Guar C-261 [Henkel/Henkel Canada/Henkel KGaA]
CATEGORY:
  Viscosity builder, thickener, stabilizer, conditioning agent
APPLICATIONS:
  Cosmetic industry preparations: conditioners (Cationic Guar C-261; Cosmedia Guar C-261); hair preparations (Cosmedia Guar C-261); personal care products (Cosmedia Guar C-261); shampoos (Cationic Guar C-261)
PROPERTIES:
Form:
  Solid (Cationic Guar C-261)
  Powder (Cosmedia Guar C-261)
Solubility:
  Disp. in water (Cosmedia Guar C-261)
Composition:
  100% conc. (Cationic Guar C-261)
Ionic Nature:
  Cationic (Cationic Guar C-261)

Hydrogenated tallow amide (CTFA)

SYNONYMS:
Amides, tallow, hydrogenated
Tallow amides, hydrogenated

STRUCTURE:

$$O$$
$$\|$$
$$RC-NH_2$$

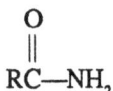

where RCO⁻ represents the hydrogenated tallow acid radicals

CAS No.:
61790-31-6

TRADENAME EQUIVALENTS:
Armid HT [Armak]

CATEGORY:
Lubricant, stabilizer, solubilizer, dispersant, thickener, intermediate, foam booster/ stabilizer, antifoam

APPLICATIONS:
Cosmetic industry preparations: hair preparations
Food applications: food packaging
Household detergents: synthetic detergents
Industrial applications: chemical synthesis; dyes and pigments; lubricating/cutting oils; paint and coatings mfg.; paper mfg.; plastics; printing inks; rust inhibitors; steam generator systems; textile/leather processing

PROPERTIES:
Form:
Flakes, powder
Color:
Gardner 7 max.
Composition:
90% amide min.
Solubility:
Only slightly sol. at R.T. in fats and fatty acids; sol. increases with increasing temps.
Only slightly sol. at R.T. in organic solvents; sol. increases with increasing temps.
Insol. in water
M.W.:
277

Sp.gr.:
   0.851 (100 C)
Fineness:
   99% –60 mesh (Armid HT, powder grade)
Visc.:
   16 cps (Brookfield #2 spindle, 10 and 20 rpm, 220 F)
M.P.:
   98–103 C
Flash Pt.:
   437 F (COC)
Fire Pt.:
   482 F
Iodine No.:
   5 max.
STD. PKGS.:
   60-lb cardboard containers (Armid HT, flakes and powder)

Hydroquinone (CTFA)

SYNONYMS:
   1,4-Benzenediol
   p-Dihydroxy benzene
   Hydrochinone
   Hydroquinol
   p-Hydroxyphenol
   Quinol
EMPIRICAL FORMULA:
   $C_6H_6O_2$
STRUCTURE:

CAS No.:
   123-31-9

Hydroquinone *(cont'd.)*

TRADENAME EQUIVALENTS:
Tecquinol [Eastman]
CATEGORY:
Antioxidant
APPLICATIONS:
Industrial applications: cellulosics; latexes; plastics; resins
PROPERTIES:
Form:
Crystals
Color:
White to gray

Hydroxyethylcellulose *(CTFA)*

SYNONYMS:
Cellulose, 2-hydroxyethyl ether
H. E. Cellulose
STRUCTURE:
Modified cellulose polymer which contains hydroxyethyl side chains
CAS No.:
9004-62-0; 9063-65-4
TRADENAME EQUIVALENTS:
Cellosize Hydroxyethyl Cellulose [Union Carbide]
Cellosize Hydroxyethyl Cellulose WP 3, WP 09 [Union Carbide]
Cellosize Hydroxyethyl Cellulose QP 40, QP 100M, QP 300, QP 4400, QP 15,000, QP 30,000, QP 52,000 [Union Carbide] (quick dispersion grades)
Natrosol, Natrosol 250 [Hercules]
Tylose H Series [Hoechst AG]
CATEGORY:
Thickener, dispersant, protective colloid, binder, stabilizer, suspending agent, plasticizer, viscosity control agent, filler
APPLICATIONS:
Cleansers: soaps and hand cleaning pastes (Tylose H Series)
Cosmetic industry preparations: (Tylose H Series)
Industrial applications: (Natrosol); adhesives (Natrosol, 250); ceramics (Natrosol, 250; Tylose H Series); latex applications (Cellosize Hydroxyethyl Cellulose; Natrosol 250); paints and coatings mfg. (Natrosol, 250); paper mfg. (Natrosol); petroleum industry (Natrosol 250); emulsion polymerization (Cellosize Hydroxy-

ethyl Cellulose; Natrosol; Tylose H Series); textile/leather processing (Natrosol); textile printing pastes/inks (Natrosol 250; Tylose H Series); tobacco industry (Tylose H Series)

Industrial cleaners: metal processing surfactants (Natrosol 250)

Pharmaceutical applications: (Natrosol)

PROPERTIES:

Form:

Powder (Cellosize Hydroxyethyl Cellulose; Tylose H Series with suffix "p")

Granules (Tylose H Series)

Free-flowing powder (Natrosol)

Free-flowing granular powder (Natrosol 250)

Color:

White (Natrosol)

White to cream (Cellosize Hydroxyethyl Cellulose)

White to light tan (Natrosol 250)

Odor:

Odorless (Natrosol)

Taste:

None (Natrosol)

Solubility:

Insol. in most organic solvents (Cellosize Hydroxyethyl Cellulose WP 3, WP 09, QP 40, QP 100M, QP 300, QP 4400, QP 15,000, QP 30,000, QP 52,000); insol. in common nonaq. organic solvents (Natrosol 250)

Swells or partially sol. in certain polar solvents, those generally miscible with water (Cellosize Hydroxyethyl Cellulose WP 3, WP 09, QP 40, QP 100M, QP 300, QP 4400, QP 15,000, QP 30,000, QP 52,000)

Readily sol. in water at any temp. (Tylose H Series); sol. (Cellosize Hydroxyethyl Cellulose WP 3, WP 09, QP 40, QP 100M, QP 300, QP 4400, QP 15,000, QP 30,000, QP 52,000); sol. in cold or hot water up to practical visc. limits (Natrosol 250); dissolves readily in hot or cold water to give clear, smooth, viscous solutions (Natrosol); delayed swelling time, insol. during first few minutes (Tylose H 4000 x)

Ionic Nature:

Nonionic (Natrosol, 250)

Sp.gr.:

1.38–1.40 (Cellosize Hydroxyethyl Cellulose)

Fineness:

90% min. thru U.S. no. 40 mesh (Natrosol 250)

Softening Pt.:

285 F (Cellosize Hydroxyethyl Cellulose)

Stability:

Unaffected by high concs. of sol. salts in sol'n.; its viscosity is little affected by mild acids and alkalies (Natrosol 250)

Hydroxyethylcellulose *(cont'd.)*

pH:
 6.0–8.5 (2% sol'n.) (Natrosol 250)
STD. PKGS.:
 Bag, drum, special tankwagon (Cellosize Hydroxyethyl Cellulose)

2-Hydroxy-4-methoxybenzophenone

SYNONYMS:
 Benzophenone-3 (CTFA)
 (2-Hydroxy-4-methoxyphenyl) phenylmethanone
 Methanone, (2-hydroxy-4-methoxyphenyl) phenyl-
 Oxybenzone
EMPIRICAL FORMULA:
 $C_{14}H_{12}O_3$
STRUCTURE:

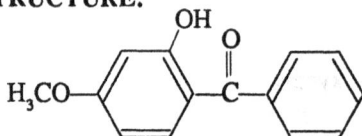

CAS No.:
 131-57-7
TRADENAME EQUIVALENTS:
 Acetorb A [Aceto]
 Anti UVA [Aceto]
 Cyasorb UV 9 [Amer. Cyanamid]
 Syntase 62 [Neville-Synthese]
 UV-Absorber 325 [Bayer AG; Mobay]
 Uvasorb MET-3 [3-V Chem.]
 Uvinul M-40 [BASF-Wyandotte]
CATEGORY:
 UV absorber, light stabilizer
APPLICATIONS:
 Cosmetic industry preparations: oil-based cosmetics (Syntase 62; Uvinul M-40)
 Food applications: food packaging (Cyasorb UV 9; Uvinul M-40)
 Industrial applications: adhesives (Cyasorb UV 9); cellulosics (Acetorb A; UV-Absorber 325; Uvasorb MET-3; Uvinul M-40); coatings (Cyasorb UV 9; Syntase 62; UV-Absorber 325; Uvinul M-40); plastics (Acetorb A; Cyasorb UV 9; Syntase 62; UV-Absorber 325; Uvasorb MET-3; Uvinul M-40)

2-Hydroxy-4-methoxybenzophenone *(cont'd.)*

Pharmaceutical applications: sunscreen (Anti UVA)

PROPERTIES:

Form:
Powder (Acetorb A; Anti UVA; Cyasorb UV 9; Syntase 62; UV-Absorber 325; Uvasorb MET-3; Uvinul M-40)

Color:
White to pale cream (Cyasorb UV 9)
Pale cream (Syntase 62)
Slightly yellowish (UV-Absorber 325)

Composition:
98% min. purity (Syntase 62)
100% active (Uvinul M-40)

Solubility:
Sol. in alcohols (Syntase 62)
Sol. in aliphatic hydrocarbons (Syntase 62)
Sol. in aromatic hydrocarbons (Syntase 62)
Sol. 56.2 g/100 g in benzene (Cyasorb UV 9)
Sol. 34.5 g/100 g in carbon tetrachloride (Cyasorb UV 9)
Sol. in esters (Syntase 62)
Sol. 5.8 g/100 g in 95% ethanol (Cyasorb UV 9)
Sol. in ketones (Syntase 62)
Sol. 59.5 g/100 g in methylene chloride (Cyasorb UV 9)
Sol. in organic solvents (UV-Absorber 325)
Sol. in paint binders (UV-Absorber 325)
Sol. in plasticizers (UV-Absorber 325)
Sol. 51.2 g/100 g in styrene (Cyasorb UV 9)
Insol. in water (Syntase 62)

M.W.:
228 (Cyasorb UV 9)

Sp.gr.:
1.324 (Cyasorb UV 9)
1.339 (Syntase 62)

Bulk Density:
3.2 lb/gal (Cyasorb UV 9)

B.P.:
150–160 C (5 mm Hg) (Cyasorb UV 9)

M.P.:
62 C min. (Syntase 62)
62–63.5 C (Acetorb A; Uvasorb MET-3)
63–64.5 C (Cyasorb UV 9)

Set Pt.:
62 C (Cyasorb UV 9)

2-Hydroxy-4-methoxybenzophenone *(cont'd.)*

Stability:
Stable to heat; not impaired by processing (UV-Absorber 325)
TOXICITY/HANDLING:
Not a skin or eye irritant (Cyasorb UV 9)
STD. PKGS.:
Drums (Anti UVA)

2-Hydroxy-4-methoxybenzophenone-5-sulfonic acid

SYNONYMS:
Benzenesulfonic acid, 5-benzoyl-4-hydroxy-2-methoxy-
Benzophenone-4 (CTFA)
5-Benzoyl-4-hydroxy-2-methoxybenzene sulfonic acid
Sulisobenzone
EMPIRICAL FORMULA:
$C_{14}H_{12}O_6S$
STRUCTURE:

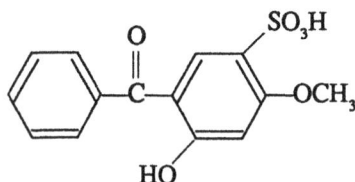

CAS No.:
4065-45-6
TRADENAME EQUIVALENTS:
Syntase 230 [Neville]
Uvinul MS-40 [BASF Wyandotte]
CATEGORY:
UV absorber, sunscreen
APPLICATIONS:
Cosmetic industry preparations: (Syntase 230; Uvinul MS-40); body creams (Syntase 230); hair preparations (Syntase 230; Uvinul MS-40); perfumery (Syntase 230; Uvinul MS-40); shampoos (Syntase 230; Uvinul MS-40); skin preparations (Uvinul MS-40)
Industrial applications: textile/leather processing (Syntase 230; Uvinul MS-40)
Pharmaceutical applications: sunscreens (Syntase 230; Uvinul MS-40)

PROPERTIES:

Form:
Solution (Uvinul MS-40)
Powder (Syntase 230; Uvinul MS-40)

Color:
Pale yellow (Syntase 230)

Composition:
99.0% min. purity (Syntase 230)
100% active (Uvinul MS-40)

Solubility:
Sol. 12% in acetone (Syntase 230)
Sol. in alcohols (Syntase 230; Uvinul MS-40)
Insol. in ethyl acetate (Syntase 230)
Insol. in n-hexane (Syntase 230)
Insol. in MEK (Syntase 230)
Sol. 59.3% in methanol (Syntase 230)
Insol. in toluene (Syntase 230)
Very sol. in water (Uvinul MS-40); sol. 38.2% (Syntase 230)

M.W.:
308 (Syntase 230)

Sp.gr.:
1.262 (Syntase 230)

M.P.:
110 C (Syntase 230)

Acid No.:
180 (Syntase 230)

pH:
0.6 (10% aq.) (Syntase 230)

Hydroxyoctacosanyl hydroxystearate (CTFA)

SYNONYMS:
12-Hydroxystearic acid, beta-hydroxyoctacosanyl ester

STRUCTURE:

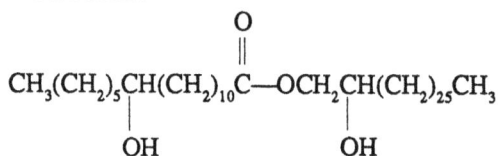

$$CH_3(CH_2)_5\underset{\underset{OH}{|}}{CH}(CH_2)_{10}C\overset{\overset{O}{\|}}{-}OCH_2\underset{\underset{OH}{|}}{CH}(CH_2)_{25}CH_3$$

Hydroxyoctacosanyl hydroxystearate *(cont'd.)*

TRADENAME EQUIVALENTS:
Elfacos C26 [Akzo Chemie BV]
CATEGORY:
Stabilizer, consistency regulator
APPLICATIONS:
Cosmetic industry preparations: decorative cosmetics ; w/o emulsions
PROPERTIES:
Form:
Waxy
Composition:
100% solids
Ionic Nature:
Nonionic

2-Hydroxy-4-(octyloxy) benzophenone

SYNONYMS:
Benzophenone-12 (CTFA)
2-Hydroxy-4-*n*-octoxybenzophenone
[2-Hydroxy-4-(octyloxy) phenyl] phenylmethanone
Methanone, [2-hydroxy-4-(octyloxy) phenyl] phenyl-
Octabenzone
2-Hydroxy-4-*n*-octoxybenzophenone
EMPIRICAL FORMULA:
$C_{21}H_{26}O_3$
STRUCTURE:

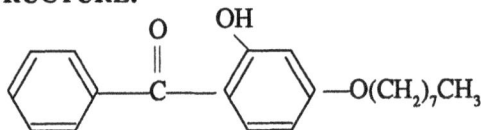

CAS No.:
1843-05-6
TRADENAME EQUIVALENTS:
Carstab 700, 705 [Cincinnati Milacron]
Cyasorb UV 531 [Amer. Cyanamid]
Mark 1413 [Argus]
Syntase 800 [Neville]
UV-Chek AM-300, AM-301 [Ferro]
Uvinul 408 [BASF Wyandotte]

2-Hydroxy-4-(octyloxy) benzophenone (cont'd.)

CATEGORY:
UV light absorber, antioxidant, stabilizer

APPLICATIONS:
Automotive products (Cyasorb UV 531; UV-Chek AM-300/301)
Cosmetic applications (Carstab 705)
Farm products: greenhouse film (UV-Chek AM-300/301); garden products (UV-Chek AM-300/301)
Food applications: food packaging materials (Carstab 700)
Industrial applications: adhesives (Cyasorb UV 531); cellulosics (Carstab 700, 705; UV-Chek AM-300/301); plastics (Carstab 700, 705; Cyasorb UV 531; Mark 1413; Syntase 800; UV-Chek AM-300/301; Uvinul 408); polymers (Carstab 700, 705; Mark 1413; Syntase 800; UV-Chek AM-300/301; Uvinul 408); protective coatings (Carstab 705; Cyasorb UV 531; Syntase 800; UV-Chek AM-300/301); resins (Syntase 800)

PROPERTIES:
Form:
Crystalline powder (Carstab 700, 705; UV-Chek AM-300)
Free-flowing crystalline powder (UV-Chek AM-301)
Powder (Cyasorb UV 531; Mark 1413; Syntase 800)
Color:
Off-white (Carstab 700, 705)
Pale yellow (Cyasorb UV 531; Syntase 800; UV-Chek AM-300, AM-301; Uvinul 408)
Light straw (Mark 1413)
Composition:
98.5% active min. (Carstab 700; Uvinul 408)
99% purity min. (Syntase 800)
Solubility:
Sol. 332 g/100 g @ 30 C in acetone (Uvinul 408); sol. 74.3 g/100 ml (Carstab 700); 74.3 g/100 g acetone (Cyasorb UV 531)
Low sol. in alcohols (UV-Chek AM-300, AM-301)
Sol. in aliphatic hydrocarbons (Carstab 705; Syntase 800; UV-Chek AM-300, AM-301)
Sol. in aromatic hydrocarbons (Carstab 705; Syntase 800; UV-Chek AM-300, AM-301)
Sol. 72.7 g/100 ml benzene (Carstab 700); 72.7 g/100 g (Cyasorb UV 531)
Sol. 20.5 g/100 g dioctyl phthalate (Cyasorb UV 531)
Sol. in esters (Carstab 705; Syntase 800)
Sol. 5.8 g/100 g @ 30 C in ethanol (Uvinul 408); insol. (Syntase 800)
Sol. 205 g/100 g @ 30 C in ethyl acetate (Uvinul 408); sol. 180 g/100 ml (Carstab 700)
Sol. 40.1 g/100 ml in hexane (Carstab 700); 40.1 g/100 g (Cyasorb UV 531); 36 g/100 g @ 30 C (Uvinul 408)
Sol. in ketones (Carstab 705; Syntase 800; UV-Chek AM-300, AM-301)

2-Hydroxy-4-(octyloxy) benzophenone (cont'd.)

Sol. 185 g/100 g @ 30 C in MEK (Uvinul 408); sol. 65.0 g/100 ml (Carstab 700)

Sol. 3.1 g/100 g @ 30 C in methanol (Uvinul 408); insol. (Syntase 800)

Sol. 69.8 g/100 g methylene chloride (Cyasorb UV 531)

Sol. 160 g/100 g @ 30 C in methyl pyrrolidone (Uvinul 408)

Sol. 49.0 g/100 ml in perchloroethylene (Carstab 700)

Sol. 333 g/100 g @ 30 C in toluene (Uvinul 408)

Insol. in water (Carstab 700, 705; Syntase 800; UV-Chek AM-300, AM-301; Uvinul 408)

M.W.:

326 (Carstab 700; UV-Chek AM-300, AM-301; Uvinul 408)

326.1 (Cyasorb UV 531)

Sp.gr.:

1.064 (Uvinul 408)

1.160 (Carstab 700; Cyasorb UV 531; Mark 1413; Syntase 800)

1.17 (UV-Chek AM-300, AM-301)

Bulk Density:

0.71 g/ml (Carstab 705)

0.96 l/kg (Uvinul 408)

3 lb/gal (Cyasorb UV 531)

F.P.:

47.4 C (Carstab 700)

M.P.:

46–48 C (Mark 1413)

47.0 C (Carstab 705; Cyasorb UV 531)

48 C (UV-Chek AM-300, AM-301)

48–49 C (Syntase 800; Uvinul 408)

Stability:

Photo and heat stable; relatively inert; unaffected by most acidic or alkaline media (Carstab 700)

Excellent light, chemical, and heat stability (Mark 1413)

TOXICITY/HANDLING:

Low oral and dermal toxicity (Carstab 700)

Considered practically nontoxic in single doses; not irritating to rabbit eyes or skin (Cyasorb UV 531)

STD. PKGS.:

51-gal (200 lb net) polyethylene-lined fiber drums (Carstab 700, 705)

Hydroxypropylcellulose (CTFA)

SYNONYMS:
Cellulose, 2-hydroxypropyl ether
CAS No.:
9004-64-2
TRADENAME EQUIVALENTS:
Klucel, 6, EF, GF, HF, J, LF, MF [Hercules]
CATEGORY:
Thickener, suspending agent, film-former, protective colloid, stabilizer, emulsification aid, whipping aid, binder
APPLICATIONS:
Cosmetic industry preparations: (Klucel GF, HF, MF); creams and lotions (Klucel GF, HF, MF); hair preparations (Klucel GF, HF, MF); toiletries (Klucel GF, HF, MF)
Food applications: (Klucel)
Industrial applications: adhesives (Klucel); industrial processing (Klucel 6); molding and extrusion operations (Klucel); polymers/polymerization (Klucel, J); printing inks (Klucel)
Pharmaceutical applications: tablet mfg. (Klucel EF, LF)
PROPERTIES:
Form:
Granular (Klucel 6)
Powder (Klucel)
Color:
White to off-white (Klucel 6)
Off-white (Klucel)
Odor:
None (Klucel)
Taste:
None (Klucel)
Solubility:
Sol. in many polar organic solvents (Klucel)
Sol. in water below 38 C; insol. above 45 C (Klucel); cold water sol., hot water insol. above 45 C (Klucel GF, J, HF); cold water sol. (Klucel EF, LF, MF)
Sp.gr.:
1.010 (2% aq. sol'n., 30 C) (Klucel)
Bulk Density:
0.5 g/ml (Klucel)
Bulking Value:
0.04 gal/lb (in sol'n.) (Klucel)
Fineness:
99% min. through 20 mesh (Klucel, 6)
Visc.:
300–600 cps (2% conc.) (Klucel 6 type G)
2000–5000 cps (2% conc.) (Klucel 6 type M)

Hydroxypropylcellulose *(cont'd.)*

Softening Pt.:
   130 C (Klucel)
Ref. Index:
   1.337 (2% aq. sol'n.) (Klucel)
pH:
   5.0–8.5 (in aq. sol'n.) (Klucel)
   7.5–10.0 (in aq. sol'n.) (Klucel 6)
Surface Tension:
   43.6 dynes/cm (0.1% aq. sol'n.) (Klucel)
STORAGE/HANDLING:
   Absorbs moisture from the atmosphere—store in tightly closed containers and in a dry atmosphere (Klucel 6)
STD. PKGS.:
   100 lb net Leverpak drums (Klucel 6)

Hydroxypropyl guar *(CTFA)*

SYNONYMS:
   Guar gum, 2-hydroxypropyl ether
CAS No.:
   39421-75-5
TRADENAME EQUIVALENTS:
   Gelcharg HP4 [Hercules]
   Progacyl COS-20, COS-70 [Lyndal]
CATEGORY:
   Thickener, stabilizer, protective colloid
APPLICATIONS:
   Cosmetic industry preparations: personal care products (Progacyl COS-20, COS-70)
   Industrial applications: (Gelcharg HP4; Progacyl COS-70); aqueous systems (Progacyl COS-20); explosives (Gelcharg HP4)
PROPERTIES:
Form:
   Viscous liquid (Gelcharg HP4)
Composition:
   10.0% moisture (Gelcharg HP4)
Solubility:
   Sol. in alcohols (Progacyl COS-20, COS-70)
   Sol. in glycols (Progacyl COS-20, COS-70)
   Sol. in water (Gelcharg HP4; Progacyl COS-20, COS-70)

Ionic Nature:
  Nonionic (Gelcharg HP4; Progacyl COS-20)
Visc.:
  15,000 cps (2% conc.) (Gelcharg HP4)
pH:
  7.0 (sol'n.) (Gelcharg HP4)

Hydroxypropyl methylcellulose (CTFA)

SYNONYMS:
  Cellulose, 2-hydroxypropyl methyl ether
  Hypromellose
  Methyl hydroxypropyl cellulose
CAS No.:
  9004-65-3
TRADENAME EQUIVALENTS:
  Methocel E, E4M, E5, E15-LV, E50-LV, F, F4M, F50-LV, J, K, K3, K4M, K15M, K35,
    K100-LV, K100M [Dow Chem.]
CATEGORY:
  Thickener, plasticizer, stabilizer, emulsifier, gelling agent, adhesive
APPLICATIONS:
  Food applications: (Methocel E4M, E5, E15-LV, E50-LV, F4M, F50-LV, K3, K4M,
    K15M, K35, K100-LV, K100M)
  Industrial applications: ceramics (Methocel E, F, J, K); paint mfg. (Methocel F, J, K)
PROPERTIES:
Form:
  Soft gel (Methocel K3, K4M, K15M, K35, K100-LV, K100M)
  Semifirm gel (Methocel E4M, E5, E15-LV, E50-LV, F4M, F50-LV)
  Powder (Methocel E, F, J, K)
Color:
  White to slightly off-white (Methocel E, F, J, K)
Odor:
  Odorless (Methocel E, F, J, K)
Taste:
  Tasteless (Methocel E, F, J, K)
Composition:
  16.5–20.0% methoxyl; 23.0–32.0% hydroxypropoxyl (Methocel J)
  19.0–24.0% methoxyl; 4.0–12.0% hydroxypropoxyl (Methocel K)
  27–30% methoxyl; 4.0–7.5% hydroxypropoxyl (Methocel F)

Hydroxypropyl methylcellulose *(cont'd.)*

Solubility:
Sol. in water (Methocel E4M, E5, E15-LV, E50-LV, F4M, F50-LV, K3, K4M, K15M, K35, K100-LV, K100M)

Ionic Nature:
Nonionic (Methocel E, F, J, K)

Sp.gr.:
1.39 (Methocel E, F, J, K)

Density:
0.25–0.70 g/cc (apparent) (Methocel E, F, J, K)
11.6 lb/gal (Methocel F, J, K)

Visc.:
3 cps (2% aq.) (Methocel K3)
5 cps (2% aq.) (Methocel E5)
15 cps (2% aq.) (Methocel E15-LV)
35 cps (2% aq.) (Methocel K35)
50 cps (2% aq.) (Methocel E50-LV, F50-LV)
100 cps (2% aq.) (Methocel K100-LV)
4000 cps (2% aq.) (Methocel E4M, F4M, K4M)
15,000 cps (2% aq.) (Methocel K15M)
100,000 cps (2% aq.) (Methocel K100M)

Gelation Temp.:
58–64 C (Methocel E4M, E5, E15-LV, E50-LV)
62–68 C (Methocel F4M, F50-LV)
70–90 C (Methocel K3, K4M, K15M, K35, K100-LV, K100M)

F.P.:
0 C (2% aq. sol'n.) (Methocel F, J, K)

Stability:
Stable over a wide pH range of 3.0–11.0 (Methocel E4M, E5, E15-LV, E50-LV, F4M, F50-LV, K3, K4M, K15M, K35, K100-LV, K100M)

Ref. Index:
1.336 (2% aq. sol'n.) (Methocel F, J, K)

pH:
7 (aq. sol'n.) (Methocel F, J, K)

Surface Tension:
44–50 dynes/cm (Methocel F)
48–52 dynes/cm (Methocel J)
50–56 dynes/cm (Methocel K)

TOXICITY/HANDLING:
Dusts could cause skin and eye irritation under extreme conditons (Methocel E, F, J, K)

STORAGE/HANDLING:
Control dusts in air to prevent possible dust explosions (Methocel E, F, J, K)

Isostearic diethanolamide

SYNONYMS:

N,N-Bis (2-hydroxyethyl) isooctadecanamide

Isooctadecanamide, N,N-bis (2-hydroxyethyl)-

Isostearamide DEA (CTFA)

EMPIRICAL FORMULA:

$C_{22}H_{45}NO_3$

CAS No.:

52794-79-3

TRADENAME EQUIVALENTS:

Monamid 150-IS [Mona] (1:1)

Schercomid ID, SI-M [Scher]

Schercomid SI [Scher] (1:1)

Standamid ID [Henkel] (1:1)

CATEGORY:

Thickener, viscosity builder, lubricant, conditioner, emulsifier, corrosion inhibitor, detergent, foam enhancer

APPLICATIONS:

Cosmetic industry preparations: (Schercomid ID; Standamid ID); conditioners (Schercomid SI-M; Standamid ID); creams and lotions (Monamid 150-IS); hair preparations (Monamid 150-IS; Standamid ID); shampoos (Schercomid SI-M)

PROPERTIES:

Form:

Liquid (Monamid 150-IS; Standamid ID)

Clear liquid (Schercomid ID, SI)

Clear liquid to soft solid (Schercomid SI-M)

Color:

Light amber (Monamid 150-IS; Schercomid ID, SI)

Amber (Schercomid SI-M; Standamid ID)

Composition:

70% min. amide (Schercomid ID)

85% min. amide (Schercomid SI-M)

87% min. amide (Schercomid SI)

100% active (Monamid 150-IS)

Solubility:

Sol. in most alcohols (Schercomid SI); sol. in SD40 alcohol (Standamid ID)

Sol. in aliphatics (lower members) (Schercomid SI)

Isostearic diethanolamide *(cont'd.)*

Sol. in aromatic hydrocarbons (Schercomid SI); @ 10% (Monamid 150-IS)
Sol. in chlorinated hydrocarbons (Schercomid SI); @ 10% (Monamid 150-IS)
Sol. @ 10% in ethanol (Monamid 150-IS)
Sol. in glycol ethers (Schercomid SI)
Sol. in glycols (Schercomid SI)
Sol. @ 10% in kerosene (Monamid 150-IS)
Sol. @10% in min. oil (Monamid 150-IS); disp. (Schercomid SI)
Sol. @ 10% in min. spirits (Monamid 150-IS)
Sol. in natural fats (Schercomid SI)
Insol. in PEG 8 (Standamid ID)
Sol. in polyols (Schercomid SI)
Insol. in propylene glycol (Standamid ID)
Sol. @ 10% in veg. oil (Monamid 150-IS)
Disp. in water (Schercomid SI); disp. @ 10% (Monamid 150-IS); insol. (Standamid ID)

Ionic Nature:
Nonionic (Monamid 150-IS; Schercomid ID, SI, SI-M)
Visc.:
1600 cps (Standamid ID)
Gel Pt.:
< 0 C (Standamid ID)
Cloud Pt.:
< −5 C (Monamid 150-IS)
Acid No.:
2 max. (Schercomid ID, SI, SI-M)
5–10 (Monamid 150-IS)
Alkali No.:
15–40 (Schercomid SI)
20–40 (Schercomid SI-M)
30–60 (Monamid 150-IS)
110–130 (Schercomid ID)
pH:
9.5 ± 0.5 (10% disp.) (Monamid 150-IS)

Isostearoamphopropionate *(CTFA)*

SYNONYMS:
1H-Imidazolium, 1-(2-carboxyethyl)-4,5-dihydro-3-(2-hydroxyethyl)-2-isoheptadecyl-, hydroxide, inner salt

EMPIRICAL FORMULA:
$C_{25}H_{50}N_2O_4 \cdot Na$
STRUCTURE:

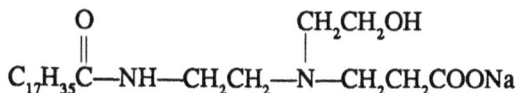

$$\begin{array}{ccc} O & & CH_2CH_2OH \\ \parallel & & | \\ C_{17}H_{35}C-NH-CH_2CH_2-N-CH_2CH_2COONa \end{array}$$

CAS No.:
68630-96-6
TRADENAME EQUIVALENTS:
Miranol ISM [Miranol]
Monateric ISA-35 [Mona]
CATEGORY:
Thickener, viscosity builder, foam booster, lubricity aid, softener, emulsifier, wetting agent, foaming agent, corrosion inhibitor, detergent, solubilizer
APPLICATIONS:
Cosmetic industry preparations: shampoos (Miranol ISM)
Household detergents: dishwashing (Miranol ISM); liquid detergents (Monateric ISA-35)
Industrial applications: aerosols (Monateric ISA-35); lubricating/cutting oils (Monateric ISA-35)
Industrial cleaners: (Monateric ISA-35); metal processing surfactants (Monateric ISA-35)
PROPERTIES:
Form:
Clear liquid (Monateric ISA-35)
Paste (Miranol ISM)
Color:
Amber (Monateric ISA-35)
Composition:
35% active (Monateric ISA-35)
Solubility:
Sol. in water (Monateric ISA-35); disp. (Miranol ISM)
Sp.gr.:
1.01 (Monateric ISA-35)
Density:
8.4 lb/gal (Monateric ISA-35)
pH:
5.4 (@ 10%) (Monateric ISA-35)
Biodegradable: (Monateric ISA-35)

Lanolin acid (CTFA)

SYNONYMS:
Fatty acids, lanolin
Lanolic acids
Lanolin fatty acids
CAS No.:
68424-43-1
TRADENAME EQUIVALENTS:
Amerlate LFA, WFA [Amerchol]
Fancor LFA [Fanning Corp.]
Nimco 1781 Lanolin Acids [Emery] (cosmetic grade)
Ritalafa [R.I.T.A.]
Skliro [Croda] (cosmetic grade)
CATEGORY:
Stabilizer, emollient, emulsifier, dispersant, suspending agent, film-former
APPLICATIONS:
Cleansers: soaps (Fancor LFA; Nimco 1781 Lanolin Acids)
Cosmetic industry preparations: (Amerlate LFA, WFA; Fancor LFA; Nimco 1781 Lanolin Acids); creams and lotions (Fancor LFA); makeup (Ritalafa); shampoos (Amerlate LFA, WFA; Fancor LFA); toiletries (Fancor LFA)
Household detergents: soap emulsions (Amerlate LFA, WFA; Fancor LFA)
Industrial applications: (Fancor LFA); dyes and pigments (Amerlate LFA); industrial processing (Skliro); water-repellent films (Skliro)
Pharmaceutical applications: (Fancor LFA)

PROPERTIES:
Form:
Wax (Amerlate LFA, WFA)
Waxy solid (Fancor LFA; Skliro)
Color:
Light (Ritalafa)
Yellow (Skliro)
Odor:
Characteristic (Skliro)
Solubility:
Sol. in isopropyl myristate (Nimco 1781 Lanolin Acids)
Sol. in min. oil (Nimco 1781 Lanolin Acids)
Sol. in oil (Ritalafa); miscible with oil (Amerlate LFA, WFA)

Lanolin acid *(cont'd.)*

M.P.:
   45–60 C (Drop) (Skliro)
   57–65 C (Fancor LFA)
Acid No.:
   135–170 (Fancor LFA)
   140–165 (Skliro)
Iodine No.:
   8–16 (Wijs) (Skliro)
   12 max. (Fancor LFA)
Saponification No.:
   155–185 (Skliro)
   158–175 (Fancor LFA)
Hydroxyl No.:
   50–70 (Skliro)

Lauramidopropyl betaine (CTFA)

SYNONYMS:
   N-(Carboxymethyl) -N,N- dimethyl -3- [(1-oxododecyl) amino] -1- propanaminium hyroxide, inner salt
   1-Propanaminium, N-(carboxymethyl)-N-N-dimethyl-3-[(1-oxododecyl)amino]-, hydroxide, inner salt

EMPIRICAL FORMULA:
   $C_{19}H_{38}N_2O_3$

STRUCTURE:

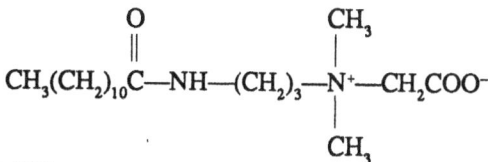

CAS No.:
   4292-10-8

TRADENAME EQUIVALENTS:
   Jortaine LMAB [Jordan]
   Mackam LMB [McIntyre]
   Mackam LMB-LS [McIntyre] (low salt)
   Mirataine BB [Miranol]
   Monateric LMAB [Mona]

Lauramidopropyl betaine *(cont'd.)*

CATEGORY:
Viscosity builder, foam booster, emulsifier, wetting agent, foaming agent, corrosion inhibitor, detergent, solubilizer

APPLICATIONS:
Bath products: bubble bath (Mackam LMB)

Cosmetic industry preparations: (Jortaine LMAB); shampoos (Mackam LMB; Mirataine BB)

Farm products: agricultural oils/sprays (Monateric LMAB)

Household detergents: dishwashing (Mackam LMB; Mirataine BB); laundry detergents (Monateric LMAB); liquid detergents (Jortaine LMAB; Monateric LMAB)

Industrial applications: (Monateric LMAB); fire fighting (Monateric LMAB); paper (Monateric LMAB); textiles (Monateric LMAB)

PROPERTIES:

Form:
Liquid (Jortaine LMAB; Mackam LMB, LMB-LS)

Clear liquid (Monateric LMAB)

Color:
Light yellow (Monateric LMAB)

Composition:
30% active (Monateric LMAB)

35% active (Jortaine LMAB; Mackam LMB, LMB-LS)

Ionic Nature:
Amphoteric (Jortaine LMAB; Mackam LMB, LMB-LS; Mirataine BB; Monateric LMAB)

Sp.gr.:
1.04 (Monateric LMAB)

Density:
8.7 lb/gal (Monateric LMAB)

pH:
8.3 (10%) (Monateric LMAB)

Biodegradable: (Monateric LMAB)

Lauric acid *(CTFA)*

SYNONYMS:
n-Dodecanoic acid

EMPIRICAL FORMULA:
$C_{12}H_{24}O_2$

STRUCTURE:
$CH_3(CH_2)_{10}COOH$

CAS No.:
143-07-7

TRADENAME EQUIVALENTS:
Emery 652 [Emery]
Hetamide LA, ML [Heterene]
Hystrene 9512 [Humko Sheffield]
   Generically sold by: [Harwick]

CATEGORY:
Foam stabilizer, emulsifier, visc. builder, activator, detergent, lubricant, plasticizer, intermediate, softener

APPLICATIONS:
Automobile cleaners: car shampoo (Hetamide LA, ML)
Cleansers: hand cleanser (Hetamide LA, ML)
Cosmetic industry preparations: (Hetamide LA, ML; Hystrene 9512); shampoos (Hystrene 9512)
Degreasers: (Hetamide LA, ML)
Industrial applications: latexes (generic—Harwick); lubricating/cutting oils (Hetamide LA, ML; Hystrene 9512); paint mfg. (Hystrene 9512); rubber processing (generic—Harwick); textile/leather processing (Hystrene 9512)
Industrial cleaners: bottle washing (Hetamide LA, ML); drycleaning compositions (Hetamide LA, ML); floor cleaners (Hetamide LA, ML); metallic soaps (Hystrene 9512)
Pharmaceutical applications: (Hystrene 9512)

PROPERTIES:
Form:
Liquid (Hetamide LA)
Solid (Hetamide ML; Hystrene 9512)
Solid or semisolid (generic—Harwick)
Color:
White (generic—Harwick)
Odor:
Typical (generic—Harwick)
Composition:
100% active (Hetamide LA, ML; Hystrene 9512)
Sp.gr.:
0.85 (generic—Harwick)
0.96 (20 C) (Hetamide ML)
1.00 (20 C) (Hetamide LA)

Lauric acid *(cont'd.)*

M.P.:
   35 C (Hetamide ML)
   40–50 C (generic—Harwick)
Acid No.:
   10–14 (Hetamide LA)
   273–283 (generic—Harwick)
   276–281 (Hystrene 9512)
Iodine No.:
   0.5 max. (generic—Harwick)
Saponification No.:
   277–282 (generic—Harwick; Hystrene 9512)
pH:
   9–10 (1% aq. sol'n.) (Hetamide LA)

Lauric diethanolamide

SYNONYMS:
   N,N-Bis (2-hydroxyethyl) dodecanamide
   N,N-Bis (2-hydroxyethyl) lauramide
   Diethanolamine lauric acid amide
   Lauramide DEA (CTFA)
   Lauric acid diethanolamide
   Lauric DEA
   Lauric fatty acid diethanolamide
   Lauroyl diethanolamide
EMPIRICAL FORMULA:
   $C_{16}H_{33}O_3$
STRUCTURE:

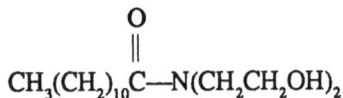

$$CH_3(CH_2)_{10}\overset{\displaystyle O}{\overset{\|}{C}}-N(CH_2CH_2OH)_2$$

CAS No.:
   120-40-1
TRADENAME EQUIVALENTS:
   Alkamide L9DE [Alkaril]
   Aminol COR-4C, LM-30-C Special [Finetex]
   Ardet DLA, LDA, LMC [Ardmore]
   Carsamide SAL-7, SAL-9, SAL-82 [Carson]
   Clindrol 200-L [Clintwood]

TRADENAME EQUIVALENTS *(cont'd.):*
  Clindrol Superamide 100L, Superamide 100LM [Clintwood]
  Comperlan LD, LDO, LDS [Henkel Canada]
  Comperlan LMD [Henkel KgaA]
  Condensate PE, PL [Continental]
  Cyclomide DL203, DL203/S, DL207/S [Cyclo]
  Emid 6510, 6511 [Emery]
  Emid 6513 [Emery]
  Emid 6519 [Emery]
  Emid 6541 [Emery]
  Empilan LDE, LDX [Albright & Wilson/Marchon]
  Ethylan MLD [Lankro]
  Hartamide LDA70, LDA90 [Hart]
  Hetamide ML [Heterene]
  Incromide L-90, LCL, LL, LM-70, LR [Croda Surfactants]
  Loramine DL 203/S [Dutton & Reinisch]
  Loropan LD [Thomas Triantaphyllou]
  Mackamide L-95, LLM, LMD [McIntyre]
  Mazamide LM-20 [Mazer]
  Mazamide LS-173, LS-196 [Mazer]
  Merpinamid KD11, LD/E, LSD/E [Kempen]
  Monamid 150-GLT, 150-LMW-C, 150-LW, 150-LWA, 716, 770 [Mona]
  Monamine ACO-100, LM-100 [Mona]
  Ninol 4821, AA-62, AA-62 Extra, P-616, P-621 [Stepan]
  Nitrene L-90 [Henkel]
  Onyxol 336, 345 [Onyx]
  Product LT 18-48 [Clintwood]
  Profan AA62 Extra [Sanyo]
  Rewomid 203/S, DL203 [Dutton & Reinisch]
  Richamide 6310 [Richardson]
  Schercomid 1214, LD, SL-Extra, SLM, SLM-C, SL-ML, SLM-LC, SLM-S [Scher]
  Sipomide 843 [Alcolac]
  Stafoam DL [Nippon Oil & Fats]
  Standamid LD, LD 80/20, LDM, LDO, LDS [Henkel]
  Steinamid 203/S [Dutton & Reinisch]
  Sterling LDEA-90 [Canada Packers]
  Super Amide L9, L9C, LL, LM [Onyx]
  Synotol L60, LM60 [PVO]
  Synotol L90 [Drew Produtos]
  Unamide J-56, LDX [Lonza]
  Varamide L-1, ML-4, SL-9 [Sherex]
  Witcamide 5195, 6310, STD-HP [Witco/Organics]

Lauric diethanolamide (cont'd.)

CATEGORY:
Detergent, surfactant, scouring aid, emulsifier, foaming agent, foam stabilizer, stabilizer, thickener, gelling agent, viscosity builder, wetting agent, penetrant, dispersant, solubilizer, coupling agent, antistat, lubricant, superfatting agent, softener, conditioner, intermediate, corrosion inhibitor

APPLICATIONS:
Automobile cleaners: (Mazamide LM-20); car shampoo (Carsamide SAL-7, SAL-9; Hetamide ML; Incromide L-90, LM-70; Standamid LDS)

Bath products: (Standamid LD, LDO); bath gels (Ninol 4821); bubble bath (Alkamide L9DE; Aminol LM-30C Special; Carsamide SAL-7, SAL-9; Clindrol 200-L; Clindrol Superamide 100L, 100LM; Comperlan LD, LDO, LDS; Condensate PE, PL; Emid 6510; Ethylan MLD; Incromide L-90, LL, LM-70; Monamid 150-LMW-C, 150-LW, 150-LWA, 716, 770; Monamine ACO-100, LM-100; Ninol 4821, AA-62 Extra, P-616, P-621; Product LT 18-48; Schercomid 1214, LD, SL-Extra, SLM, SL-ML; Sipomide 843; Standamid LD 80/20, LDM, LDS; Super Amide L9, L9C, LM; Synotol L60, LM60; Unamide LDX); shower soaps (Ninol 4821)

Cleansers: (Synotol L60, LM60); germicidal skin cleanser (Synotol L60, LM60); hand cleanser (Hetamide ML; Mazamide LM-20; Monamid 150-LMW-C, 150-LW, 150-LWA, 716, 770; Monamine ACO-100, LM-100; Ninol 4821; Sipomide 843; Standamid LDO, LDS)

Cosmetic industry preparations: (Clindrol 200-L; Comperlan LMD; Condensate PE; Cyclomide DL203/S, DL207/S; Ethylan MLD; Hetamide ML; Loropan LD; Mazamide LS-173, LS-196; Monamid 150-GLT, 150-LMW-C, 150-LW, 150-LWA, 716, 770; Monamine ACO-100, LM-100; Onyxol 345; Schercomid 1214, LD, SLM-S; Stafoam DL; Sterling LDEA-90; Super Amide L9, L9C, LM; Synotol L90; Unamide LDX; Witcamide 5195, 6310, STD-HP); conditioners (Alkamide L9DE; Monamid 150-LMW-C, 150-LW, 150-LWA, 716, 770; Monamine ACO-100, LM-100); creams and lotions (Carsamide SAL-7, SAL-9; Comperlan LD, LDO, LDS; Incromide L-90, LM-70; Schercomid SL-ML; Standamid LD 80/20); personal care products (Ninol 4821; Onyxol 336; Sipomide 843); perfumery (Rewomid DL203; Standamid LD); shampoos (Alkamide L9DE; Aminol LM-30C Special; Carsamide SAL-7, SAL-9; Clindrol 200-L; Clindrol Superamide 100L, 100LM; Comperlan LD, LDO, LDS; Condensate PE, PL; Cyclomide DL203; Emid 6510, 6541; Empilan LDE, LDX; Ethylan MLD; Incromide L-90, LL, LM-70; Loramine DL203/S; Loropan LD; Merpinamid KD11, LD/E, LSD/E; Monamid 150-LMW-C, 150-LW, 150-LWA, 716, 770; Monamine ACO-100, LM-100; Profan AA62 Extra; Ninol 4821, AA-62, AA-62 Extra, P-616, P-621; Product LT 18-48; Rewomid 203/S; Richamide 6310; Schercomid 1214, LD, SL-Extra, SLM, SL-ML; Stafoam DL; Sipomide 843; Standamid LD, LD 80/20, LDM, LDO, LDS; Steinamid 203/S; Super Amide L9, L9C, LM; Synotol L60, LM60; Unamide LDX; Varamide L-1, ML-4, SL-9); shaving preparations (Schercomid SLM, SL-ML; Synotol L60, LM60); toiletries (Alkamide L9DE; Empilan LDX; Ethylan MLD;

Mazamide LS-173, LS-196; Monamid 150-GLT; Schercomid SLM-S; Unamide LDX; Witcamide 5195, 6310, STD-HP)

Degreasers: (Hetamide ML; Sipomide 843)

Farm products: agricultural oils/sprays (Monamid 150-LMW-C, 150-LW, 150-LWA, 716, 770; Monamine ACO-100, LM-100)

Food applications: fruit/vegetable washing (Condensate PL)

Household detergents: (Alkamide L9DE; Carsamide SAL-7, SAL-9; Clindrol 200-L; Condensate PE; Incromide L-90, LM-70; Loramine DL203/S; Rewomid 203/S; Sipomide 843; Steinamid 203/S; Super Amide L9, L9C, LM; Synotol L90); all-purpose cleaner (Alkamide L9DE; Emid 6510, 6541; Schercomid SL-Extra); built detergents (Emid 6511); carpet & upholstery shampoos (Alkamide L9DE; Carsamide SAL-7, SAL-9; Incromide L-90, LM-70; Monamid 150-LMW-C, 150-LW, 150-LWA, 716, 770; Monamine ACO-100, LM-100; Sipomide 843); detergent base (Ardet DLA; Loropan LD); dishwashing (Clindrol 200-L; Clindrol Super-amide 100L, 100LM; Condensate PE, PL; Mazamide LM-20; Monamid 150-LMW-C, 150-LW, 150-LWA, 716, 770; Monamine ACO-100, LM-100; Ninol AA-62 Extra, P-616, P-621; Product LT 18-48; Richamide 6310; Schercomid SL-Extra; Stafoam DL; Standamid LD 80/20, LDS; Super Amide L9, L9C, LM; Varamide ML-4, SL-9); hard surface cleaner (Clindrol 200-L; Mazamide LM-20; Richamide 6310; Schercomid 1214, LD; Sipomide 843; Unamide J-56; Varamide SL-9); heavy-duty cleaner (Emid 6511); laundry detergent (Schercomid SL-Extra; Stafoam DL; Varamide SL-9); light-duty cleaners (Standamid LDS; Unamide J-56, LDX; Varamide SL-9); liquid detergents (Cyclomide DL203; Empilan LDE, LDX; Hartamide LDA70, LDA90; Incromide LL; Merpinamid KD11, LD/E, LSD/E; Ninol AA-62; Nitrene L-90; Schercomid SL-Extra); powdered detergents (Hartamide LDA70, LDA90; Monamid 150-LMW-C, 150-LW, 150-LWA, 716, 770; Monamine ACO-100, LM-100)

Industrial applications: (Witcamide 5195); aerosols (Witcamide 5195); dyes and pigments (Monamid 150-LMW-C, 150-LW, 150-LWA, 716, 770; Monamine ACO-100, LM-100; Stafoam DL); lubricating/cutting oils (Condensate PL; Hetamide ML; Monamid 150-LMW-C, 150-LW, 150-LWA, 716, 770; Monamine ACO-100, LM-100); metalworking (Mazamide LM-20; Monamid 150-LMW-C, 150-LW, 150-LWA, 716, 770; Monamine ACO-100, LM-100; Witcamide 5195); petroleum industry (Monamid 150-LMW-C, 150-LW, 150-LWA, 716, 770; Monamine ACO-100, LM-100); plastics (Empilan LDX; Ethylan MLD); polishes and waxes (Condensate PE, PL; Monamid 150-LMW-C, 150-LW, 150-LWA, 716, 770; Monamine ACO-100, LM-100); textile/leather processing (Condensate PE, PL; Monamid 150-LMW-C, 150-LW, 150-LWA, 716, 770; Monamine ACO-100, LM-100; Stafoam DL; Witcamide 5195)

Industrial cleaners: (Carsamide SAL-7, SAL-9; Condensate PE; Hetamide ML; Incromide L-90, LM-70; Merpinamid KD11, LD/E, LSD/E; Sipomide 843; Super Amide L9, L9C, LM; Unamide J-56; Witcamide 6310); bottle cleaners (Condensate

Lauric diethanolamide *(cont'd.)*

PE; Hetamide ML); drycleaning compositions (Aminol COR-4C; Hetamide ML; Monamid 150-LMW-C, 150-LW, 150-LWA, 716, 770; Monamine ACO-100, LM-100; Stafoam DL); institutional cleaners (Carsamide SAL-7, SAL-9; Incromide L-90, LM-70); lime soap dispersant (Synotol L60, LM60); metal processing surfactants (Monamid 150-LMW-C, 150-LW, 150-LWA, 716, 770; Monamine ACO-100, LM-100); textile cleaning (Monamid 150-LMW-C, 150-LW, 150-LWA, 716, 770; Monamine ACO-100, LM-100)

PROPERTIES:

Form:

Liquid (Aminol COR-4C, LM-30C Special; Ardet LMC; Comperlan LMD; Condensate PL; Emid 6519; Incromide LCL, LR; Mackamide LLM; Mazamide LM-20; Monamid 770; Monamine LM-100; Onyxol 336, 345; Product LT 18-48; Schercomid SLM; Standamid LDO, LDS; Super Amide LL; Synotol L60, LM60; Unamide J-56)

Clear liquid (Carsamide SAL-82; Monamid 150-GLT, 716; Ninol 4821; Schercomid 1214, SLM-C, SL-ML, SLM-LC; Standamid LD 80/20); (95 F) (Varamide SL-9)

Clear, slightly viscous liquid (Standamid LDM)

Viscous liquid (Comperlan LDO, LDS; Merpinamid KD11, LD/E, LSD/E; Schercomid LD)

Liquid to paste (Cyclomide DL203, DL207/S; Incromide LL)

Paste (Clindrol 200-L; Clindrol Superamide 100LM; Emid 6541; Monamine ACO-100; Ninol AA-62, P-621; Richamide 6310; Standamid LD; Super Amide L9, L9C, LM; Varamide L-1, ML-4; Witcamide 6310)

Liquid to wax (Ardet DLA)

Solid (Carsamide SAL-7, SAL-9; Clindrol Superamide 100L, 100LM; Condensate PE; Emid 6510, 6511, 6513; Hartamide LDA70, LDA90; Incromide L-90, LM-70; Loramine DL203/S; Loropan LD; Mackamide L-95, LMD; Mazamide LS-173, LS-196; Monamid 150-LMW-C, 150-LW, 150-LWA; Nitrene L-90; Profan AA-62 Extra; Rewomid 203/S; Stafoam DL; Sipomide 843; Steinamid 203/S; Sterling LDEA-90; Synotol L90; Unamide LDX; Witcamide STD-HP); (@ 20 C) (Empilan LDE, LDX; Hetamide ML)

Crystalline solid (Cyclomide DL203/S; Schercomid SL-Extra, SLM-S)

Wax (Ardet LDA; Ninol AA-62 Extra, P-616; Rewomid DL203)

Waxy flake (Ethylan MLD)

Waxy solid (Comperlan LD)

Waxy solid, liquid > 25 C (Alkamide L9DE)

Color:

Clear (Hartamide LDA70, LDA90)

White (Alkamide L9DE; Clindrol 200-L; Clindrol Superamide 100L, 100LM; Condensate PE; Ninol AA-62 Extra, P-616, P-621; Rewomid DL203; Schercomid SL-Extra, SLM-S; Stafoam DL; Unamide LDX)

White to off-white (Cyclomide DL203, DL207/S)

Off-white (Cyclomide DL203/S; Loramine DL203/S; Rewomid 203/S; Steinamid 203/S)

Pale cream (Ethylan MLD)

Cream (@ 20 C) (Empilan LDE, LDX)

Light amber (Condensate PL; Monamid 716; Standamid LDS)

Amber (Comperlan LDO, LDS; Monamid 150-GLT; Ninol 4821; Schercomid SLM-C, SL-ML, SLM-LC; Standamid LD 80/20, LDO; Unamide J-56)

Light yellow (Schercomid SLM)

Yellow (Carsamide SAL-82; Incromide LCL; Merpinamid KD11, LD/E, LSD/E; Product LT 18-48; Schercomid 1214, LD)

Gardner 2 (Carsamide SAL-7, SAL-9; Emid 6510, 6511, 6541; Varamide SL-9)

Gardner 2 max. (Incromide L-90, LM-70)

Gardner 4 max. (Incromide LL)

Gardner 5 max. (Incromide LR)

GVCS-33: 3 max. (Monamid 150-LMW-C, 150-LW, 150-LWA)

GVCS-33: 4 max. (Monamine LM-100, 770)

GVCS-33: 5 max. (Monamine ACO-100)

Odor:

Negligible (Ethylan MLD)

Low (Alkamide L9DE)

Bland (Incromide LL, LR)

Mild (Schercomid SLM, SL-ML)

Mild, fatty (Carsamide SAL-7, SAL-9; Incromide L-90)

Mild fruity (Schercomid SL-Extra)

Typical (Condensate PE, PL)

Typical, slight (Loramine DL203/S; Rewomid 203/S; Schercomid 1214, LD; Steinamid 203/S)

Typical alkanolamide (Clindrol 200-L; Clindrol Superamide 100L)

Composition:

50% amide (Synotol L60, LM60)

80% conc. (Standamid LD 80/20)

85% active (Monamid 770)

85% amide content min. (Schercomid SLM-C, SLM-LC)

87% amide content min. (Schercomid SLM-S)

88% conc. (Synotol L90)

90% amide content min. (Product LT 18-48; Standamid LD)

90% conc. (Loropan LD; Super Amide L9)

90% nonionic content min. (Ninol 4821)

90–95% amide content (Condensate PE)

92% active (Alkamide L9DE)

95% active (Ethylan MLD)

97% conc. (Onyxol 336, 345; Super Amide L9C)

Lauric diethanolamide *(cont'd.)*

100% active (Clindrol 200-L; Clindrol Superamide 100L, 100LM; Condensate PL; Cyclomide DL203, DL203/S, DL207/S; Emid 6510, 6511, 6541; Hartamide LDA70, LDA90; Incromide L-90, LL, LM-70, LR; Loramine DL203/S; Monamid 150-GLT, 150-LMW-C, 150-LW, 150-LWA, 716; Monamine ACO-100, LM-100; Ninol AA-62; Rewomid 203/S, DL203; Schercomid 1214, LD, SL-Extra, SLM, SL-ML; Sipomide 843; Steinamid 203/S; Sterling LDEA-90; Unamide J-56, LDX; Varamide L-1, ML-4)

100% conc. (Aminol COR-4C, LM-30C Special; Ardet DLA, LDA, LMC; Comperlan LMD; Emid 6513, 6519; Mackamide L-95, LLM, LMD; Ninol AA-62 Extra, P-616, P-621, P-616; Nitrene L-90; Profan AA-62 Extra; Standamid LDM, LDS)

Solubility:

Sol. in alcohols (Carsamide SAL-7, SAL-9; Clindrol 200-L; Clindrol Superamide 100L, 100LM; Incromide L-90, LM-70; Product LT 18-48; Schercomid 1214, LD, SL-Extra, SLM, SL-ML); sol. in lower alcohols (Standamid LDO, LDS)

Sol. in aliphatic hydrocarbons (Schercomid SL-ML); sol. in lower aliphatics (Schercomid SL-Extra, SLM); disp. in aliphatic hydrocarbons (Product LT 18-48; Schercomid 1214, LD)

Completely sol. in anionic or nonionic surfactants (Condensate PE)

Sol. in aromatic hydrocarbons (Alkamide L9DE; Carsamide SAL-7, SAL-9; Clindrol 200-L; Clindrol Superamide 100L, 100LM; Incromide L-90, LM-70; Product LT 18-48; Schercomid 1214, LD, SL-Extra, SLM, SL-ML); (@ 10%) (Monamid 150-LMW-C, 150-LW, 150-LWA, 716, 770; Monamine ACO-100, LM-100)

Sol. in benzene (> 2%) (Stafoam DL)

Disp. in butyl stearate (Emid 6510, 6511)

Sol. in chlorinated aliphatic solvents (Alkamide L9DE)

Sol. in chlorinated hydrocarbons (Carsamide SAL-7, SAL-9; Clindrol Superamide 100L, 100LM; Incromide L-90, LM-70; Product LT 18-48; Schercomid 1214, LD, SL-Extra, SLM, SL-ML); (@ 10%) (Monamid 150-LMW-C, 150-LW, 150-LWA, 716, 770; Monamine ACO-100, LM-100)

Sol. in detergent systems (Alkamide L9DE); disperses readily into liquid dishwashing detergent formulations and shampoo systems (Varamide SL-9)

Sol. in diethylene glycol (> 2%) (Stafoam DL)

Sol. in esters (Product LT 18-48)

Sol. in ethyl alcohol (@ 10%) (Monamid 150-LMW-C, 150-LW, 150-LWA, 716, 770; Monamine ACO-100, LM-100); (> 2%) (Stafoam DL)

Sol. in natural fats (Schercomid SL-Extra)

Sol. in natural fats and oils (Schercomid SLM, SL-ML); sol. (@ 10%) (Monamid 716, 770); disp. (Incromide L-90, LM-70; Schercomid 1214, LD); disp. (@ 10%) (Monamid 150-LMW-C, 150-LW, 150-LWA)

Disp. in glycerol trioleate (Emid 6510, 6511)

Sol. in glycol ethers (Schercomid 1214, LD, SL-Extra, SLM, SL-ML)

Sol. in glycols (Clindrol 200-L; Product LT 18-48; Schercomid 1214, LD, SL-Extra,

SLM, SL-ML)
Sol. in isopropyl alcohol (>2%) (Stafoam DL)
Disp. in kerosene (Incromide L-90, LM-70); (@ 10%) (Monamid 150-LMW-C, 150-LW, 150-LWA)
Sol. in ketones (Clindrol 200-L; Clindrol Superamide 100L, 100LM; Product LT 18-48)
Sol. in min. oils (Alkamide L9DE); disp. (Emid 6510, 6511; Incromide L-90, LM-70; Schercomid 1214, LD, SL-Extra, SLM, SL-ML)
Sol. in min. spirits (@ 10%) (Monamid 150-LMW-C); disp. (@ 10%) (Monamid 150-LW, 150-LWA)
Insol. in most oils (Standamid LDS)
Sol. in organic solvents (Schercomid 1214, LD, SL-ML)
Sol. in perchloroethylene (Emid 6510, 6511, 6541)
Sol. in petroleum solvents (Alkamide L9DE)
Sol. in polyethylene glycol (Standamid LDO, LDS)
Sol. in polyols (Schercomid 1214, LD, SL-Extra, SL-ML)
Sol. in propylene glycol (Standamid LDO, LDS); (> 2%) (Stafoam DL)
Sol. in Stoddard solvent (Emid 6510, 6511, 6541)
Sol. in tetrachloroethylene (> 2%) (Stafoam DL)
Sol. in veg. oil (Product LT 18-48)
Sol. in water (Clindrol 200-L; Emid 6541; Ninol AA-62; Schercomid 1214, LD; Witcamide 6310, STD-HP); sol. (@ 10%) (Monamid 716, 770; Monamine ACO-100, LM-100); sol. hazy (< 10%) (Unamide LDX); disp. (Emid 6510, 6511; Ethylan MLD; Incromide L-90, LM-70; Product LT 18-48; Schercomid SL-Extra, SLM, SL-ML; Standamid LDS); disp. in low concs., forms gels at higher levels (Condensate PE); gels (@ 10%) (Monamid 150-LMW-C, 150-LW, 150-LWA); insol. (Standamid LDO)
Disp. in white min. oil (@ 10%) (Monamid 150-LMW-C, 150-LW, 150-LWA)
Sol. in xylene (> 2%) (Stafoam DL)
Ionic Nature:
Nonionic (Alkamide L9DE; Carsamide SAL-7, SAL-9; Clindrol 200L, Superamide 100L, 100LM; Condensate PL; Emid 6510, 6511, 6541; Empilan LDE, LDX; Ethylan MLD; Hartamide LDA70, LDA90; Hetamide ML; Incromide L-90, LL, LM-70, LR; Loramine DL203/S; Mazamide LS-173, LS-196; Monamid 150-GLT. 150-LMW-C, 150-LW, 150-LWA, 716, 770; Ninol 4821, AA-62, AA-62 Extra, P-616, P-621; Onyxol 336, 345; Rewomid DL203, 203/S; Richamide 6310; Schercomid SL-Extra, SLM, SLM-C, SL-ML, SLM-LC, SLM-S; Standamid LD, LD 80/20, LDM, LDS; Steinamid 203/S; Sterling LDEA-90; Super Amide L9, L9C, LL, LM; Synotol L60, LM60; Unamide J-56, LDX; Varamide L-1, ML-4; Witcamide 5195)
Nonionic/anionic (Alkamide 2124; Mazamide LM-20; Monamine ACO-100, LM-100; Schercomid 1214, LD)

Lauric diethanolamide *(cont'd.)*

Sp.gr.:
   0.96 (25/20 C) (Super Amide L9C); (20 C) (Hetamide ML)
   0.969 (50 C) (Stafoam DL)
   0.97 (Schercomid SL-ML)
   0.97 ± 0.01 (45 C) (Schercomid SL-Extra, SLM)
   0.98 (Condensate PE; Standamid LDS); (25/20 C) (Super Amide L9, LL, LM); (20 C)
      (Monamid 716); (40 C) (Monamid 150-LMW-C, 150-LW, 150-LWA)
   0.98–1.00 (Mazamide LS-173, LS-196)
   0.985 (Standamid LDO)
   0.99 (Clindrol Superamide 100L, 100LM; Product LT 18-48; Unamide LDX); (20 C)
      (Monamid 770)
   0.99–1.01 (Mazamide LM-20)
   1.00 (Rewomid DL203); (25/20 C) (Onyxol 345)
   1.01 (Schercomid 1214); (25/20 C) (Onyxol 336); (20 C) (Monamine ACO-100)
   1.01 ± 0.01 (Schercomid LD)
   1.02 (Clindrol 200-L; Condensate PL); (20 C) (Monamine LM-100)
   1.04 (Alkamide L9DE)
Density:
   0.97 g/cc (20 C) (Rewomid DL203)
   1.0 g/cc (20 C) (Empilan LDE, LDX)
   8.0 lb/gal (Schercomid SL-ML)
   8.1 lb/gal (Carsamide SAL-9; Richamide 6310); (45 C) (Schercomid SL-Extra, SLM)
   8.2 lb/gal (Carsamide SAL-7; Monamid 150-LMW-C, 150-LW, 150-LWA, 716;
      Ninol 4821, AA-62 Extra, P-616, P-621; Standamid LDS; Unamide LDX)
   8.25 lb/gal (Monamid 770)
   8.3 lb/gal (Emid 6541; Nitrene L-90; Varamide SL-9)
   8.34 lb/gal (Ninol AA-62)
   8.4 lb/gal (Monamine ACO-100; Schercomid 1214, LD)
   8.5 lb/gal (Monamine LM-100)
Visc.:
   130 cs (60 C) (Ethylan MLD)
   179 cps (50 C) (Stafoam DL)
   500 cps max. (Carsamide SAL-82)
   1000 cps min. (Schercomid LD)
   1800 cps min. (10% aq. disp.) (Schercomid SL-Extra, SLM)
F.P.:
   25 C (Alkamide L9DE)
M.P.:
   25–30 C (Clindrol 200-L)
   29–33 C (Clindrol Superamide 100LM)
   35 C (Hetamide ML)
   40 C (Emid 6511; Loramine DL203/S; Rewomid 203/S; Steinamid 203/S)

42 C (Emid 6510)
42 ± 3 C (Schercomid SL-Extra)
42–46 C (Clindrol Superamide 100L)
42–47 C (Stafoam DL)
97.5–99.0 F (Condensate PE)
105 F (Standamid LD)
104–114 F (Carsamide SAL-7)
Pour Pt.:
40 C (Ethylan MLD)
Set Pt.:
31 C (Empilan LDX)
34 C (Empilan LDE)
Congeal Pt.:
–2 C (Schercomid 1214)
8 C (Product LT 18-48)
16 C (Schercomid SL-ML)
37 ± 3 C (Schercomid SLM)
Solidification Pt.:
84 F (Varamide SL-9)
Flash Pt.:
> 170 C (OC) (Schercomid 1214, LD, SLM)
> 180 C (OC) (Schercomid SL-ML)
148 F (Super Amide L9C)
178 F (Super Amide L9, LL, LM)
> 200 F (Onyxol 336, 345)
> 300 F (COC) (Ethylan MLD)
Cloud Pt.:
0 C (Standamid LD 80/20, LDM)
< 1 C (Monamid 716)
Acid No.:
0.5 max. (Standamid LD 80/20)
1.0 max. (Incromide L-90, LCL, LL, LM-70, LR; Monamid 150-GLT, 150-LMW-C,
150-LW, 150-LWA, 770; Schercomid SL-Extra, SLM-C, SLM-LC, SLM-S;
Standamid LD)
1.5 max. (Schercomid SL-ML)
2.0 max. (Standamid LDO; Super Amide L9C, LL)
3.0 (Standamid LDM)
3.0 max. (Monamid 716; Standamid LDS; Unamide LDX)
8–13 (Onyxol 345)
9.0 max. (Super Amide L9)
10 max. (Super Amide LM)
10–14 (Monamine ACO-100)

Lauric diethanolamide *(cont'd.)*

    12–16 (Schercomid 1214)
    18–23 (Monamine LM-100)
    18–24 (Onyxol 336)
    18–25 (Unamide J-56)
    20–26 (Schercomid LD)
    20–40 (Schercomid SLM)
Alkali No.:
    16 max. (Super Amide L9C)
    20–40 (Schercomid SL-ML, SLM-S)
    25–35 (Standamid LD 80/20)
    27 max. (Super Amide LM)
    30–40 (Standamid LD)
    30–47 (Condensate PE)
    30–50 (Schercomid SLM-C, SLM-LC)
    35–50 (Standamid LDO)
    37 max. (Super Amide LL)
    45 max. (Super Amide L9)
    45–60 (Standamid LDS)
    130–165 (Onyxol 336, 345)
    150–170 (Schercomid 1214)
    175–205 (Condensate PL)
Iodine No.:
    30–40 (Schercomid SL-ML)
Amine No.:
    0–16 (Unamide LDX)
    155–170 (Unamide J-56)
Stability:
    Stable (Clindrol Superamide 100L, 100LM)
    Good (Schercomid LD)
    Good hard water stability (Condensate PL)
    Good in presence of hard water and electrolytes (Ethylan MLD)
    Stable over a wide pH range (Incromide LL)
    Chemically stable at pH 5–12 (Standamid LD)
    Crystallizes on aging (Schercomid SLM)
    May become hazy or crystallize at cold temps. (Monamid 150-GLT)
Storage Stability:
    1 yr min. shelf life; alkali value may decrease slightly with aging, but does not affect product performance (Schercomid 1214, SLM, SL-ML)
    Product viscosity may increase with aging which does not affect performance (Schercomid SL-ML)
pH:
    8.0–9.0 (1% aq.) (Ethylan MLD)

9.0–9.7 (1% disp.) (Condensate PE)
9.0–10.0 (1% sol'n.) (Condensate PL)
9.0–10.5 (Mazamide LS-173, LS-196); (5% sol'n.) (Unamide J-56, LDX); (5% aq.) (Ninol AA-62); (1% sol'n.) (Monamid 150-GLT); (1% disp.) (Product LT 18-48)
9.0–11.0 (1.0% sol'n.) (Standamid LDO, LDS)
9.2–10.0 (Mazamide LM-20)
9.2–10.2 (10% sol'n.) (Monamid 770)
9.5–10.5 (10% sol'n.) (Monamid 150-LWA; Monamine ACO-100, LM-100; Standamid LD, LD 80/20)
9.6–10.6 (1% aq.) (Stafoam DL)
9.7 ± 0.5 (10% sol'n.) (Schercomid LD)
9.7–10.3 (Incromide LL)
10.0–11.0 (10% sol'n.) (Monamid 150-LW, 716)
10.2 ± 0.5 (10% aq. disp.) (Schercomid SL-Extra, SLM)
10.2–11.2 (10% sol'n.) (Monamid 150-LMW-C)
12.5 (Varamide SL-9)

Surface Tension:
5 dynes/cm (0.1%) (Stafoam DL)
33 dynes/cm (0.025 % aq. disp., 22 C) (Varamide SL-9)

Biodegradable: (Carsamide SAL-7, SAL-9; Mazamide LM-20; Mazamide LS-173, LS-196; Monamid 150-LMW-C, 150-LW, 150-LWA, 716, 770; Monamine ACO-100, LM-100; Standamid LDS; Unamide LDX); rapidly biodegradable (Incromide L-90, LM-70); presumed biodegradable (Varamide SL-9)

TOXICITY/HANDLING:
Low eye irritation (Onyxol 345)
Skin and eye irritant (Clindrol 200-L; Clindrol Superamide 100L, 100LM)
Irritating to skin and eyes in conc. form (Standamid LDS)
Defatting to skin (Mazamide LS-173, LS-196)
Avoid prolonged contact with skin (Incromide LCL, LL, LM-70)
Avoid contact with skin and eyes (Hartamide LDA70, LDA90)
Avoid prolonged contact with skin and eyes (Incromide L-90)
Avoid prolonged contact with conc. form; spillages may be slippery (Ethylan MLD)
Not believed hazardous; however, avoid contact with eyes and prolonged contact with skin (Ninol 4821)

STORAGE/HANDLING:
Store in cool, dry place (Incromide L-90)
Store in a dry place at 50–120 F (Incromide LCL, LL, LM-70)
Store in closed containers between 20–40 C (Standamid LDS)
Easily pumpable at temps. as low as 20 C and can be stored at lower temps. without degradation; do not store above 40 C (Standamid LDO)
Do not mix with strong oxidizing or reducing agents (potentially explosive) (Ethylan MLD)

Lauric diethanolamide *(cont'd.)*

STD. PKGS.:
Mild steel drums (Rewomid 203/S; Steinamid 203/S)
Drums/T/L (Hartamide LDA70, LDA90)
17-kg C/N or bulk (Stafoam DL)
200-kg net steel drums (Merpinamid KD11, LD/E, LSD/E)
55-gal steel drums (Schercomid LD, SL-Extra)
55-gal open-head steel drums (Carsamide SAL-7, SAL-9)
55-gal lined steel or fiber drums (Standamid LDO)
55-gal (425 lb net) unlined open-head drums (Incromide LR)
55-gal (425 lb net) open-head steel drums (Clindrol 200-L; Clindrol Superamide 100LM)
55-gal (435 lb net) open-head steel drums, 40 lb net cubitainers (Clindrol Superamide 100L)
55-gal (440 lb net) drums (Incromide LCL)
55-gal (440 lb net) unlined open-head steel drums (Incromide LL, LM-70)
55-gal (450 lb net) lined open-head steel drums (Schercomid SLM)
55-gal (450 lb net) open-head steel drums, tank wagon, tank car (Varamide SL-9)
55-gal (450 lb net) open-head drums (Incromide L-90)
55-gal (450 lb net) open-head Liquipak drums (Schercomid 1214, SL-ML)
55-gal (450 lb net) fiber drums, bulk, tank wagons, rail cars (Standamid LDS)
450 lb net open-head steel drums (Monamid 150-LMW-C, 150-LW, 150-LWA, 770; Monamine ACO-100, LM-100)
450 lb net tight-head steel drums (Monamid 716)

Lauric monoethanolamide

SYNONYMS:
Dodecanamide, N-(2-hydroxyethyl)-
N-(2-Hydroxyethyl) dodecanamide
N-(2-hydroxyethyl) lauramide
Lauramide MEA (CTFA)
Lauric acid monoethanolamide
Lauroyl monoethanolamide
Monoethanolamine lauric acid amide
EMPIRICAL FORMULA:
$C_{14}H_{29}NO_2$
STRUCTURE:

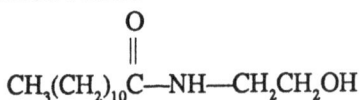

CAS No.:
142-78-9

TRADENAME EQUIVALENTS:
Alkamide L9ME [Alkaril]
Ardet LEA [Ardmore]
Cedemide MX [Domtar/CDC]
Comperlan LM [Henkel/Canada]
Cyclomide L203 [Cyclo]
Empilan LME [Albright & Wilson/Detergents]
Hartamide LMEA-90 [Hart]
Lauridit LM [Akzo Chemie]
Loramine L203 [Dutton & Reinisch]
Mackamide LMM [McIntyre]
Monamid LMA, LMMA [Mona]
Rewomid L203 [Dutton & Reinisch]
Standamid LM [Henkel]
Steinamid L203 [Dutton & Reinisch]

CATEGORY:
Detergent, emulsifier, foaming agent, foam stabilizer, wetting agent, viscosity builder, thickener, intermediate, superfatting agent, corrosion inhibitor, lubricant, dispersant

APPLICATIONS:
Bath products: (Comperlan LM; Standamid LM); bubble bath (Hartamide LMEA-90; Lauridit LM; Monamid LMA, LMMA)

Cleansers: (Comperlan LM); hand cleanser (Alkamide L9ME; Monamid LMA, LMMA; Standamid LM)

Cosmetic industry preparations: (Alkamide L9ME; Lauridit LM; Monamid LMA, LMMA); conditioners (Monamid LMA, LMMA); creams and lotions (Alkamide L9ME); perfumery (Loramine L203; Rewomid L203; Steinamid L203); shampoos (Comperlan LM; Empilan LME; Lauridit LM; Monamid LMA, LMMA; Standamid LM)

Farm products: agricultural oils/sprays (Monamid LMA, LMMA)

Household detergents: (Alkamide L9ME; Lauridit LM; Loramine L203; Rewomid L203; Steinamid L203); carpet & upholstery shampoos (Empilan LME; Monamid LMA, LMMA); dishwashing (Monamid LMA, LMMA); heavy-duty cleaner (Lauridit LM); laundry detergent (Alkamide L9ME); liquid detergents (Cedemide MX; Empilan LME); paste detergents (Loramine L203; Rewomid L203; Steinamid L203); powdered detergents (Empilan LME; Hartamide LMEA-90; Loramine L203; Monamid LMA, LMMA; Rewomid L203)

Industrial applications: dyes and pigments (Monamid LMA, LMMA); lubricating/cutting oils (Monamid LMA, LMMA); metalworking (Monamid LMA, LMMA); petroleum industry (Monamid LMA, LMMA); polishes and waxes (Monamid

Lauric monoethanolamide (cont'd.)

LMA, LMMA); textile/leather processing (Monamid LMA, LMMA)

Industrial cleaners: (Alkamide L9ME); drycleaning compositions (Monamid LMA, LMMA); institutional cleaners (Alkamide L9ME); metal processing surfactants (Monamid LMA, LMMA); textile cleaning (Lauridit LM; Monamid LMA, LMMA)

Pharmaceutical applications: (Alkamide L9ME)

PROPERTIES:

Form:

Solid (Cedemide MX; Comperlan LM; Hartamide LMEA-90; Mackamide LMM)

Flake (Ardet LEA; Cyclomide L203; Lauridit LM; Loramine L203; Rewomid L203; Standamid LM; Steinamid L203)

Beads (Standamid LM)

Granular (Monamid LMA, LMMA)

Waxy flake (@ 20 C) (Empilan LME)

Waxy solid (Alkamide L9ME)

Color:

White (Alkamide L9ME; Hartamide LMEA-90; Loramine L203; Rewomid L203; Steinamid L203)

Off-white (Cyclomide L203)

Cream (20 C) (Empilan LME)

Yellowish (Lauridit LM)

Tan (Monamid LMA, LMMA)

Odor:

Low (Alkamide L9ME)

Typical, slight (Loramine L203; Rewomid L203; Steinamid L203)

Composition:

93% active (Lauridit LM)

95% active (Alkamide L9ME)

95% amide (Hartamide LMEA-90)

100% active (Cyclomide L203; Loramine L203; Monamid LMA, LMMA; Rewomid L203; Steinamid L203)

100% conc. (Ardet LEA; Comperlan LM; Cedemide MX; Mackamide LMM)

Solubility:

Sol. in lower alcohols (Standamid LM)

Sol. in aromatic hydrocarbons (Alkamide L9ME); disp. (@ 10%) (Monamid LMA, LMMA)

Sol. in chlorinated aliphatic solvents (Alkamide L9ME)

Sol. in detergent systems (Alkamide L9ME)

Sol. in ethanol (@ 10%) (Monamid LMA, LMMA)

Disp. in natural fats and oils (@ 10%) (Monamid LMA, LMMA)

Disp. in kerosene (@ 10%) (Monamid LMA, LMMA)

Sol. in min. oils (Alkamide L9ME)

Sol. in petroleum solvents (Alkamide L9ME)

Sol. in polyethylene glycols (Standamid LM)

Sol. in propylene glycol (Standamid LM)

Poor sol. in water (Lauridit LM); disp. (@ 10%) (Monamid LMA, LMMA); insol. (Standamid LM)

Disp. in white min. oil (@ 10%) (Monamid LMA, LMMA)

Ionic Nature:

Nonionic (Ardet LEA; Cedemide MX; Comperlan LM; Empilan LME; Loramine L203; Mackamide LMM; Monamid LMA, LMMA; Rewomid L203; Steinamid L203)

Anionic (Hartamide LMEA-90)

Sp.gr.:

0.915 (Standamid LM)

1.01 (80 C) (Lauridit LM)

Density:

0.4 g/cc (20 C) (Empilan LME)

M.P.:

60 C (Alkamide L9ME)

80 C (Standamid LM)

80–84 C (Lauridit LM)

84 C (Loramine L203; Rewomid L203; Steinamid L203)

85 C (Empilan LME)

Solidification Pt.:

80 ± 2 C (Monamid LMA, LMMA)

Acid No.:

0–1 (Monamid LMA, LMMA)

2.0 max. (Standamid LM)

4.0 max. (Lauridit LM)

Alkali No.:

12 max. (Standamid LM)

Saponification No.:

20 max. (Lauridit LM)

Storage Stability:

Storable over a wide temperature range (0–45 C) with no product degradation (Standamid LM)

pH:

9.0 (Lauridit LM)

9.0–11.0 (1.0% sol'n.) (Standamid LM)

9.7–10.7 (10% sol'n.) (Monamid LMMA)

10.0–11.0 (10% sol'n.) (Monamid LMA)

Biodegradable: (Monamid LMA, LMMA)

STORAGE/HANDLING:

Lauric monoethanolamide *(cont'd.)*

Protect from high temperature (Lauridit LM)
STD. PKGS.:
Paper sacks (Loramine L203; Rewomid L203; Steinamid L203)
25-kg paper bags (Lauridit LM)
200 lb net fiberboard drums (Monamid LMA, LMMA)
200 lb fiber drums, 50 lb boxes (Standamid LM)

Lauric monoisopropanolamide

SYNONYMS:
Dodecanamide, N-(2-hydroxyproyl)-
N-(2-Hydroxypropyl) dodecanamide
Lauramide MIPA (CTFA)
Lauric fatty acid monoisopropanolamide
Lauric isopropanolamide
Lauric MIPA
Lauroyl isopropanolamide
Monoisopropanolamine lauric acid amide
EMPIRICAL FORMULA:
$C_{15}H_{31}NO_2$
STRUCTURE:

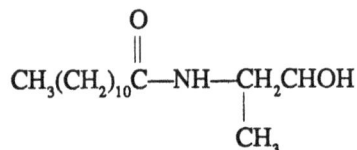

$$CH_3(CH_2)_{10}\overset{\overset{\displaystyle O}{\|}}{C}-NH-CH_2CHOH$$
$$|$$
$$CH_3$$

CAS No.:
142-54-1
TRADENAME EQUIVALENTS:
Alkamide LIPA [Alkaril]
Ardet LIPA [Ardmore]
Clindrol 101LI [Clintwood]
Comperlan LP [Henkel]
Cyclomide LIPA [Cyclo]
Empilan LIS [Albright&Wilson/Marchon]
Incromide LI [Croda Surfactants]
Loramine IPL203 [Dutton & Reinisch]
Merpinamid LMIPA [Kempen]
Monamid LIPA [Mona]

Lauric monoisopropanolamide (cont'd.)

TRADENAME EQUIVALENTS *(cont'd.):*
  Monamid LMIPA [Mona]
  Rewomid IPL203 [Dutton & Reinisch]
  Steinamid IPL203 [Dutton & Reinisch]
CATEGORY:
  Foam booster, foam stabilizer, stabilizer, thickener, viscosity builder, intermediate,
    emulsifier, detergent, wetting agent, corrosion inhibitor, lubricant, dispersant,
    superfatting agent
APPLICATIONS:
  Bath products: bubble bath (Monamid LIPA)
  Cleansers: hand cleanser (Clindrol 101LI; Monamid LIPA)
  Cosmetic industry preparations: (Monamid LIPA, LMIPA); conditioners (Monamid
    LIPA); shampoos (Clindrol 101LI; Empilan LIS; Merpinamid LMIPA; Monamid
    LIPA)
  Farm products: agricultural oils/sprays (Monamid LIPA)
  Household detergents: (Merpinamid LMIPA; Monamid LMIPA); carpet & upholstery
    shampoos (Clindrol 101LI; Empilan LIS; Monamid LIPA); dishwashing (Empilan
    LIS; Monamid LIPA); laundry detergent (Alkamide LIPA); liquid detergents
    (Alkamide LIPA; Empilan LIS; Merpinamid LMIPA); powdered detergents
    (Alkamide LIPA; Empilan LIS; Monamid LIPA)
  Industrial applications: dyes and pigments (Monamid LIPA); lubricating/cutting oils
    (Monamid LIPA); petroleum industry (Monamid LIPA); polishes and waxes
    (Monamid LIPA); textile/leather processing (Monamid LIPA)
  Industrial cleaners: (Merpinamid LMIPA; Monamid LMIPA); drycleaning composi-
    tions (Monamid LIPA); metal processing surfactants (Monamid LIPA); textile
    cleaning (Monamid LIPA)
PROPERTIES:
Form:
  Solid (Alkamide LIPA; Comperlan LP; Monamid LMIPA)
  Flake (Ardet LIPA; Clindrol 101LI; Cyclomide LIPA; Incromide LI; Loramine
    IPL203; Rewomid IPL203; Steinamid IPL203)
  Granular (Monamid LIPA)
  Wax (Ardet LIPA; Merpinamid LMIPA)
  Waxy flake (Empilan LIS)
Color:
  White (Incromide LI; Loramine IPL203; Merpinamid LMIPA; Rewomid IPL203;
    Steinamid IPL203)
  Cream (Empilan LIS)
  Light yellow (Alkamide LIPA)
  Tan (Clindrol 101LI; Monamid LIPA)
Odor:
  Typical, slight (Loramine IPL203; Rewomid IPL203; Steinamid IPL203)

Lauric monoisopropanolamide (cont'd.)

Composition:
　87.5% active (Empilan LIS)
　95% amide (Alkamide LIPA)
　100% active (Clindrol 101LI; Incromide LI; Loramine IPL203; Monamid LIPA;
　　Rewomid IPL203; Steinamid IPL203)
　100% conc. (Ardet LIPA; Comperlan LP; Cyclomide LIPA; Monamid LMIPA)
Solubility:
　Sol. in alcohols (Clindrol 101LI)
　Sol. in aromatic hydrocarbons (Clindrol 101LI)
　Sol. in chlorinated hydrocarbons (Clindrol 101LI)
　Sol. in esters (Clindrol 101LI)
　Sol. in ethanol (@ 10%) (Monamid LIPA)
　Disp. in natural fats and oils (@ 10%) (Monamid LIPA)
　Sol. in glycols (Clindrol 101LI)
　Disp. in kerosene (@ 10%) (Monamid LIPA)
　Sol. in ketones (Clindrol 101LI)
　Disp. in min. oil (@ 10%) (Alkamide LIPA)
　Disp. in min. spirits (@ 10%) (Alkamide LIPA)
　Disp. in perchloroethylene (@ 10%) (Alkamide LIPA)
　Disp. in water (@ 10%) (Alkamide LIPA; Monamid LIPA); disp. in hot water (Empilan
　　LIS)
　Disp. in white min. oil (@ 10%) (Monamid LIPA)
Ionic Nature:
　Nonionic (Ardet LIPA; Clindrol 101LI; Comperlan LP; Cyclomide LIPA; Empilan
　　LIS; Incromide LI; Monamid LIPA, LMIPA; Rewomid IPL203)
Density:
　0.4 g/cc (Empilan LIS)
M.P.:
　53–60 C (Clindrol 101LI)
　54 C (Incromide LI)
　55 C (Loramine IPL203; Rewomid IPL203; Steinamid IPL203)
　63–68 C (Alkamide LIPA)
Solidification Pt.:
　55 ± 3 C (Monamid LIPA)
Acid No.:
　1 max. (Incromide LI; Monamid LIPA)
Stability:
　Stable (Clindrol 101LI)
　Stable to acids and alkalis under certain conditions (Empilan LIS)
pH:
　8.0–11.0 (1% DW) (Alkamide LIPA)
　10.3–11.3 (10% sol'n.) (Monamid LIPA)

Lauric monoisopropanolamide (cont'd.)

Biodegradable: (Monamid LIPA)
TOXICITY/HANDLING:
  Avoid prolonged contact with skin (Incromide LI)
  Skin and eye irritant (Clindrol 101LI)
STORAGE/HANDLING:
  Store in a cool, dry place (Incromide LI)
STD. PKGS.:
  Paper sacks (Loramine IPL203; Rewomid IPL203; Steinamid IPL203)
  200-kg net iron drums (Merpinamid LMIPA)
  44-gal (150 lb net) Leverpak with polyliner (Incromide LI)
  55-gal (200 lb net) open-head fiber drums (Clindrol 101LI)
  175 lb net fiberboard drums (Monamid LIPA)

Lauroyl sarcosine (CTFA)

SYNONYMS:
  Glycine, N-methyl-N-(1-oxododecyl)-
  N-Methyl-N-(1-oxododecyl) glycine
EMPIRICAL FORMULA:
  $C_{15}H_{29}NO_3$
STRUCTURE:

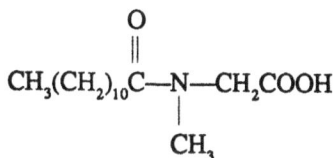

CAS No.:
  97-78-9
TRADENAME EQUIVALENTS:
  Hamposyl L [W.R. Grace]
  Sarkosyl L, NL-30 [Ciba-Geigy]
CATEGORY:
  Foam stabilizer, foaming agent, detergent, emulsifier, corrosion inhibitor, wetting
    agent, conditioner, shampoo base, lubricant
APPLICATIONS:
  Bath products: bath oils (Sarkosyl L)
  Cleansers: hand cleanser (Sarkosyl L); skin cleanser (Hamposyl L)
  Cosmetic industry preparations: (Hamposyl L); shampoos (Hamposyl L; Sarkosyl L,
    NL-30); toilet soaps (Sarkosyl L)

Lauroyl sarcosine (cont'd.)

Degreasers: solvent degreasing (Sarkosyl L)

Household detergents: (Sarkosyl L); carpet & upholstery shampoos (Hamposyl L; Sarkosyl L, NL-30); dishwashing (Sarkosyl L); laundry detergent (Sarkosyl L, NL-30); window cleaners (Sarkosyl L, NL-30)

Industrial applications: lubricating/cutting oils (Sarkosyl L); metal processing (Sarkosyl L)

Pharmaceutical applications: (Sarkosyl L); antiperspirant/deodorant (Sarkosyl L); antidandruff shampoos (Sarkosyl L); balms (Sarkosyl L); dental preparations (Sarkosyl L, NL-30); mouthwash (Sarkosyl L)

PROPERTIES:
Form:
Liquid (Sarkosyl NL-30)
Solid (Hamposyl L)
Powder (Sarkosyl L)
Color:
Colorless (Sarkosyl NL-30)
Gardner 2 (Hamposyl L)
Composition:
30% active (Sarkosyl NL-30)
94% active (Hamposyl L)
94% min. purity (Sarkosyl L)
Solubility:
Sol. in aliphatic hydrocarbons (Sarkosyl L)
Sol. in glycerin (Sarkosyl L)
Sol. in glycols (Sarkosyl L)
Sol. in most organic solvents (Hamposyl L; Sarkosyl L)
Sol. in phosphate esters (Sarkosyl L)
Sol. in silicones (Sarkosyl L)
Insol. in water (Sarkosyl L)
Ionic Nature:
Anionic (Hamposyl L; Sarkosyl L, NL-30)
M.W.:
264–285 (Sarkosyl L)
Sp.gr.:
0.969 (Sarkosyl L)
0.97–0.99 (Hamposyl L)
M.P.:
34–37 C (Hamposyl L)
35–37 C (Sarkosyl L)
Stability:
Good at elevated temps. (Sarkosyl L)

pH:
   7.5–8.5 (1% sol'n.) (Sarkosyl NL-30)
Biodegradable: (Hamposyl L)
TOXICITY/HANDLING:
   Skin irritant on prolonged contact with conc. acid (Sarkosyl L)
STD. PKGS.:
   55-gal drums (Hamposyl L)

Linoleoyl diethanolamide

SYNONYMS:
   N,N-Bis (2-hydroxyethyl) linoleamide
   N,N-Bis (2-hydroxyethyl)-9,12-octadecadienamide
   Diethanolamine linoleic acid amide
   Linoleamide DEA (CTFA)
   9,12-Octadecadienamide, N,N-bis (2-hydroxyethyl)-
EMPIRICAL FORMULA:
   $C_{22}H_{41}NO_3$
STRUCTURE:

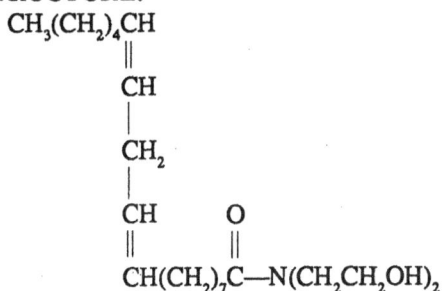

CAS No.:
   56863-02-6
TRADENAME EQUIVALENTS:
   Aminol LNO [Finetex]
   Clindrol LT15-73-1 [Clintwood] (1:1)
   Comperlan F [Henkel KGaA]
   Comperlan VOD [Henkel (Canada)] (1:1)
   Cyclomide DIN 295/S [Cyclo] (1:1)
   Cyclomide DOTS [Cyclo] (1:1)
   Emid 6540 [Emery]
   Foamole A [Van Dyk]

Linoleoyl diethanolamide (cont'd.)

TRADENAME EQUIVALENTS *(cont'd.):*
Incromide LA [Croda Surfactants]
Jordamide LLD [Jordan]
Monamid 15-70W, 150-ADY [Mona] (1:1)
Monamine ADY-100 [Mona] (1:2)
Product LT 15-73-1 [Clintwood]
Rewomid F [Rewo Chemische Werke GmbH]
Schercomid SLE, SLS [Scher]
Standamid SOD [Henkel] (1:1)
Standamid SOMD [Henkel]

CATEGORY:
Thickener, viscosity builder, stabilizer, foam stabilizer, foam booster, conditioner, emulsifier, dispersant, emollient, solubilizer, superfatting agent, lubricant, corrosion inhibitor, suspending agent, detergent, wetting agent

APPLICATIONS:
Bath products: (Comperlan VOC); bubble bath (Incromide LA; Monamid 15-70W, 150-ADY; Monamine ADY-100; Standamid SOMD)

Cleansers: (Comperlan VOC); hand cleanser (Clindrol LT15-73-1; Monamid 15-70W, 150-ADY; Monamine ADY-100; Standamid SOMD)

Cosmetic industry preparations: (Jordamide LLD; Monamid 15-70W, 150-ADY; Monamine ADY-100; ; Schercomid SLE); conditioners (Foamole A; Monamid 15-70W, 150-ADY; Monamine ADY-100; ; Schercomid SLE); hair preparations (Foamole A; Schercomid SLS); shampoos (Aminol LNO; Comperlan F, VOD; Foamole A; Incromide LA; Monamid 15-70W, 150-ADY; Monamine ADY-100; Rewomid F; Schercomid SLE, SLS; Standamid SOMD); skin preparations (Clindrol LT15-73-1; Schercomid SLS)

Farm products: agricultural oils/sprays (Monamid 15-70W, 150-ADY; Monamine ADY-100)

Household detergents: (Aminol LNO; Clindrol LT15-73-1; Cyclomide DIN 295/S; Incromide LA; Jordamide LLD; Standamid SOD); carpet & upholstery shampoos (Monamid 15-70W, 150-ADY; Monamine ADY-100); dishwashing (Monamid 15-70W, 150-ADY; Monamine ADY-100); liquid soaps (Foamole A); powdered soaps (Monamid 15-70W, 150-ADY; Monamine ADY-100)

Industrial applications: (Jordamide LLD); dyes and pigments (Cyclomide DOTS; Schercomid SLS); industrial processing (Schercomid SLS); lubricating/cutting oils (Monamid 15-70W, 150-ADY; Monamine ADY-100; Standamid SOD); metalworking (Monamid 15-70W, 150-ADY; Monamine ADY-100); petroleum industry (Monamid 15-70W, 150-ADY; Monamine ADY-100); textile/leather processing (Monamid 15-70W, 150-ADY; Monamine ADY-100)

Industrial cleaners: (Standamid SOD); drycleaning compositions (Monamid 15-70W, 150-ADY; Monamine ADY-100); institutional cleaners (Standamid SOD); metal processing surfactants (Monamid 15-70W, 150-ADY; Monamine ADY-100);

textile scouring (Monamid 15-70W, 150-ADY; Monamine ADY-100)
Pharmaceutical applications: vitamin E substitute (Comperlan F)

PROPERTIES:

Form:
Liquid (Clindrol LT15-73-1; Comperlan F, VOD; Cyclomide DOTS; Emid 6540; Incromide LA; Jordamide LLD; Monamid 15-70W, 150-ADY; Monamine ADY-100; Product LT 15-73-1; Standamid SOD)
Clear liquid (Schercomid SLE, SLS)
Med.-visc. liquid (Rewomid F)
Paste/solid (Standamid SOMD)

Color:
Straw to amber (Clindrol LT15-73-1; Product LT 15-73-1)
Amber (Cyclomide DOTS; Jordamide LLD; Schercomid SLS)
Dark amber (Schercomid SLE; Standamid SOD)
Gardner 5 max. (Incromide LA)
GVCS-33: 9 max. (Monamine ADY-100)
GVCS-33: 10 max. (Monamid 150-ADY)
GVCS-33: 11 max. (Monamid 15-70W)

Odor:
Bland (Incromide LA)
Mild, typical (Schercomid SLE)
Mild, fruity (Schercomid SLS)

Composition:
100% conc. (Comperlan F, VOD; Emid 6540; Rewomid F; Standamid SOD)
100% active (Incromide LA; Monamid 15-70W, 15-ADY; Monamine ADY-100)
100% active; 80% amide content min. (Schercomid SLS)
100% active; 83% amide content min. (Clindrol LT15-73-1; Product LT 15-73-1)
100% active; 87% amide content min. (Schercomid SLE)
100% active; ≈ 95% amide content (Cyclomide DOTS)

Solubility:
Sol. in acetone (Standamid SOD)
Sol. in alcohol ethers (Clindrol LT15-73-1; Product LT 15-73-1)
Sol. in alcohols (Clindrol LT15-73-1; Product LT 15-73-1; Schercomid SLE, SLS); sol. in lower alcohols (Standamid SOD, SOMD)
Sol. in aliphatic hydrocarbons (Clindrol LT15-73-1; Product LT 15-73-1; Schercomid SLE, SLS); disp. (Standamid SOD)
Sol. in aromatic hydrocarbons (Clindrol LT15-73-1; Monamid 15-70W, 150-ADY; Product LT 15-73-1; Schercomid SLE, SLS); disp. (Monamine ADY-100)
Sol. in chlorinated hydrocarbons (Monamid 15-70W, 150-ADY; Schercomid SLE, SLS); disp. (Monamine ADY-100)
Sol. in esters (Clindrol LT15-73-1; Product LT 15-73-1); disp. (Standamid SOD)
Sol. in ethanol (Monamid 15-70W, 150-ADY; Monamine ADY-100); sol. 70%

ethanol (Foamole A)
Sol. in ethers (Clindrol LT15-73-1; Product LT 15-73-1)
Sol. in glycol ethers (Schercomid SLE, SLS)
Sol. in glycols (Schercomid SLE, SLS; Standamid SOD)
Sol. in kerosene (Monamid 15-70W, 150-ADY; Monamine ADY-100)
Disp. in higher ketones (Standamid SOD)
Sol. in min. oil (Schercomid SLE, SLS); disp. (Standamid SOD); disp. white min. oil (Monamid 15-70W, 150-ADY)
Sol. in min. spirits (Monamid 15-70W, 150-ADY; Monamine ADY-100)
Sol. in natural oils and fats (Monamid 15-70W); disp. in natural oils and fats (Monamid 150-ADY)
Sol. in most organic solvents (Schercomid SLE, SLS)
Sol. in peanut oil (Foamole A)
Sol. in propylene glycol (Foamole A); insol. (Standamid SOMD)
Sol. in toluene (Standamid SOD)
Sol. in veg. oil (Schercomid SLE, SLS)
Disp. in water (Clindrol LT15-73-1; Monamine ADY-100; Product LT 15-73-1; Schercomid SLE, SLS; Standamid SOD); gels in water (Foamole A; Monamid 15-70W, 150-ADY); insol. (Standamid SOMD)
Sol. in xylene (Standamid SOD)

Ionic Nature:
Nonionic (Comperlan F, VOD; Emid 6540; Incromide LA; Monamid 15-70W, 150-ADY; Monamine ADY-100; Schercomid SLE, SLS; Standamid SOD)

Sp.gr.:
0.935 (Standamid SOMD)
0.96 (Monamid 15-70W)
0.965 (Schercomid SLE)
0.97 (20 C) (Monamid 150-ADY)
0.98 (Clindrol LT15-73-1; Product LT 15-73-1; Schercomid SLS); (20 C) (Monamine ADY-100)

Density:
8.0 lb/gal (Monamid 15-70W; Schercomid SLE, SLS)
8.10 lb/gal (Monamid 150-ADY)
8.20 lb/gal (Monamine ADY-100; Standamid SOD)

Congeal Pt.:
5 C min. (Schercomid SLE)
≈ 10 C (Clindrol LT15-73-1; Product LT 15-73-1)

Flash Pt.:
> 180 C (OC) (Schercomid SLE, SLS)

Acid No.:
0–1 (Monamid 15-70W, 150-ADY)
0–2 (Monamine ADY-100)

1.0 max. (Incromide LA; Schercomid SLE; Standamid SOD)
2.0 max. (Schercomid SLS; Standamid SOMD)

Iodine No.:
100 (Schercomid SLE)

Alkali No.:
20–40 (Schercomid SLE, SLS)
25–40 (Monamid 15-70W)
30–40 (Standamid SOD)
30–45 (Monamid 150-ADY)
30–50 (Standamid SOMD)
110–130 (Monamine ADY-100)

Storage Stability:
1 yr min shelf life; alkali value decreases with aging which does not affect performance (Schercomid SLE)
1 yr min. shelf life; alkali value decreases and acid value increases slightly with aging which does not affect performance (Schercomid SLS)

pH:
8.5–11.0 (1% disp.) (Clindrol LT15-73-1; Product LT 15-73-1)
10–11 (10% sol'n.) (Monamid 15-70W, 150-ADY)
10.5–11.5 (10% sol'n.) (Monamine ADY-100)
9–11 (1% sol'n.) (Standamid SOD, SOMD)

Biodegradable: (Monamid 15-70W, 150-ADY; Monamine ADY-100; Standamid SOD)

TOXICITY/HANDLING:
Avoid prolonged contact with skin (Incromide LA)
Irritating to skin and eyes in conc. form (Standamid SOD)

STORAGE/HANDLING:
Store in a dry place at temps. not lower than 50 F (Incromide LA)
Store in closed containers @ 20–40 C (Standamid SOD)
Do not store at temps. above 45 C (Standamid SOMD)

STD. PKGS.:
55-gal lined steel drums (Standamid SOMD)
55-gal (440 lb net) lined, bung-head, steel drums (Schercomid SLE)
55-gal (440 lb net) lined, bung-head, steel drums or open-head Liquipak drums (Schercomid SLS)
55-gal (440 lb net) unlined, open-head drums (Incromide LA)
55-gal (440 lb net) tighthead steel drums (Monamid 15-70W)
55-gal (450 lb net) open-head steel drums (Monamid 150-ADY; Monamine ADY-100)
450 lb net fiber drums, bulk, tank wagons, rail cars (Standamid SOD)

Locust bean gum (CTFA)

SYNONYMS:
Algaroba
Carob bean gum
Carob flour
Ceratonia
St. John's bread
CAS No.:
9000-40-2
TRADENAME EQUIVALENTS:
Locust Bean Gum FL 50-40, FL 50-50, FL 70-50 [Hercules]
CATEGORY:
Stabilizer, thickener, binder
APPLICATIONS:
Food applications: food additives (Locust Bean Gum FL 50-40, FL 50-50, FL 70-50)
PROPERTIES:
Form:
Gum (Locust Bean Gum FL 50-40, FL 50-50, FL 70-50)
Color:
White (Locust Bean Gum FL 70-50)
Odor:
Bland (Locust Bean Gum FL 70-50)
Composition:
13% max. moisture (Locust Bean Gum FL 50-40, FL 50-50, FL 70-50)
Fineness:
15–35% through 200 mesh (Locust Bean Gum FL 50-40)
40–60% through 200 mesh (Locust Bean Gum FL 50-50, FL 70-50)
Visc.:
2400–3200 cps (1% sol'n., Brookfield RVF, #3 spindle, 20 rpm) (Locust Bean Gum FL 50-40, FL 50-50)
3200–3600 cps (1% sol'n., Brookfield RVF, #3 spindle, 20 rpm) (Locust Bean Gum FL 70-50)
pH:
5.4–7.0 (Locust Bean Gum FL 50-40, FL 50-50, FL 70-50)
STORAGE/HANDLING:
Store in a dry place, away from heat, and out of the sun (Locust Bean Gum FL 50-40, FL 50-50, FL 70-50)
STD. PKGS.:
50-lb net multiwall paper bags with a polyethylene barrier (Locust Bean Gum FL 50-40, FL 50-50, FL 70-50)

Magnesium aluminum silicate (CTFA)

SYNONYMS:
  Aluminosilicic acid, magnesium salt
  Aluminum magnesium silicate
CAS No.:
  1327-43-1
TRADENAME EQUIVALENTS:
  Macaloid [NL Chem.]
  Van Gel [R.T. Vanderbilt]
  Veegum, F, HS, HV, K, Neutral, S-728, WG [R.T. Vanderbilt]
  Veegum T [R.T. Vanderbilt] (tech.)
CATEGORY:
  Stabilizer, thickener, visc. modifier, suspending agent, thixotrope, binder, dispersant
APPLICATIONS:
  Cosmetic industry preparations: (Veegum, F, HS, HV, S-728); creams and lotions (Macaloid; Veegum, F, HS, HV, S-728); toiletries (Veegum, F, HS, HV, S-728)
  Industrial applications: (Van Gel; Veegum T); chemical specialties (Veegum, F, HS, HV, S-728, T); coatings (Van Gel); dyes and pigments (Veegum, F, HS, S-728, T); paint mfg. (Veegum, F, HS, S-728, T); textile/leather processing (Veegum, F, HS, S-728, T)
  Pharmaceutical applications: (Veegum, F, HS, HV, K, Neutral, S-728, WG); dental preparations (Veegum, F, HS, HV, K, S-728); ointments (Veegum, F, HS, HV, K, S-728, WG); tablet mfg. (Veegum, F, HS, HV, K, S-728, WG)

PROPERTIES:
Form:
  Flake (Van Gel; Veegum HS, HV, K, Neutral, S-728, T)
  Small flakes (Veegum)
  50 mesh powder (Veegum WG)
  Microfine powder (Veegum F)
Color:
  White (Veegum, F, HS, HV, K, Neutral, S-728, WG)
  Near white (Veegum T)
  Off-white (Van Gel)
Odor:
  None (Van Gel; Veegum, F, HS, HV, K, Neutral, S-728, T, WG)
Taste:
  None (Veegum, F, HS, HV, K, Neutral, S-728, T, WG)

Magnesium aluminum silicate *(cont'd.)*

Composition:
< 8% moisture (Veegum, F, HS, HV, K, S-728, T, WG)
12% max. moisture (Van Gel)
Solubility:
Insol. in alcohols (Veegum, F, HS, HV, K, Neutral, S-728, T, WG)
Disp. in water (Macaloid); disp. and swells in water (Van Gel); insol. in water, swells in water to form colloidal dispersions (Veegum, F, HS, HV, K, Neutral, S-728, T, WG)
Density:
2.6 mg/m³ (Van Gel; Veegum)
Fineness:
7 min. (4% disp., Hegman) (Van Gel)
Visc.:
100 cps ± 50% (7% disp.) (Veegum Neutral)
150 cps ± 50% (5.5% disp.) (Veegum HS)
200 cps ± 50% (4% disp.) (Veegum T)
220 cps ± 25% (5.5% disp.) (Veegum K)
250 cps ± 25% (5% aq. disp.) (Veegum, F, S-728)
800 ± 400 cps (4% disp., RTV, #3 spindle, 60 rpm) (Van Gel)
Storage Stability:
Indefinite storage life in original containers (Veegum, F, HS, HV, K, Neutral, S-728, T, WG)
pH:
6.5–7.5 (5% aq. disp.) (Veegum Neutral)
≈ 9 (5% aq. disp.) (Veegum, F, HS, HV, K, S-728, T, WG)
9.5 ± 0.5 (4% disp.) (Van Gel)
STORAGE/HANDLING:
Store under dry conditions to prevent absorption of moisture (Veegum, F, HS, HV, K, Neutral, S-728, T, WG)
STD. PKGS.:
100 lb fiber drums with moisture barriers (Veegum, F, HS, HV, K, Neutral, S-728, T, WG)

Methylcellulose *(CTFA)*

SYNONYMS:
Cellulose, methyl ether
CAS No.:
9004-67-5
TRADENAME EQUIVALENTS:
Methocel A, A4C, A4M, A15-LV [Dow Chem.]

Methylcellulose (cont'd.)

CATEGORY:
Plasticizer, thickener, stabilizer, emulsifier, adhesive, gelling agent, binder

APPLICATIONS:
Food applications: (Methocel A4C, A4M, A15-LV)
Industrial applications: ceramics (Methocel A); paint mfg. (Methocel A)

PROPERTIES:
Form:
Firm gel (Methocel A4C, A4M, A15-LV)
Powder (Methocel A)
Color:
White to slightly off-white (Methocel A)
Odor:
Odorless (Methocel A)
Taste:
Tasteless (Methocel A)
Composition:
27.5–31.5% methoxyl (Methocel A)
Solubility:
Sol. in water (Methocel A4C, A4M, A15-LV)
Ionic Nature:
Nonionic (Methocel A)
Sp.gr.:
1.39 (Methocel A)
Density:
11.6 lb/gal (Methocel A)
Apparent Density:
0.25–0.70 g/cc (Methocel A)
Visc.:
15 cps (2% aq.) (Methocel A15-LV)
400 cps (2% aq.) (Methocel A4C)
4000 cps (2% aq.) (Methocel A4M)
Gelation Temp.:
50–55 C (Methocel A4C, A4M, A15-LV)
F.P.:
0 C (2% aq. sol'n.) (Methocel A)
Stability:
Stable over wide pH range of 3.0–11.0 (Methocel A4C, A4M, A15-LV)
Ref. Index:
1.336 (2% aq. sol'n.) (Methocel A)
pH:
7 (aq. sol'n.) (Methocel A)

215

Methylcellulose (cont'd.)

Surface Tension:
47–53 dynes/cm (aq. sol'n.) (Methocel A)
TOXICITY/HANDLING:
Dusts could cause skin and eye irritation under extreme conditons (Methocel A)
STORAGE/HANDLING:
Control dusts in air to prevent possible dust explosions (Methocel A)

Methyl oleate (CTFA)

SYNONYMS:
Methyl 9-octadecenoate
9-Octadecenoic acid, methyl ester
EMPIRICAL FORMULA:
$C_{19}H_{36}O_2$
STRUCTURE:

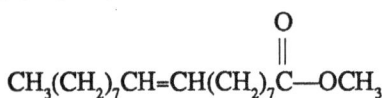

CAS No.:
112-62-9
TRADENAME EQUIVALENTS:
Emerest 2301 [Emery]
Emery Methyl Oleate [Emery/Oleochem Group]
Grocor 8002, 8008 [A. Gross]
Kemester 104, 105, 115, 205 [Humko]
Radia 7060 [Oleofina S.A.]
Stepan C68 [Stepan]
CATEGORY:
Wetting agent, lubricant, plasticizer, softener, intermediate, emulsifier, emollient, extrusion aid, solvent
APPLICATIONS:
Automotive applications: lubricants (Emery Methyl Oleate)
Cosmetic industry preparations: (Grocor 8002, 8008; Kemester 104)
Household detergents: (Stepan C68)
Industrial applications: chemical specialties (Grocor 8002, 8008; Radia 7060; Stepan C68); defoamers (Emerest 2301); dyes and pigments (Grocor 8002, 8008); lubricating/cutting oils (Emerest 2301; Radia 7060); metalworking (Emery Methyl Oleate; Kemester 105, 115, 205); mold release (Emerest 2301); plastics (Emerest 2301; Grocor 8002, 8008); resins (Stepan C68); rubber (Kemester 105); textile/leather

216

processing (Emery Methyl Oleate; Grocor 8002, 8008; Kemester 104; Radia 7060; Stepan C68)

PROPERTIES:

Form:

  Liquid (Emerest 2301; Grocor 8002, 8008)

Color:

  Light yellow (Grocor 8002, 8008)

  Gardner 1 max. (Kemester 205)

  Gardner 2 max. (Kemester 105; Stepan C68)

  Gardner 4 max. (Kemester 115)

  Gardner < 6 (Emerest 2301)

Odor:

  Mild (Grocor 8002)

Composition:

  98% min. active (Grocor 8002, 8008)

Solubility:

  Sol. in isopropanol (Grocor 8002, 8008); (5%) in isopropanol (Emerest 2301)

  Sol. in min. oil (Radia 7060); (5%) in min. oil (Emerest 2301)

  Sol. in most solvents (Radia 7060)

  Sol. (5%) in toluene (Emerest 2301)

  Sol. in veg. oil (Radia 7060)

  Insol. in water (Emerest 2301)

  Sol. (5%) in xylene (Emerest 2301)

Ionic Nature:

  Nonionic (Grocor 8002, 8008)

M.W.:

  286 (avg.) (Radia 7060)

Sp.gr.:

  0.862 (37.8 C) (Radia 7060)

  0.872 (Grocor 8002, 8008)

Density:

  7.3 lb/gal (Emerest 2301)

Visc.:

  4.10 cps (37.8 C) (Radia 7060)

  5 cSt (100 F) (Emerest 2301)

M.P.:

  16 C (Stepan C68)

Pour Pt.:

  −16 C (Emerest 2301)

Flash Pt.:

  350 F (Emerest 2301)

  178 C (COC) (Radia 7060)

Methyl oleate *(cont'd.)*

Cloud Pt.:
  −11 C (Radia 7060)
  0 C (Grocor 8002, 8008)
Acid No.:
  2.0 max. (Stepan C68)
  4.0 max. (Kemester 105, 115, 205)
Iodine No.:
  65–75 (Stepan C68)
  80–90 (Kemester 105, 205)
  90 max. (Kemester 115)
Saponification No.:
  185–205 (Kemester 115)
  191–197 (Stepan C68)
  194–203 (Kemester 105, 205)
Ref. Index:
  1.4513 (Radia 7060)
pH:
  Neutral (Grocor 8002, 8008)
Surface Tension:
  31.50 dynes/cm (Radia 7060)
STD. PKGS.:
  Bulk or drums (Grocor 8002, 8008)
  Drums (Kemester 105)

Microcrystalline cellulose *(CTFA)*

SYNONYMS:
  Isolated, colloidal crystallite portion of cellulose fibers
CAS No.:
  9004-34-6
TRADENAME EQUIVALENTS:
  Avicel PH-101, PH-102, PH-103, PH-105 [FMC] (NF grades)
  Avicel RC-591 [FMC]
CATEGORY:
  Stabilizer, suspending agent, binder, carrier, hardening agent, absorbent, disintegrant, flow aid, filler, peptizing agent, foam stabilizer
APPLICATIONS:
  Cosmetic industry preparations: makeup (Avicel PH-101, PH-102, PH-105); perfumery (Avicel PH-105)

Farm products: animal health products (Avicel PH-101, PH-102, PH-103, PH-105)
Pharmaceutical applications: dry antiperspirant sticks (Avicel PH-101); tablet mfg.
(Avicel PH-101, PH-102, PH-103, PH-105)

PROPERTIES:

Form:

Powder (Avicel PH-101, PH-102, PH-103, PH-105, RC-591)

Color:

White (Avicel PH-101, PH-102, PH-103, PH-105, RC-591)

Odor:

Odorless (Avicel PH-101, PH-102, PH-103, PH-105)

Taste:

Tasteless (Avicel PH-101, PH-102, PH-103, PH-105)

Composition:

active

Solubility:

Insol. in dilute acids (Avicel PH-101, PH-102, PH-103, PH-105)
Swells in dilute alkali (Avicel PH-101, PH-102, PH-103, PH-105)
Insol. in organic solvents (Avicel PH-101, PH-102, PH-103, PH-105)
Disp. in water (Avicel PH-101, PH-102, PH-103, PH-105, RC-591)

pH:

5.0–7.0 (Avicel PH-105)
5.5–7.0 (Avicel PH-101, PH-102, PH-103)

Myristic monoethanolamide

SYNONYMS:

N-(2-Hydroxyethyl) tetradecanamide
Monoethanolamine myristic acid condensate
Myristamide MEA (CTFA)
Myristoyl monoethanolamide
Tetradecanamide, N-(2-hydroxyethyl)-

EMPIRICAL FORMULA:

$C_{16}H_{33}NO_2$

STRUCTURE:

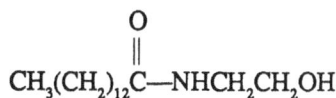

$$CH_3(CH_2)_{12}\overset{\overset{\displaystyle O}{\|}}{C}—NHCH_2CH_2OH$$

CAS No.:

142-58-5

Myristic monoethanolamide (cont'd.)

TRADENAME EQUIVALENTS:
Schercomid MME [Scher]
Witcamide MM [Witco]

CATEGORY:
Thickener, gelling agent, emulsifier, pearling agent, opacifier, foam booster, conditioner, lubricant, mold release agent, binder

APPLICATIONS:
Cosmetic industry preparations: (Schercomid MME; Witcamide MM); creams and lotions (Schercomid MME); shampoos (Schercomid MME)
Household detergents: dishwashing (Schercomid MME); powdered detergents (Schercomid MME)
Pharmaceutical applications: (Schercomid MME)

PROPERTIES:

Form:
Wax (Schercomid MME; Witcamide MM)

Color:
Pale (Schercomid MME)

Odor:
Slight ammoniacal (Schercomid MME)

Composition:
100% active; 90% amide content min. (Schercomid MME)

Solubility:
Sol. in alcohols (Schercomid MME)
Sol. in aliphatic hydrocarbons (Schercomid MME)
Sol. in aromatic hydrocarbons (Schercomid MME)
Sol. in chlorinated hydrocarbons (Schercomid MME)
Partly sol. in glycerol (Schercomid MME)
Sol. in glycol ethers (Schercomid MME)
Sol. in glycols (Schercomid MME)
Partly sol. in triols (Schercomid MME)
Disp. in water (Schercomid MME)

Ionic Nature:
Nonionic (Schercomid MME; Witcamide MM)

M.W.:
271 (Schercomid MME)

M.P.:
88 ± 4.0 C (Schercomid MME)

Flash Pt.:
> 170 C (OC) (Schercomid MME)

Acid No.:
2.0 max. (Schercomid MME)

Alkali No.:
15 max. (Schercomid MME)

SYNONYMS:
Dimethyl myristyl amine oxide
N,N-Dimethyl-1-tetradecanamine-N-oxide
Myristamine oxide (CTFA)
Tetradecanamine, N,N-dimethyl-, N-oxide
Tetradecyl dimethyl amine oxide

EMPIRICAL FORMULA:
$C_{16}H_{35}NO$

STRUCTURE:

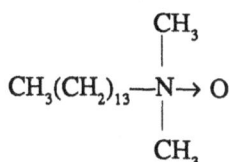

$$CH_3(CH_2)_{13}-\overset{\overset{\displaystyle CH_3}{|}}{\underset{\underset{\displaystyle CH_3}{|}}{N}} \rightarrow O$$

CAS No.:
3332-27-2

TRADENAME EQUIVALENTS:
Ammonyx MCO, MO [Onyx]
Aromox DM14D-W [Akzo Chemie]
Barlox 14 [Lonza]
Chemadox M [Richardson]
Conco XA-M [Continental]
Emcol M [Witco/Organics]
Empigen OH [Albright & Wilson/Detergents]
Incromine Oxide M [Croda Surfactants]
Jordamox MDA [Jordan]
Ninox M [Stepan]
Schercamox DMA, DMM [Scher]

CATEGORY:
Detergent, surfactant, wetting agent, foaming agent, foam stabilizer, stabilizer, thickener, viscosity builder, emollient, antistat

APPLICATIONS:
Bath products: bubble bath (Chemadox M; Empigen OH; Ninox M; Schercamox DMA, DMM)
Cosmetic industry preparations: (Ammonyx MCO, MO; Aromox DM14D-W; Chemadox M; Conco XA-M; Incromine Oxide M; Jordamox MDA); conditioners (Incromine Oxide M; Ninox M; Schercamox DMM); creams and lotions (Chemadox M; Incromine Oxide M); shampoos (Chemadox M; Empigen OH; Ninox M; Schercamox DMA, DMM); shaving preparations (Ninox M)
Household detergents: (Ammonyx MCO, MO; Aromox DM14D-W; Conco XA-M; Empigen OH; Jordamox MDA); dishwashing (Schercamox DMA); light-duty cleaners (Ninox M); liquid detergents (Empigen OH)

Myristyl dimethyl amine oxide *(cont'd.)*

Industrial applications: dyes and pigments (Conco XA-M); electroplating (Conco XA-M); photography (Conco XA-M); polymers/polymerization (Conco XA-M); textile/leather processing (Conco XA-M)

Industrial cleaners: (Conco XA-M; Jordamox MDA); janitorial cleaners (Ammonyx MCO, MO)

PROPERTIES:
Form:
Liquid (Ammonyx MCO, MO; Aromox DM14D-W; Barlox 14; Emcol M; Jordamox MDA); (@ 20 C) (Empigen OH)
Clear liquid (Ninox M; Schercamox DMA, DMM)
Viscous liquid (Incromine Oxide M)

Color:
Colorless (Incromine Oxide M)
Water-white (Schercamox DMA, DMM)
Pale straw (Empigen OH)
Gardner 1 max. (Aromox DM14D-W)
Klett 50 max. (Chemadox M)

Odor:
Mild (Schercamox DMA, DMM)

Composition:
29–31% conc. (Jordamox MDA)
29.5–30.5% amine oxide (Incromine Oxide M)
30% active (Conco XA-M; Ninox M; Schercamox DMA, DMM)
30% active min. in water (Aromox DM14D-W)
30% amine oxide (Ammonyx MCO, MO)
30% conc. (Barlox 14)
30% solids (Chemadox M)
30 ± 1.5% active (Empigen OH)

Solubility:
Sol. in alcohols (Schercamox DMA, DMM)
Sol. in ethanol (Conco XA-M)
Sol. in glycol ethers (Schercamox DMA, DMM)
Sol. in glycols (Schercamox DMA, DMM)
Sol. in hexylene glycol (Conco XA-M)
Sol. in isopropanol (Conco XA-M)
Sol. in polyols (Schercamox DMA, DMM)
Sol. in triols (Schercamox DMA, DMM)
Sol. in water (Conco XA-M; Schercamox DMA, DMM)

Ionic Nature:
Nonionic (Ammonyx MCO, MO; Barlox 14; Emcol M; Jordamox MDA; Incromine Oxide M)
Nonionic/cationic (Ninox M; Schercamox DMA, DMM)

Cationic/amphoteric (Conco XA-M)
M.W.:
256 avg. (Schercamox DMA)
263 avg. (Schercamox DMM)
Sp.gr.:
0.96 (Aromox DM14D-W); (25/20 C) (Ammonyx MCO, MO)
0.98 ± 0.05 (Schercamox DMM)
Density:
0.99 g/cc (20 C) (Empigen OH)
7.99 lb/gal (Chemadox M)
Visc.:
900 cs (20 C) (Empigen OH)
M.P.:
−3 C (Aromox DM14D-W)
Flash Pt.:
Nonflammable (Aromox DM14D-W)
> 200 F (Ammonyx MCO, MO)
Stability:
Good (Schercamox DMA, DMM)
pH:
6.7 (Chemadox M)
7.0 ± 1.0 (Schercamox DMA, DMM)
7.0–8.0 (Aromox DM14D-W); (5%) (Incromine Oxide M)
7.5 ± 0.5 (5% aq. sol'n.) (Empigen OH)
Biodegradable: (Conco XA-M; Ninox M)
TOXICITY/HANDLING:
Avoid prolonged contact with skin and eyes (Incromine Oxide M)
STORAGE/HANDLING:
Avoid contact with heavy metals, salts of Fe, Cu, Ni, etc. (Aromox DM14D-W)
Store in a cool, dry place (Incromine Oxide M)
STD. PKGS.:
200-kg net bung-type polythene drum (Aromox DM14D-W)
55-gal poly-lined drums (Schercamox DMA, DMM)
55-gal (450 lb net) polyethylene-lined Leverpak (Incromine Oxide M)

223

Naphthenic oil

SYNONYMS:
Naphthenic petroleum oil ASTM D2226 Type 103

TRADENAME EQUIVALENTS:
Flexon 580, 641, 680 [Exxon]
Naphthenic Oil 100 SUS, 150 SUS, 200 SUS, 1300 SUS, 2400 SUS, 6000 SUS [R.E. Carroll]

CATEGORY:
Plasticizer, softener, processing aid

APPLICATIONS:
Industrial applications: resins (Flexon 580); rubber (Flexon 580, 641, 680; Naphthenic Oil 100 SUS, 150 SUS, 200 SUS, 1300 SUS, 2400 SUS, 6000 SUS)

PROPERTIES:
Color:
ASTM 1.0 (Flexon 641)
ASTM 2.5 (Flexon 680)
ASTM 3.0 (Flexon 580)
Odor:
Petroleum oil (Flexon 580, 641, 680)
Composition:
35% aromatics (Flexon 641)
41% aromatics (Flexon 680)
42.9% aromatics (Naphthenic Oil 150 SUS)
44% aromatics (Naphthenic Oil 100 SUS)
45.8% aromatics (Naphthenic Oil 200 SUS)
47% aromatics (Flexon 580; Naphthenic Oil 1300 SUS)
47.5% aromatics (Naphthenic Oil 2400 SUS)
48.9% aromatics (Naphthenic Oil 6000 SUS)
Sp.gr.:
0.897 (60/60 F) (Flexon 641)
0.919 (Naphthenic Oil 150 SUS); (60/60 F) (Flexon 680)
0.921 (Naphthenic Oil 100 SUS)
0.928 (Naphthenic Oil 200 SUS)
0.935 (60/60 F) (Flexon 580)
0.9440 (Naphthenic Oil 1300 SUS)
0.9490 (Naphthenic Oil 2400 SUS)
0.9509 (Naphthenic Oil 6000 SUS)

Naphthenic oil (cont'd.)

Visc.:
39 SUS (210 F) (Flexon 641)
79 SUS (210 F) (Flexon 580)
79.2 SUS (210 F) (Flexon 680)
110 SUS (100 F) (Naphthenic Oil 100 SUS)
156 SUS (100 F) (Naphthenic Oil 150 SUS)
209 SUS (100 F) (Naphthenic Oil 200 SUS)
1250 SUS (100 F) (Naphthenic Oil 1300 SUS)
2525 SUS (100 F) (Naphthenic Oil 2400 SUS)
5945 SUS (100 F) (Naphthenic Oil 6000 SUS)

Flash Pt.:
325 F (Naphthenic Oil 100 SUS)
330 F (Naphthenic Oil 150 SUS)
345 F (Naphthenic Oil 200 SUS)
405 F (Naphthenic Oil 1300 SUS)
445 F (Naphthenic Oil 2400 SUS)
495 F (Naphthenic Oil 6000 SUS)

Aniline Pt.:
144 F (Naphthenic Oil 100 SUS)
151 F (Naphthenic Oil 200 SUS)
156 F (Naphthenic Oil 150 SUS)
166 F (Naphthenic Oil 1300 SUS)
172 F (Flexon 641)
173 F (Naphthenic Oil 2400 SUS)
182 F (Flexon 580)
183 F (Naphthenic Oil 6000 SUS)
200 F (Flexon 680

225

n-Octyl-n-decyl adipate

SYNONYMS:
NODA
TRADENAME EQUIVALENTS:
Monoplex NODA [C.P. Hall; Rohm & Haas]
Plasthall NODA [C.P. Hall]
Staflex NODA [Reichhold]
CATEGORY:
Plasticizer
APPLICATIONS:
Industrial applications: plastics (Monoplex NODA; Staflex NODA); rubber (Mono-
plex NODA); wire and cable (Staflex NODA)
PROPERTIES:
Form:
Clear liquid (Monoplex NODA; Plasthall NODA)
Color:
APHA 10 (Staflex NODA)
APHA 20 (Monoplex NODA)
M.W.:
390 (Plasthall NODA)
399 (Staflex NODA)
400 (Monoplex NODA)
Sp.gr.:
0.913 (25/15 C) (Monoplex NODA)
0.918 (23 C) (Staflex NODA)
Density:
7.6 lb/gal (Monoplex NODA; Staflex NODA)
Visc.:
16 cps (Monoplex NODA)
60 cSt (kinematic, 0 C) (Staflex NODA)
F.P.:
32 F (Monoplex NODA)
M.P.:
0 C (Monoplex NODA)
Pour Pt.:
26 F (Staflex NODA)
Flash Pt.:
430 F (Staflex NODA)

446 F (Monoplex NODA)
Cloud Pt.:
 < 26 F (Staflex NODA)
Acid No.:
 0.08 (Monoplex NODA)
Saponification No.:
 280 (Monoplex NODA)
Ref. Index:
 1.4458 (23 C) (Staflex NODA)
 1.446 (Monoplex NODA

Oleamidopropyl betaine (CTFA)

SYNONYMS:
 N-(Carboxymethyl)-N,N-dimethyl-3-[(1-oxooctadecenyl) amino]-1-propanaminium
  hydroxide, inner salt
 Oleamido betaine
 Oleamidopropyl dimethyl glycine
 Oleyl amido betaine
 Oleyl amidopropyl betaine
 1-Propanaminium, N-(carboxymethyl)-N,N-dimethyl-3-[(1-oxooctadecenyl)
  amino]-hydroxide, inner salt
EMPIRICAL FORMULA:
 $C_{25}H_{48}N_2O_3$
STRUCTURE:

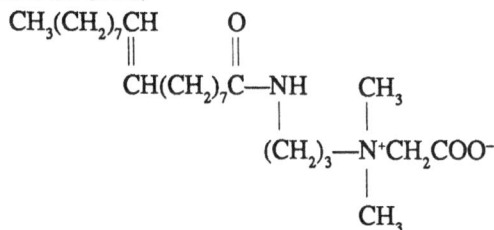

CAS No.:
 25054-76-6
TRADENAME EQUIVALENTS:
 Cycloteric BET O-30 [Cyclo]
 Incronam OP-30 [Croda Surfactants]
 Mackam HV [McIntyre]
 Schercotaine OAB [Scher]

Oleamidopropyl betaine (cont'd.)

CATEGORY:
Thickener, surfactant, foaming agent, foam stabilizer, conditioner, softener, lubricant
APPLICATIONS:
Cosmetic industry preparations: (Incronam OP-30); conditioners (Incronam OP-30); creams and lotions (Incronam OP-30); hair preparations (Cycloteric BET O-30); shampoos (Incronam OP-30); skin preparations (Cycloteric BET O-30)
Industrial applications: textile/leather processing (Schercotaine OAB)
PROPERTIES:
Form:
Liquid (Cycloteric BET O-30; Mackam HV; Schercotaine OAB)
Clear liquid (Incronam OP-30)
Color:
Yellow (Cycloteric BET O-30; Incronam OP-30)
Composition:
29–31% active (Cycloteric BET O-30)
30% active (Incronam OP-30)
35% conc. (Mackam HV; Schercotaine OAB)
Ionic Nature:
Amphoteric (Cycloteric BET O-30; Incronam OP-30; Mackam HV; Schercotaine OAB)
pH:
5.5–6.5 (Cycloteric BET O-30)
6.0–7.2 (5% sol'n.) (Incronam OP-30)
TOXICITY/HANDLING:
Avoid prolonged contact with skin (Incronam OP-30)
STD. PKGS.:
55-gal (450 lb net) lined drums (Incronam OP-30)

Oleic acid (CTFA)

SYNONYMS:
9-Octadecenoic acid
EMPIRICAL FORMULA:
$C_{18}H_{34}O_2$
STRUCTURE:
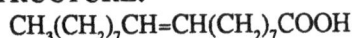
$CH_3(CH_2)_7CH=CH(CH_2)_7COOH$
CAS No.:
112-80-1

TRADENAME EQUIVALENTS:
Emersol 210, 220, 233LL[Emery]
Emersol 6321 [Emery] (food grade)
Groco 2 Oleic Acid, 4 Oleic Acid [A. Gross]
Industrene 105 [Humko]
Generically sold by: [Unichema]
CATEGORY:
Activator, plasticizer, softener, emulsifier, stabilizer (soap form), frothing agent
APPLICATIONS:
Industrial applications: latexes (generic—Unichema; Emersol 210, 220, 233 LL, 6321); polymers/polymerization (Industrene 105); rubber (generic—Unichema; Groco 2 Oleic Acid, 4 Oleic Acid; Industrene 105)
PROPERTIES:
Form:
Oily liquid (generic—Unichema)
Color:
Light yellow (generic—Unichema)
Gardner 6 (Industrene 105)
1″ Lovibond 1.0R–5.0Y max. (Groco 2 Oleic Acid, Groco 4 Oleic Acid)
Composition:
70% oleic acid (Groco 2 Oleic Acid, 4 Oleic Acid)
Sp.gr.:
0.89–0.91 (generic—Unichema)
M.P.:
0–8 C (generic—Unichema)
Flash Pt.:
≈ 365 F (generic—Unichema)
Acid No.:
197–203 (generic—Unichema)
198–204 (Groco 2 Oleic Acid, 4 Oleic Acid)
Iodine No.:
85–95 (Industrene 105)
90–95 (generic—Unichema)
95 max. (Groco 4 Oleic Acid)
96 max. (Groco 2 Oleic Acid)
STD. PKGS.:
Bulk or drums (Groco 2 Oleic Acid, 4 Oleic Acid)
Steel drums (Industrene 105)

Oleic diethanolamide

SYNONYMS:
N,N-Bis (2-hydroxyethyl)-9-octadecenamide
N,N-Bis (2-hydroxyethyl) oleamide
Diethanolamine oleic acid amide
9-Octadecenamide, N,N-bis (2-hydroxyethyl)-
Oleamide DEA (CTFA)
Oleic acid diethanolamide
Oleic fatty acid diethanolamide

EMPIRICAL FORMULA:
$C_{22}H_{43}NO_3$

STRUCTURE:

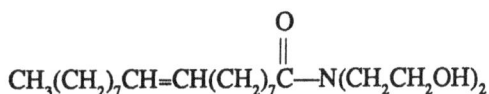

$$CH_3(CH_2)_7CH=CH(CH_2)_7\overset{\displaystyle O}{\overset{\|}{C}}-N(CH_2CH_2OH)_2$$

CAS No.:
93-83-4

TRADENAME EQUIVALENTS:
Alkamide SDO [Alkaril]
Alrosol O [Ciba-Geigy]
Aminol OF [Finetex]
Calamide O [Pilot]
Chimipal OLD [La Tessilchimica]
Clindrol 200-O [Clintwood]
Comperlan OD [Henkel KGaA]
Crillon ODE [Croda Chem. Ltd.]
Cyclomide DO280 [Cyclo] (2:1)
Cyclomide DO280/S [Cyclo] (1:1)
Emid 6545 [Emery]
Hartamide 9137 [Hart Chem. Ltd.]
Incromide OPD [Croda Surfactants]
Jordamide 201 [Jordan] (2:1)
Lauridit OD [Akzo Chemie]
Loramine DO280/SE [Dutton & Reinisch]
Loropan OD [Thomas Triantaphyllou]
Mackamide O [McIntyre]
Marlamid D1885 [Chemische Werke Huls AG]
Mazamide O-20 [Mazer] (2:1)
Merpinamid OD [Elektrochemische Fabrik Kempen]
Ninol 201 [Stepan]
Product LT 10-8-1 [Clintwood]
Rewomid DO280SE [Dutton & Reinisch]
Richamide 5085 [Richardson] (2:1)

Oleic diethanolamide (cont'd.)

TRADENAME EQUIVALENTS *(cont'd.):*
Schercomid ODA [Scher]
Schercomid SO-A [Scher] (1:1)
Steinamid DO280/SE [Dutton & Reinisch]
Varamide A-7 [Sherex]
CATEGORY:
Emulsifier, wetting agent, detergent, foam booster, foam stabilizer, foam suppressant, thickener, viscosity builder, corrosion inhibitor, emollient, lubricant, conditioning agent, superfatting agent, solubilizer, coupling agent, dispersant
APPLICATIONS:
Bath products: bubble bath (Alkamide SDO; Calamide O; Chimipal OLD; Clindrol 200-O; Comperlan OD; Crillon ODE; Incromide OPD; Lauridit OD; Loropan OD)
Cleansers: hand cleanser (Crillon ODE)
Cosmetic industry preparations: (Alkamide SDO; Clindrol 200-O; Lauridit OD; Loramine DO280/SE; Rewomid DO280/SE; Steinamid DO280/SE); conditioners (Calamide O; Schercomid ODA); creams and lotions (Incromide OPD); hair color preparations (Aminol OF); personal care products (Ninol 201); shampoos (Alkamide SDO; Calamide O; Chimipal OLD; Clindrol 200-O; Comperlan OD; Crillon ODE; Cyclomide DO280/S; Incromide OPD; Lauridit OD; Loropan OD; Merpinamid OD; Richamide 5085; Schercomid ODA)
Degreasers: (Calamide O; Clindrol 200-O; Incromide OPD; Mackamide O)
Household detergents: (Alkamide SDO; Clindrol 200-O; Crillon ODE; Lauridit OD; Marlamid D1885; Merpinamid OD); dishwashing (Lauridit OD; Richamide 5085); hard surface cleaner (Richamide 5085); heavy-duty cleaner (Jordamide 201); laundry detergent (Alkamide SDO); light-duty cleaners (Jordamide 201); liquid detergents (Chimipal OLD; Lauridit OD; Merpinamid OD; Ninol 201)
Industrial applications: (Calamide O); dyes and pigments (Alrosol O; Emid 6545; Schercomid ODA); lubricating/cutting oils (Alkamide SDO; Alrosol O; Calamide O; Clindrol 200-O; Incromide OPD; Mackamide O; Ninol 201); metalworking (Calamide O); petroleum industry (Clindrol 200-O); textile/leather processing (Alkamide SDO; Emid 6545)
Industrial cleaners: (Marlamid D1885; Merpinamid OD; Schercomid ODA); institutional cleaners (Alkamide SDO)
PROPERTIES:
Form:
Liquid (Aminol OF; Chimipal OLD; Clindrol 200-O; Comperlan OD; Crillon ODE; Cyclomide DO280, DO280/S; Emid 6545; Hartamide 9137; Jordamide 201; Lauridit OD; Loramine DO280/SE; Loropan OD; Mackamide O; Marlamid D1885; Mazamide O-20; Ninol 201; Rewomid DO280SE; Richamide 5085; Schercomid ODA; Steinamid DO280/SE)
Clear liquid (Calamide O; Schercomid SO-A)
Viscous liquid (Alkamide SDO; Alrosol O; Merpinamid OD; Product LT 10-8-1)

231

Oleic diethanolamide (cont'd.)

Paste/liquid (Incromide OPD)

Color:

Honey (Loramine DO280/SE; Rewomid DO280/SE; Steinamid DO280/SE)

Light amber (Schercomid SO-A)

Amber (Alkamide SDO; Alrosol O; Calamide O; Clindrol 200-O; Product LT 10-8-1; Schercomid ODA)

Yellow (Incromide OPD; Merpinamid OD)

Yellow-brown (Lauridit OD)

Gardner 7 (Emid 6545)

Odor:

Low (Alkamide SDO)

Bland (Incromide OPD)

Slight, typical oleic (Schercomid SO-A)

Mild, characteristic (Schercomid ODA)

Composition:

80% active (Alkamide SDO)

80% amide (Hartamide 9137)

85% active (Lauridit OD)

90% conc. (Comperlan OD; Loropan OD)

100% active (Clindrol 200-O; Cyclomide DO280/S; Incromide OPD; Loramine DO280/SE; Marlamid D1885; Rewomid DO280SE; Schercomid ODA, SO-A; Steinamid DO280/SE; Varamide A-7)

100% conc. (Aminol OF; Chimipal OLD; Cyclomide DO280; Jordamide 201; Mackamide O)

Solubility:

Sol. in alcohols (Clindrol 200-O; Schercomid ODA, SO-A)

Sol. in aliphatic hydrocarbons (Clindrol 200-O; Product LT 10-8-1; Schercomid ODA, SO-A)

Sol. in aromatic hydrocarbons (Alkamide SDO; Clindrol 200-O; Product LT 10-8-1; Schercomid SO-A); partly sol. (Schercomid ODA)

Sol. in butyl stearate (@ 5%) (Emid 6545)

Sol. in chlorinated aliphatic solvents (Alkamide SDO)

Sol. in chlorinated hydrocarbons (Clindrol 200-O; Schercomid ODA, SO-A)

Sol. in detergent systems (Alkamide SDO)

Sol. in esters (Product LT 10-8-1)

Sol. in ethanol (Product LT 10-8-1)

Sol. in natural fats (Schercomid SO-A)

Sol. in glycerol (Schercomid SO-A)

Sol. in glycerol trioleate (@ 5%) (Emid 6545)

Sol. in glycol ethers (Schercomid ODA, SO-A)

Sol. in glycols (Clindrol 200-O; Product LT 10-8-1; Schercomid ODA, SO-A)

Sol. in isopropanol (Product LT 10-8-1)

Sol. in ketones (Clindrol 200-O; Product LT 10-8-1)

Sol. in methanol (Product LT 10-8-1)

Sol. in min. oils (Alkamide SDO; Schercomid SO-A); (@ 5%) (Emid 6545)

Sol. in oils (Ninol 201); readily disp. (Calamide O)

Sol. in most organic solvents (Schercomid SO-A)

Sol. in petroleum solvents (Alkamide SDO)

Sol. in Stoddard solvent (@ 5%) (Emid 6545)

Sol. in water (100 g/l @ 20 C) (Lauridit OD); readily disp. (Calamide O); disp. (Ninol 201; Schercomid ODA, SO-A); disp. (@ 5%) (Emid 6545); insol. (Product LT 10-8-1)

Sol. in xylene (@ 5%) (Emid 6545)

Ionic Nature:

Nonionic (Aminol OF; Chimipal OLD; Clindrol 200-O; Crillon ODE; Cyclomide DO280; Incromide OPD; Jordamide 201; Lauridit OD; Loramine DO280/SE; Loropan OD; Marlamid D1885; Mackamide O; Mazamide O-20; Merpinamid OD; Ninol 201; Rewomid DO280SE; Varamide A-7)

Nonionic/anionic (Schercomid ODA, SO-A)

Anionic (Hartamide 9137)

Sp.gr.:

0.950 ± 0.01 (Schercomid ODA, SO-A)

0.963 (Product LT 10-8-1)

0.97 (Lauridit OD)

0.99 (Clindrol 200-O)

0.99–1.01 (Mazamide O-20)

Density:

0.96 g/cm^3 (20 C) (Loramine DO280/SE; Rewomid DO280/SE; Steinamid DO280/SE)

7.7 lb/gal (Emid 6545)

7.9 lb/gal (Schercomid ODA, SO-A)

8.0 lb/gal (Product LT 10-8-1)

8.23 lb/gal (Ninol 201)

8.3 lb/gal (Richamide 5085)

Visc.:

61.5 s (Saybolt, 210 F) (Product LT 10-8-1)

290 cSt (100 F) (Emid 6545)

450 cps min. (Schercomid SO-A)

1200 cps min. (Schercomid ODA)

F.P.:

–8 C (Clindrol 200-O)

0 C (Lauridit OD)

10 C max. (Schercomid ODA)

Oleic diethanolamide (cont'd.)

Pour Pt.:
   −3 C (Emid 6545)
Flash Pt.:
   100 C (PMCC) (Lauridit OD)
   > 170 C (OC) (Schercomid ODA)
   280 F (COC) (Product LT 10-8-1)
   475 F (Emid 6545)
Cloud Pt.:
   < 25 C (Emid 6545)
Acid No.:
   1 max. (Incromide OPD)
   4 max. (Lauridit OD)
   8 max. (Schercomid SO-A)
   12–16 (Schercomid ODA)
pH:
   9.0 (Lauridit OD)
   9.0–10.0 (Mazamide O-20)
   9.5–10.5 (Ninol 201)
   9.8 ± 0.5 (10% disp.) (Schercomid SO-A)
   9.9 ± 0.5 (10% disp.) (Schercomid ODA)
   10.0 (1% aq.) (Calamide O)
Biodegradable: (Mazamide O-20); rapidly biodegradable (Incromide OPD)
TOXICITY/HANDLING:
   Defatting to skin (Mazamide O-20)
   Skin and eye irritant (Clindrol 200-O)
   Avoid prolonged contact with skin (Incromide OPD)
STORAGE/HANDLING:
   Store in a dry place at 50–120 F (Incromide OPD)
STD. PKGS.:
   20 tons; 125-kg drums (Lauridit OD)
   200-kg net iron drums (Merpinamid OD)
   55-gal steel drums (Schercomid ODA)
   55-gal (450 lb net) lined steel drums (Calamide O)
   55-gal (450 lb net) tight-head steel drums (Clindrol 200-O)
   55-gal (450 lb net) polyethylene-lined Leverpak (Incromide OPD)

SYNONYMS:
N-(2-Hydroxypropyl)-9-octadecenamide
Monoisopropanolamine oleic acid amide
9-Octadecenamide, N-2-(hydroxypropyl)-
Oleamide MIPA (CTFA)
Oleic isopropanolamide

EMPIRICAL FORMULA:
$C_{21}H_{41}NO_2$

STRUCTURE:

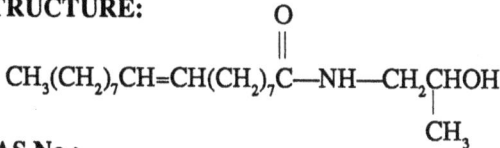

$$CH_3(CH_2)_7CH=CH(CH_2)_7C-NH-CH_2CHOH$$
$$CH_3$$

CAS No.:
111-05-7

TRADENAME EQUIVALENTS:
Loramine IPE 280 [Dutton & Reinisch]
Rewomid IPE 280 [Dutton & Reinisch]
Schercomid OMI [Scher]
Steinamid IPE 280 [Dutton & Reinisch]
Witcamide 61 [Witco]

CATEGORY:
Thickener, detergent, superfatting agent, foam stabilizer, conditioning agent, emulsifier, lubricant, emollient

APPLICATIONS:
Cosmetic industry preparations: (Loramine IPE 280; Rewomid IPE 280; Steinamid IPE 280; Witcamide 61); conditioners (Schercomid OMI); creams and lotions (Schercomid OMI); shampoos (Schercomid OMI); toiletries (Witcamide 61)
Household detergents: (Loramine IPE 280; Rewomid IPE 280; Steinamid IPE 280)

PROPERTIES:

Form:
Liquid to soft solid (Schercomid OMI)
Semisolid (Loramine IPE 280; Rewomid IPE 280; Steinamid IPE 280)

Color:
Amber (Schercomid OMI)
Light yellow (Loramine IPE 280; Rewomid IPE 280; Steinamid IPE 280)

Odor:
Mild ammoniacal (Schercomid OMI)
Typical, slight (Loramine IPE 280; Rewomid IPE 280; Steinamid IPE 280)

Composition:
100% active (Loramine IPE 280; Rewomid IPE 280; Steinamid IPE 280)
100% active; 85% amide min. (Schercomid OMI)

Oleic monoisopropanolamide (cont'd.)

Solubility:

Sol. in alcohols (Schercomid OMI)

Sol. in aliphatic hydrocarbons (Schercomid OMI)

Sol. in aromatic hydrocarbons (Schercomid OMI)

Sol. in chlorinated hydrocarbons (Schercomid OMI)

Sol. in esters (Schercomid OMI)

Sol. in glycol ethers (Schercomid OMI)

Partly sol. in glycols (Schercomid OMI)

Sol. in min. oil (Schercomid OMI)

Sol. in most organic solvents (Schercomid OMI)

Disp. in poylols (Schercomid OMI)

Disp. in triols (Schercomid OMI)

Sol. in veg. oil (Schercomid OMI)

Disp. in water (Schercomid OMI)

Ionic Nature:

Nonionic (Loramine IPE 280; Rewomid IPE 280; Schercomid OMI; Steinamid IPE 280)

Sp.gr.:

0.90 ± 0.01 (Schercomid OMI)

Density:

7.5 lb/gal (Schercomid OMI)

Congeal Pt.:

23–26 C (Schercomid OMI)

Flash Pt.:

180 C (OC) (Schercomid OMI)

Acid No.:

10 max. (Schercomid OMI)

Alkali No.:

20 max. (Schercomid OMI)

Storage Stability:

1 yr min. shelf life; alkali value decreases slightly with aging which does not affect product performance (Schercomid OMI)

pH:

Alkaline (Loramine IPE 280; Rewomid IPE 280; Steinamid IPE 280)

STD. PKGS.:

Lacquered drums (Loramine IPE 280; Rewomid IPE 280; Steinamid IPE 280)

55-gal (450 lb net) bung-head steel drums (Schercomid OMI)

Oleyl hydroxyethyl imidazoline (CTFA)

SYNONYMS:
2-(8-Heptadecenyl)-4,5-dihydro-1H-imidazole-1-ethanol
1H-Imidazole-1-ethanol, 2-(8-heptadecenyl)-4,5-dihydro-
Oleyl imidazoline

EMPIRICAL FORMULA:
$C_{22}H_{42}N_2O$

STRUCTURE:

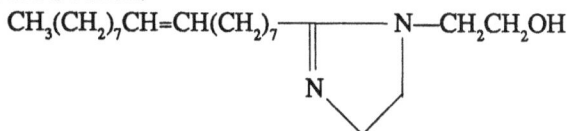

CAS No.:
95-38-5; 21652-27-7; 27136-73-8

TRADENAME EQUIVALENTS:
Amine O [Ciba-Geigy]
Mackazoline O [McIntyre]
Miramine OC [Miranol]
Monazoline O [Mona]
Schercozoline O [Scher]
Unamine O [Lonza]

CATEGORY:
Corrosion inhibitor, antistat, emulsifier, thickener, wetting agent, detergent, softener, bactericide, microbicide, dispersant

APPLICATIONS:
Automobile cleaners: (Monazoline O)
Cosmetic industry preparations: (Schercozoline O)
Farm products: agricultural oils/sprays (Monazoline O; Schercozoline O); herbicides (Monazoline O)
Industrial applications: adhesives and sealants (Monazoline O); cement/concrete (Schercozoline O); dyes and pigments (Monazoline O; Schercozoline O); lubricating/cutting oils (Miramine OC; Schercozoline O); metalworking (Miramine OC; Schercozoline O); ore flotation (Monazoline O); paint mfg. (Monazoline O; Schercozoline O); paper mfg. (Schercozoline O); petroleum industry (Monazoline O); plastics (Monazoline O); polishes and waxes (Schercozoline O); printing inks (Monazoline O); surface treatment (Schercozoline O); textile/leather processing (Monazoline O; Schercozoline O)
Industrial cleaners: (Miramine OC; Monazoline O; Schercozoline O); acid cleaners (Amine O; Monazoline O; Schercozoline O); dairy cleaners (Monazoline O)

PROPERTIES:
Form:
Liquid (Amine O; Mackazoline O; Schercozoline O; Unamine O)

Oleyl hydroxyethyl imidazoline (cont'd.)

Liquid; may crystallize on aging (Monazoline O)
Liquid to pasty solid (Miramine OC)
Color:
Amber (Miramine OC; Monazoline O)
Dark amber (Schercozoline O)
Yellow (Amine O)
Odor:
Ammoniacal (Miramine OC)
Composition:
90% min. imidazoline (Monazoline O; Schercozoline O)
99% active (Miramine OC)
100% conc. (Mackazoline O)
Solubility:
Sol. in acidic sol'ns. (Unamine O)
Sol. @ 10% in chlorinated hydrocarbons (Monazoline O)
Sol. @ 10% in ethanol (Monazoline O)
Sol. @ 10% in kerosene (Monazoline O)
Sol. @ 10% in min. oil (Monazoline O)
Sol. @ 10% in min. spirits (Monazoline O)
Sol. in oil (Miramine OC)
Sol. @ 10% in toluene (Monazoline O)
Sol. @ 10% in veg. oil (Monazoline O)
Disp. in water (Miramine OC); disp. @ 10% (Monazoline O)
Ionic Nature:
Nonionic (Mackazoline O)
Cationic (Miramine OC; Monazoline O; Schercozoline O)
M.W.:
345 (Monazoline O)
350 (Schercozoline O)
Sp.gr.:
0.92 (Monazoline O)
0.930 (Miramine OC)
Density:
7.66 lb/gal (Monazoline O)
Acid No.:
1 max. (Monazoline O)
Alkali No.:
160–170 (Schercozoline O)
167 (Miramine OC)
pH:
10.0–11.5 (10% disp.) (Monazoline O)
Biodegradable: (Miramine OC)

Oleyl hydroxyethyl imidazoline *(cont'd.)*

TOXICITY/HANDLING:

Very severe primary skin irritant; produces conjunctival and corneal eye irritation; handle with protective gloves and goggles (Miramine OC)

Fairly strong organic base—handle with caution; wear protective gloves and goggles (Monazoline O)

STORAGE/HANDLING:

Drums should be kept closed when not in use; crystallization may take place upon aging, exposure to air, or to low temps.; mild heat and agitation will restore it to its original state (Miramine OC)

STD. PKGS.:

55-gal closed head steel drums (Monazoline O)

Oleyl sarcosine

SYNONYMS:

Glycine, N-methyl-N-(1-oxo-9-octadecenyl)-
N-Methyl-N-(1-oxo-9-octadecenyl) glycine
Oleoyl sarcosine (CTFA)
Oleyl methylaminoethanoic acid
Oleyl N-methylaminoacetic acid
Oleyl N-methylglycine

EMPIRICAL FORMULA:

$C_{21}H_{39}NO_3$

STRUCTURE:

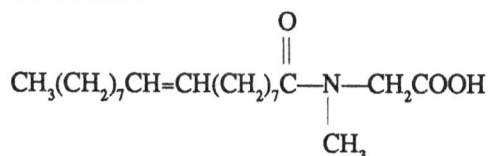

CAS No.:

110-25-8

TRADENAME EQUIVALENTS:

Hamposyl O [W.R. Grace]
Sarkosyl O [Ciba Geigy]

CATEGORY:

Shampoo base, foam stabilizer, foam booster, detergent, emulsifier, corrosion inhibitor, conditioner, wetting agent, lubricant

APPLICATIONS:

Bath products: bath oils (Sarkosyl O)

Oleyl sarcosine (cont'd.)

Cleansers: hand cleanser (Sarkosyl O); skin cleanser (Hamposyl O)

Cosmetic industry preparations: (Hamposyl O); shampoos (Hamposyl O; Sarkosyl O); toilet soaps (Sarkosyl O)

Degreasers: solvent degreasing (Sarkosyl O)

Household detergents: (Sarkosyl O); carpet & upholstery shampoos (Hamposyl O; Sarkosyl O); dishwashing (Sarkosyl O); laundry detergent (Sarkosyl O); window cleaners (Sarkosyl O)

Industrial applications: lubricating/cutting oils (Sarkosyl O); metalworking (Sarkosyl O)

Pharmaceutical applications: (Sarkosyl O); antidandruff shampoos (Sarkosyl O); antiperspirant/deodorant (Sarkosyl O); balms (Sarkosyl O); dental preparations (Sarkosyl O); mouthwash (Sarkosyl O)

PROPERTIES:

Form:

Liquid (Hamposyl O)

Powder (Sarkosyl O)

Composition:

94% min. purity (Sarkosyl O)

100% conc. (Hamposyl O)

Solubility:

Sol. in aliphatic hydrocarbons (Sarkosyl O)

Sol. in glycerin (Sarkosyl O)

Sol. in glycols (Sarkosyl O)

Sol. in most organic solvents (Sarkosyl O)

Sol. in phosphate esters (Sarkosyl O)

Sol. in silicones (Sarkosyl O)

Insol. in water (Sarkosyl O)

Ionic Nature:

Anionic (Hamposyl O; Sarkosyl O)

M.W.:

340–360 (Sarkosyl O)

Sp.gr.:

0.948 (Sarkosyl O)

Biodegradable: (Hamposyl O)

TOXICITY/HANDLING:

Skin irritant on prolonged contact with conc. acid (Sarkosyl O)

Palmitamidopropyl betaine (CTFA)

SYNONYMS:

N-(Carboxymethyl)-N,N-dimethyl-3-[(1-oxohexadecyl) amino]-1-propanaminium hydroxide, inner salt

1-Propanaminium, N-(carboxymethyl)-N,N-dimethyl-3-[(1-oxohexadecyl)amino]-, hydroxide, inner salt

EMPIRICAL FORMULA:

$C_{23}H_{46}N_2O_3$

STRUCTURE:

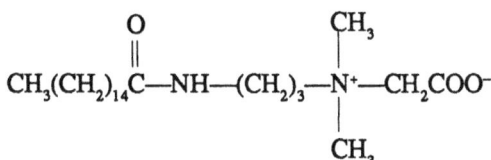

CAS No.:

32954-43-1

TRADENAME EQUIVALENTS:

Incronam P-30 [Croda Surfactants]

Schercotaine PAB [Scher]

CATEGORY:

Thickener, conditioner

APPLICATIONS:

Cosmetic industry preparations: conditioners (Incronam P-30; Schercotaine PAB); creams and lotions (Schercotaine PAB); hair preparations (Incronam P-30; Schercotaine PAB); shampoos (Incronam P-30); skin preparations (Incronam P-30; Schercotaine PAB)

PROPERTIES:

Form:

Viscous liquid to slurry (Incronam P-30)

Soft gel (Schercotaine PAB)

Color:

Light yellow (Schercotaine PAB)

Yellow (Incronam P-30)

Composition:

30% min. active (Incronam P-30)

35% min. dry solids (Schercotaine PAB)

Ionic Nature:

Amphoteric (Incronam P-30; Schercotaine PAB)

Palmitamidopropyl betaine *(cont'd.)*

pH:
   5.0–7.0 (Schercotaine PAB)
   6.0–7.2 (5%) (Incronam P-30)
TOXICITY/HANDLING:
   Avoid prolonged contact with skin (Incronam P-30)
STORAGE/HANDLING:
   Store in a dry place (Incronam P-30)
STD. PKGS.:
   55 gal (450 lb net) polyethylene-lined Leverpak (Incronam P-30)

Palm kernel oil acid diethanolamide

SYNONYMS:
   N,N-Bis (2-hydroxyethyl) palm kernel oil acid amide
   Diethanolamine palm kernel oil acid amide
   Palm kernelamide DEA (CTFA)
   Palm kernel oil acid amide, N,N-bis (2-hydroxyethyl)-
STRUCTURE:

$$R-\overset{\displaystyle O}{\overset{\displaystyle \|}{C}}-N(CH_2CH_2OH)_2$$

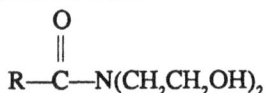

   where RCO⁻ represents the palm kernel oil acid radical
CAS No.:
   RD No. 977069-44-5
TRADENAME EQUIVALENTS:
   Cyclomide DP 240/S [Cyclo] (1:1)
   Mackamide PK [McIntyre]
CATEGORY:
   Viscosity builder, foam stabilizer
PROPERTIES:
Form:
   Liquid (Cyclomide DP 240/S; Mackamide PK)
Composition:
   100% active; 85% amide content (Cyclomide DP 240/S)
   100% conc. (Mackamide PK)
Solubility:
   Insol. in min. oil (Cyclomide DP 240/S)
   Disp. in water (Cyclomide DP 240/S)
Ionic Nature:
   Nonionic (Mackamide PK)

Palm oil glycerides (CTFA)

SYNONYMS:
Glycerides, palm oil mono-, di-, and tri-
TRADENAME EQUIVALENTS:
Imwitor 940 [Dynamit-Nobel]
Monomuls 60-30 [Chemische Fabrik Grunau GmbH]
CATEGORY:
Stabilizer, dispersant, emfulsifier, opacifier
APPLICATIONS:
Cosmetic industry preparations: (Monomuls 60-30)
Food applications: (Monomuls 60-30)
Industrial applications: asphalt emulsions (Imwitor 940); clay systems (Imwitor 940)
Pharmaceutical applications: (Monomuls 60-30)
PROPERTIES:
Form:
Solid (Monomuls 60-30)
Composition:
100% conc. (Monomuls 60-30)
Solubility:
Sol. in fats (Imwitor 940)
Sol. in oils (Imwitor 940)
Sol. in waxes (Imwitor 940)
Ionic Nature:
Anionic (Monomuls 60-30)

Pectin (CTFA)

SYNONYMS:
Citrus pectin
CAS No.:
9000-69-5
TRADENAME EQUIVALENTS:
Genu 04CB or 04CG, 100–120 Grade [Hercules] (low methoxyl)
Genu 12CB or 12CG, 85–100 Grade [Hercules] (low methoxyl)
Genu 15AB or 104AS, 100–120 Grade [Hercules] (low methoxyl)
Genu 18CB or 18CG, 70–90 Grade [Hercules] (low methoxyl)
Genu 20AB or 20AS, 100 Grade [Hercules] (low methoxyl)
Genu 21AB or 21AS [Hercules] (low methoxyl)
Genu 102AS Buffered, 100 Grade [Hercules] (low methoxyl)
Genu AA Medium-Rapid Set, 150 Grade [Hercules] (high methoxyl)

Pectin *(cont'd.)*

TRADENAME EQUIVALENTS *(cont'd.):*
Genu BA-King, 150 Grade [Hercules] (high methoxyl)
Genu BB Rapid Set, 150 Grade [Hercules] (high methoxyl)
Genu DD Extra-Slow Set, 150 Grade [Hercules] (high methoxyl)
Genu DD Extra-Slow Set-C, 150 Grade [Hercules] (high methoxyl)
Genu DD Slow Set, 150 Grade [Hercules] (high methoxyl)
Genu JM [Hercules] (high methoxyl)

CATEGORY:
Gelling agent, thickener, stabilizer, texturizer, foam stabilizer

APPLICATIONS:
Food applications: citrus drink processing (Genu BB Rapid Set); food additives (Genu 04CB, 04CG, 12CB, 12CG, 15AB, 104AS, 18CB, 18CG, 20AB, 20AS, 21AB, 21AS, 102AS Buffered, AA Medium-Rapid Set, BA-King, BB Rapid Set, DD Extra-Slow Set, DD Extra-Slow Set-C, DD Slow Set); milk products (Genu JM)

PROPERTIES:
Composition:
27–33% methoxylation (Genu 15AB or 104AS)
27–35% methoxylation (Genu 12CB, 12CG)
33–40% methoxylation (Genu 20AB, 20AS, 21AB, 21AS)
34–45% methoxylation (Genu 04CG, 04 CB)
35–45% methoxylation (Genu 18CB, 18CG)
60–63% methoxylation (Genu DD Extra-Slow Set-C)
61–64% methoxylation (Genu DD Extra-Slow Set)
63–66% methoxylation (Genu DD Slow Set)
67–70% methoxylation (Genu AA Medium-Rapid Set)
68–71% methoxylation (Genu BA-King)
70–75% methoxylation (Genu JM)
71–75% methoxylation (Genu BB Rapid Set)

Setting Time:
110–175 s (Genu BA-King)
130–200 s (Genu AA Medium-Rapid Set)
140 s max. (Genu BB Rapid Set)
220–300 s (Genu DD Slow Set)
250 s min. (Genu DD Extra-Slow Set, DD Extra-Slow Set-C)

Setting Temp.:
≤ 60 C (Genu DD Extra-Slow Set, DD Extra-Slow Set-C)
60–70 C (Genu DD Slow Set)
75–80 C (Genu AA Medium-Rapid Set)
80 C (Genu BA-King)
85–95 C (Genu BB Rapid Set)

pH:
2.7–3.2 (4% sol'n.) (Genu BA-King, JM)

244

Pectin (cont'd.)

3.2–4.0 (1% sol'n.) (Genu 18CB, 18CG)
3.4–4.2 (4% sol'n.) (Genu AA Medium-Rapid Set, BB Rapid Set, DD Extra-Slow Set, DD Slow Set)
3.8–4.4 (1% sol'n.) (Genu 12CB, 12CG, 15AB, 104AS, 20AB, 20AS, 21AB, 21AS)
4.8–5.2 (1% sol'n.) (Genu 04CG, 04 CB)

POE (6) coconut amide

SYNONYMS:
Coconut amide polyglycolether (6 moles EO)
Coconut monoethanolamide ethoxylate (6 moles EO)
PEG-6 cocamide (CTFA)
PEG 300 coconut amide
POE 6 coco monoethanolamide

STRUCTURE:

$$RC\overset{\displaystyle O}{\overset{\|}{-}}NH-(CH_2CH_2O)_nH$$

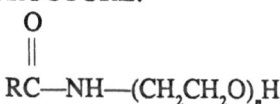

    where   RCO⁻ represents the coconut acid radical and
                avg. $n = 6$

CAS No.:
61791-08-0 (generic)
RD No.: 977061-47-4

TRADENAME EQUIVALENTS:
Alkamidox CME-5 [Alkaril] (5 EO)
Amidox C-5 [Stepan]
Empilan MAA [Albright & Wilson]
Mazamide C-5 [Mazer]
Rewopal C-6 [Dutton & Reinisch]
Unamide C-5 [Lonza]

CATEGORY:
Foam stabilizer, foaming agent, thickener, viscosity builder, detergent, surfactant, solubilizer, emulsifier, rust inhibitor, dispersant, wetting agent, solvent

APPLICATIONS:
Bath products: bubble bath (Rewopal C-6; Unamide C-5)
Cosmetic industry preparations: (Alkamidox CME-5); perfumery (Rewopal C-6); personal care products (Unamide C-5); shampoos (Amidox C-5; Rewopal C-6; Unamide C-5)
Household detergents: (Amidox C-5; Mazamide C-5; Rewopal C-6; Unamide C-5);

245

POE (6) coconut amide (cont'd.)

dishwashing (Alkamidox CME-5); hard surface cleaner (Unamide C-5); heavy-duty cleaner (Alkamidox CME-5); laundry detergent (Alkamidox CME-5); light-duty cleaners (Unamide C-5); liquid detergents (Alkamidox CME-5; Empilan MAA; Unamide C-5)

Industrial applications: (Unamide C-5)

Industrial cleaners: buffing compounds (Mazamide C-5)

PROPERTIES:

Form:

Liquid (Alkamidox CME-5; Amidox C-5; Mazamide C-5; Rewopal C-6; Unamide C-5)

Soft paste (Empilan MAA)

Color:

Straw (Alkamidox CME-5)

Amber (Rewopal C-6)

Creamy yellow (Empilan MAA)

Light yellow (Unamide C-5)

Gardner 5 max. (Amidox C-5)

Odor:

Faint characteristic (Empilan MAA)

Composition:

100% active (Alkamidox CME-5; Empilan MAA; Rewopal C-6; Unamide C-5)

Solubility:

Sol. in ethanol (Alkamidox CME-5); (@ 5%) (Amidox C-5)

Insol. in kerosene (Alkamidox CME-5); (@ 5%) (Amidox C-5)

Insol. in min. oil (Alkamidox CME-5); (@ 5%) (Amidox C-5)

Sol. in water (Alkamidox CME-5); (@ 5%) (Amidox C-5)

Sol. in xylene (Alkamidox CME-5); (@ 5%) (Amidox C-5)

Ionic Nature:

Nonionic (Empilan MAA; Mazamide C-5; Rewopal C-6; Unamide C-5)

Sp.gr.:

1.0–1.1 (Mazamide C-5)

Congeal Pt.:

55–59 F (Amidox C-5)

Cloud Pt.:

80–90 C (2% in dist. water) (Rewopal C-6)

Saponification No.:

12 max. (Unamide C-5)

15 max. (Empilan MAA)

Hydroxyl No.:

145–165 (Unamide C-5)

Stability:

Stable to acids and alkalies (Empilan MAA)

Stable in dilute acid and alkaline sol'ns. (Rewopal C-6)
Unstable in aq. systems; stable to strong alkalies (Alkamidox CME-5)
More stable than diethanolamides in aq. systems under alkaline conditions (Amidox C-5)

pH:
Slightly alkaline (Empilan MAA)
8.0–10.0 (1% solids) (Rewopal C-6)
9.0–10.0 (5% sol'n.) (Unamide C-5)
9.5–10.5 (Mazamide C-5)
10.0–11.5 (1% aq.) (Amidox C-5)

POE (20) dilaurate

SYNONYMS:
PEG-20 dilaurate (CTFA)
PEG 1000 dilaurate

STRUCTURE:

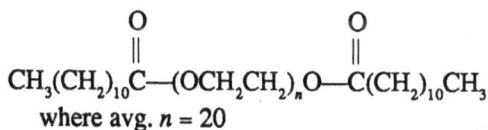

$$CH_3(CH_2)_{10}C\!\!-\!\!(OCH_2CH_2)_nO\!\!-\!\!C(CH_2)_{10}CH_3$$
where avg. $n = 20$

CAS No.:
9005-02-1 (generic)
RD No.: 977063-47-0

TRADENAME EQUIVALENTS:
Cithrol 10DL [Croda]
Kessco PEG 1000 Dilaurate [Armak]

CATEGORY:
Thickener, surfactant, solubilizer, dispersant, emulsifier, wetting agent, cosolvent

APPLICATIONS:
Bath products: bath oils (Kessco PEG 1000 Dilaurate)
Cosmetic industry preparations: (Cithrol 10DL; Kessco PEG 1000 Dilaurate); hair preparations (Kessco PEG 1000 Dilaurate); perfumery (Kessco PEG 1000 Dilaurate); shampoos (Kessco PEG 1000 Dilaurate)
Farm products: (Kessco PEG 1000 Dilaurate)
Food applications: (Kessco PEG 1000 Dilaurate)
Industrial applications: (Cithrol 10DL; Kessco PEG 1000 Dilaurate); plastics (Kessco PEG 1000 Dilaurate)
Pharmaceutical applications: (Kessco PEG 1000 Dilaurate)

POE (20) dilaurate *(cont'd.)*

PROPERTIES:
Form:
  Paste (Cithrol 10DL)
  Soft solid (Kessco PEG 1000 Dilaurate)
Color:
  Cream (Kessco PEG 1000 Dilaurate)
Composition:
  97% conc. (Cithrol 10DL)
Solubility:
  Sol. in acetone (Kessco PEG 1000 Dilaurate)
  Sol. in carbon tetrachloride (Kessco PEG 1000 Dilaurate)
  Sol. in ethyl acetate (Kessco PEG 1000 Dilaurate)
  Sol. in isopropanol (Kessco PEG 1000 Dilaurate)
  Sol. in isopropyl myristate (Kessco PEG 1000 Dilaurate)
  Sol. in water (Kessco PEG 1000 Dilaurate)
Ionic Nature:
  Nonionic (Cithrol 10DL; Kessco PEG 1000 Dilaurate)
Sp.gr.:
  1.015 (65 C) (Kessco PEG 1000 Dilaurate)
F.P.:
  38 C (Kessco PEG 1000 Dilaurate)
Flash Pt.:
  475 F (COC) (Kessco PEG 1000 Dilaurate)
Fire Pt.:
  495 F (Kessco PEG 1000 Dilaurate)
HLB:
  14.5 (Kessco PEG 1000 Dilaurate)
Acid No.:
  10.0 max. (Kessco PEG 1000 Dilaurate)
Iodine No.:
  5.5 max. (Kessco PEG 1000 Dilaurate)
Saponification No.:
  68–78 (Kessco PEG 1000 Dilaurate)

POE (32) dilaurate

SYNONYMS:
  PEG-32 dilaurate (CTFA)
  PEG 1540 dilaurate

248

STRUCTURE:

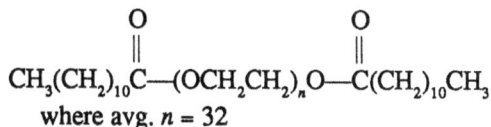

$$CH_3(CH_2)_{10}C-(OCH_2CH_2)_nO-C(CH_2)_{10}CH_3$$

where avg. $n = 32$

CAS No.:

9005-02-1 (generic)

RD No.: 977065-53-4

TRADENAME EQUIVALENTS:

Kessco PEG 1540 Dilaurate [Armak]

CATEGORY:

Thickener, surfactant, solubilizer

APPLICATIONS:

Bath products: bath oils

Cosmetic industry preparations: hair preparations; perfumery; shampoos

Farm products

Food applications

Industrial applications: plastics

Pharmaceutical applications

PROPERTIES:

Form:

Wax

Color:

Cream

Solubility:

Sol. hot in acetone

Sol. hot in carbon tetrachloride

Sol. hot in ethyl acetate

Sol. hot in isopropanol

Sol. in 5% Na_2SO_4

Sol. hot in propylene glycol

Sol. hot in toluol

Sol. in water

Ionic Nature:

Nonionic

Sp.gr.:

1.04 (65 C)

F.P.:

42 C

Flash Pt.:

450 F (COC)

POE (32) dilaurate *(cont'd.)*

Fire Pt.:
515 F
HLB:
15.7
Acid No.:
10 max.
Iodine No.:
4.0 max.
Saponification No.:
48–56
pH:
4.5 (3% disp., 20 C)

POE (75) dilaurate

SYNONYMS:
PEG-75 dilaurate (CTFA)
PEG 4000 dilaurate
STRUCTURE:

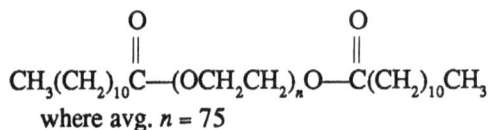

where avg. $n = 75$
CAS No.:
9005-02-1 (generic)
RD No.: 977065-82-9

TRADENAME EQUIVALENTS:
Kessco PEG 4000 Dilaurate [Armak]

CATEGORY:
Thickener, surfactant, solubilizer

APPLICATIONS:
Bath products: bath oils
Cosmetic industry preparations: hair preparations; perfumery; shampoos
Farm products
Food applications
Industrial applications: plastics
Pharmaceutical applications

PROPERTIES:
Form:
  Wax
Color:
  Cream
Solubility:
  Sol. hot in acetone
  Sol. hot in carbon tetrachloride
  Sol. hot in ethyl acetate
  Sol. hot in isopropanol
  Sol. in 5% Na_2SO_4
  Sol. hot in propylene glycol
  Sol. hot in toluol
  Sol. in water
Ionic Nature:
  Nonionic
Sp.gr.:
  1.065 (65 C)
F.P.:
  52 C
Flash Pt.:
  495 F (COC)
Fire Pt.:
  515 F
HLB:
  17.6
Acid No.:
  5.0 max.
Iodine No.:
  2.0 max.
Saponification No.:
  20–30
pH:
  4.5 (3% disp., 20 C)

POE (150) dilaurate

SYNONYMS:
  PEG-150 dilaurate (CTFA)
  PEG 6000 dilaurate

STRUCTURE:

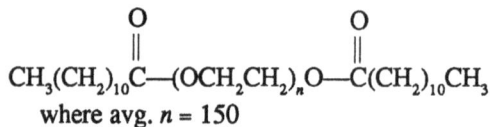

where avg. $n = 150$

CAS No.:

9005-02-1 (generic)

RD No.: 977055-05-2

TRADENAME EQUIVALENTS:

Kessco PEG 6000 Dilaurate [Armak]

CATEGORY:

Thickener, surfactant, solubilizer

APPLICATIONS:

Bath products: bath oils

Cosmetic industry preparations: hair preparations; perfumery; shampoos

Farm products

Food applications

Industrial applications: plastics

Pharmaceutical applications

PROPERTIES:

Form:

Wax

Color:

Cream

Solubility:

Sol. hot in acetone

Sol. hot in carbon tetrachloride

Sol. hot in ethyl acetate

Sol. in 5% Na_2SO_4

Sol. hot in propylene glycol

Sol. hot in toluol

Sol. in water

Ionic Nature:

Nonionic

Sp.gr.:

1.077 (65 C)

F.P.:

57 C

Flash Pt.:

435 F (COC)

Fire Pt.:

495 F

HLB:
　18.7
Acid No.:
　9.0 max.
Iodine No.:
　1.5 max.
Saponification No.:
　12–20
pH:
　4.5 (3% disp., 20 C)

POE (150) distearate

SYNONYMS:
　PEG-150 distearate (CTFA)
　PEG 6000 distearate
STRUCTURE:

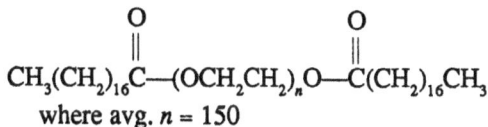

where avg. $n = 150$

CAS No.:
　9005-08-7 (generic)
　RD No.: 977055-11-0
TRADENAME EQUIVALENTS:
　Alkamuls 6000-DS [Alkaril]
　Emulvis [C.P. Hall Co.]
　Kessco PEG 6000 Distearate [Armak]
　Lipopeg 6000-DS [Lipo]
　Mapeg 6000DS [Mazer]
　Pegosperse 6000-DS [Glyco]
　Plurest DS150 [BASF Wyandotte]
　Rewopal PEG 6000 DS [Rewo Chemische Werke GmbH]
CATEGORY:
　Emulsifier, viscosity builder, thickener, solubility retarder, antiblock agent, spreading agent, dispersant, surfactant, lubricant, emollient, softener, conditioner, opacifier, antistat, solubilizer, stabilizer, moisturizer, foam modifier, plasticizer, wetting agent

POE (150) distearate (cont'd.)

APPLICATIONS:
Bath products: bath oils (Kessco PEG 6000 Distearate; Lipopeg 6000-DS)
Cleansers: skin cleanser (Emulvis)
Cosmetic industry preparations: (Alkamuls 6000-DS; Emulvis; Kessco PEG 6000 Distearate; Mapeg 6000DS; Pegosperse 6000-DS); conditioners (Pegosperse 6000-DS); creams and lotions (Lipopeg 6000-DS); hair preparations (Emulvis; Kessco PEG 6000 Distearate; Pegosperse 6000-DS); makeup (Pegosperse 6000-DS); perfumery (Kessco PEG 6000 Distearate); shampoos (Pegosperse 6000-DS); toiletries (Pegosperse 6000-DS; Rewopal PEG 6000 DS)
Farm applications: (Kessco PEG 6000 Distearate); ponds (Emulvis)
Food applications: (Kessco PEG 6000 Distearate)
Household detergents: (Pegosperse 6000-DS)
Industrial applications: inks (Emulvis); lubricating/cutting oils (Mapeg 6000DS); metalworking (Mapeg 6000DS); plastics (Kessco PEG 6000 Distearate); textile/ leather processing (Mapeg 6000DS)
Pet shampoos: (Emulvis)
Pharmaceutical applications: (Kessco PEG 6000 Distearate; Mapeg 6000DS; Pegosperse 6000-DS); suppositories (Emulvis); tablet mfg. (Emulvis)
PROPERTIES:
Form:
Solid (Alkamuls 6000-DS; Mapeg 6000DS)
Flake (Emulvis; Lipopeg 6000-DS; Mapeg 6000DS; Pegosperse 6000-DS; Rewopal PEG 6000 DS)
Wax (Kessco PEG 6000 Distearate)
Waxy flakes (Plurest DS150)
Color:
White (Mapeg 6000DS; Pegosperse 6000-DS)
White to off-white (Plurest DS150)
Off-white (Lipopeg 6000-DS)
White to cream (Emulvis)
Cream (Alkamuls 6000-DS; Kessco PEG 6000 Distearate)
Odor:
Bland (Emulvis)
Composition:
100% active (Lipopeg 6000-DS)
100% conc. (Mapeg 6000DS; Rewopal PEG 6000 DS)
Solubility:
Sol. in acetone (Kessco PEG 6000 Distearate)
Disp. in aromatic solvent (@ 10%) (Alkamuls 6000-DS)
Sol. in carbon tetrachloride (Kessco PEG 6000 Distearate)
Sol. in ethanol (Pegosperse 6000-DS)
Sol. in ethyl acetate (Kessco PEG 6000 Distearate)

Sol. in isopropanol (Kessco PEG 6000 Distearate; Mapeg 6000DS); sol. 50% in 70% IPA (Emulvis)

Disp. in min. oil (@ 10%) (Alkamuls 6000-DS); insol. (Pegosperse 6000-DS)

Disp. in min. spirits (@ 10%) (Alkamuls 6000-DS)

Disp. in perchloroethylene (@ 10%) (Alkamuls 6000-DS)

Sol. in propylene glycol (Kessco PEG 6000 Distearate; Mapeg 6000DS)

Sol. in toluol (Kessco PEG 6000 Distearate; Mapeg 6000DS)

Insol. in veg. oil (Pegosperse 6000-DS)

Sol. in water (Mapeg 6000DS; Pegosperse 6000-DS); (@ 10%) (Alkamuls 6000-DS); partly sol. (Kessco PEG 6000 Distearate); 3% sol. with greater water sol. at high temps. (Emulvis)

Ionic Nature:

Nonionic (Emulvis; Kessco PEG 6000 Distearate; Mapeg 6000DS; Pegosperse 6000-DS)

Sp.gr.:

1.075 (65 C) (Kessco PEG 6000 Distearate)

M.P.:

53–58 C (Emulvis)

55 C (Kessco PEG 6000 Distearate; Mapeg 6000DS; Pegosperse 6000-DS; Plurest DS150)

Flash Pt.:

475 F (COC) (Kessco PEG 6000 Distearate)

Fire Pt.:

525 F (Kessco PEG 6000 Distearate)

HLB:

18.0 ± 1 (Pegosperse 6000-DS)

18.4 (Alkamuls 6000-DS; Kessco PEG 6000 Distearate; Mapeg 6000DS; Plurest DS150)

18.4 ± 1 (Lipopeg 6000-DS)

Acid No.:

9 (Pegosperse 6000-DS)

9 max. (Kessco PEG 6000 Distearate; Mapeg 6000DS; Plurest DS150)

10 max. (Lipopeg 6000-DS)

Iodine No.:

0.1 (Pegosperse 6000-DS)

0.1 max. (Kessco PEG 6000 Distearate; Mapeg 6000DS)

Saponification No.:

12–20 (Lipopeg 6000-DS)

14–20 (Alkamuls 6000-DS; Kessco PEG 6000 Distearate; Mapeg 6000DS; Plurest DS150)

17 (Pegosperse 6000-DS)

POE (150) distearate *(cont'd.)*

pH:
    4.0–6.0 (3% aq.) (Emulvis)
    5.0 (3% disp.) (Kessco PEG 6000 Distearate)
STD. PKGS.:
    55-gal (225 lb net) fiber drums (Plurest DS150)

POE (6) lauryl amide

SYNONYMS:
    PEG-6 lauramide (CTFA)
    PEG 300 lauryl amide
STRUCTURE:

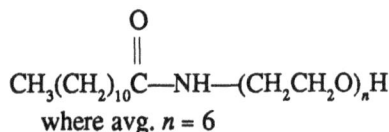

$$CH_3(CH_2)_{10}C\!\!-\!\!NH\!\!-\!\!(CH_2CH_2O)_nH$$
    where avg. $n = 6$
CAS No.:
    26635-75-6 (generic)
    RD No. 977061-48-5
TRADENAME EQUIVALENTS:
    Mazamide L-5 [Mazer]
    Unamide L-5 [Lonza]
CATEGORY:
    Thickener, visc. builder, emulsifier, foam booster, foam stabilizer, rust inhibitor, detergent
APPLICATIONS:
    Bath products: bubble bath (Unamide L-5)
    Cosmetic industry preparations: personal care products (Unamide L-5); shampoos (Unamide L-5)
    Household products: (Unamide L-5); hard surface cleaner (Unamide L-5); light-duty cleaners (Unamide L-5)
    Industrial applications: (Unamide L-5)
PROPERTIES:
Form:
    Solid (Mazamide L-5)
    Soft wax (Unamide L-5)
Color:
    White (Unamide L-5)

Composition:
    100% active (Unamide L-5)
Ionic Nature:
    Nonionic (Mazamide L-5; Unamide L-5)
Sp.gr.:
    1.0–1.1 (Mazamide L-5)
Saponification No.:
    12 max. (Unamide L-5)
Hydroxyl No.:
    145–165 (Unamide L-5)
pH:
    9.0–10.0 (5% sol'n.) (Unamide L-5)
    9.5–10.5 (Mazamide L-5)

POE (120) methyl glucose dioleate

SYNONYMS:
    PEG-120 methyl glucose dioleate (CTFA)
    PEG (120) methyl glucose dioleate
    POE (120) methyl glucoside dioleate
TRADENAME EQUIVALENTS:
    Glucamate DOE-120 [Amerchol]
CATEGORY:
    Solubilizer, thickening agent, emulsifier
APPLICATIONS:
    Cosmetic industry preparations: shampoos
PROPERTIES:
Form:
    Waxy solid
Solubility:
    Sol. in water
Ionic Nature:
    Nonionic

POE (20) monolaurate

SYNONYMS:
 PEG-20 laurate (CTFA)
 PEG 1000 monolaurate
STRUCTURE:

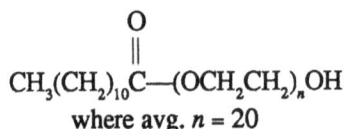

 where avg. $n = 20$
CAS No.:
 9004-81-3 (generic)
 RD No.: 977063-49-2
TRADENAME EQUIVALENTS:
 Chemax E-1000ML [Chemax]
 Cithrol 10ML [Croda]
 Ethylan L10 [Lankro]
 Kessco PEG 1000 Monolaurate [Armak]

CATEGORY:
 Emulsifier, dispersant, detergent, lubricant, penetrant, wetting agent, thickener, solubilizer, softening agent, antistat

APPLICATIONS:
 Bath products: bath oils (Kessco PEG 1000 Monolaurate)
 Cosmetic industry preparations: (Cithrol 10ML; Ethylan L10; Kessco PEG 1000 Monolaurate); hair preparations (Kessco PEG 1000 Monolaurate); perfumery (Kessco PEG 1000 Monolaurate); shampoos (Kessco PEG 1000 Monolaurate)
 Farm products: (Kessco PEG 1000 Monolaurate)
 Food applications: (Kessco PEG 1000 Monolaurate)
 Household detergents: mechanical washing (Cithrol 10ML)
 Industrial applications: glass fiber (Cithrol 10ML); lubricating/cutting oils (Chemax E-1000ML); metal treatment (Cithrol 10ML); plastics (Kessco PEG 1000 Monolaurate); textile/leather processing (Cithrol 10ML)
 Pharmaceutical applications: (Ethylan L10; Kessco PEG 1000 Monolaurate)

PROPERTIES:
Form:
 Liquid (Chemax E-1000ML)
 Soft solid (Kessco PEG 1000 Monolaurate)
 Solid (Ethylan L10)
Color:
 White (Ethylan L10)
 Cream (Kessco PEG 1000 Monolaurate)
Composition:
 100% conc. (Chemax E-1000ML)

POE (20) monolaurate *(cont'd.)*

Solubility:
Sol. in acetone (Kessco PEG 1000 Monolaurate)
Sol. in alcohols (Cithrol 10ML)
Sol. in carbon tetrachloride (Kessco PEG 1000 Monolaurate)
Sol. in ethyl acetate (Kessco PEG 1000 Monolaurate)
Sol. in isopropanol (Kessco PEG 1000 Monolaurate)
Sol. in Na_2SO_4 (Kessco PEG 1000 Monolaurate)
Sol. in polar solvents (Cithrol 10ML)
Sol. hot in propylene glycol (Kessco PEG 1000 Monolaurate)
Sol. in toluol (Kessco PEG 1000 Monolaurate)
Sol. in water (Cithrol 10ML; Kessco PEG 1000 Monolaurate)
Ionic Nature:
Nonionic (Chemax E-1000ML; Kessco PEG 1000 Monolaurate)
M.W.:
1000 (Cithrol 10ML)
Sp.gr.:
1.035 (65 C) (Kessco PEG 1000 Monolaurate)
1.056 (Ethylan L10)
Visc.:
76 cs (40 C) (Ethylan L10)
F.P.:
40 C (Kessco PEG 1000 Monolaurate)
Flash Pt.:
490 F (COC) (Kessco PEG 1000 Monolaurate)
Fire Pt.:
505 F (Kessco PEG 1000 Monolaurate)
HLB:
16.5 (Kessco PEG 1000 Monolaurate)
18 (Cithrol 10ML)
Acid No.:
5.0 max. (Kessco PEG 1000 Monolaurate)
Iodine No.:
3.5 max. (Kessco PEG 1000 Monolaurate)
Saponification No.:
41–51 (Kessco PEG 1000 Monolaurate)
51–55 (Cithrol 10ML)
STD. PKGS.:
450-lb steel drums (Kessco PEG 1000 Monolaurate)

POE (30) nonyl phenyl ether

SYNONYMS:
Nonoxynol-30 (CTFA)
Nonyl phenol ethoxylate (30 moles EO)
Nonyl phenol 30 polyglycol ether
Nonylphenoxy polyethoxy ethanol (30 moles EO)
Nonylphenoxy poly(ethyleneoxy) ethanol (30 moles EO)
PEG-30 nonyl phenyl ether
PEG (30) nonyl phenyl ether
POE (30) nonyl phenol

STRUCTURE:
$C_9H_{19}C_6H_4(OCH_2CH_2)_nOH$
where avg. $n = 30$

CAS No.:
9016-45-9 (generic); 26027-38-3 (generic); 37205-87-1 (generic)
RD No.: 977006-96-4

TRADENAME EQUIVALENTS:
Ablunol NP30, NP30 70% [Taiwan Surfactant]
Alkasurf NP-30 70% [Alkaril]
Arkopal N-300 [Hoechst AG]
Carsonon N-30, N-30 70% [Carson]
Cedepal CO-880, CO-887 [Domtar]
Chemax NP-30, NP-30/70 [Chemax]
Conco NI-187 [Continental Chem.]
Ethylan N30 [Lankro]
Iconol NP-30, NP-30-70% [BASF Wyandotte]
Igepal CO-880, CO-887 [GAF]
Macol NP-30, NP-30(70) [Mazer]
Makon 30 [Stepan]
Merpoxen NO300 [Kempen]
Polystep F-9 [Stepan]
Renex 650 [ICI United States]
Rexol 25/30, 25/307 [Hart Chem. Ltd.]
Serdox NNP30/70 [Servo B.V.]
Sermul EN30/70, EN145 [Servo B.V.]
Surfonic N-300, NB-5 [Jefferson]
Synperonic NP30 [ICI Petrochem. Div.]
T-Det N-30, N-307 [Thompson-Hayward]
Teric N30 [ICI Australia Ltd.]
Triton N-302 [Rohm & Haas]
Trycol NP-30, NP-307 [Emery]

POE (30) nonyl phenyl ether (cont'd.)

CATEGORY:

Stabilizer, solubilizer, dispersant, wetting agent, rewetting agent, detergent, emulsifier, coemulsifier, surfactant, demulsifier, intermediate, penetrant, foaming agent, foam stabilizer, dyeing assistant

APPLICATIONS:

Cosmetic industry preparations: (Conco NI-187; Makon 30); essential oils/perfumery (Carsonon N-30, N-30 70%; Ethylan N30; Igepal CO-880; Makon 30; Rexol 25/30, 25/307; Teric N30); shampoos (Makon 30)

Farm products: agricultural oils/sprays (Makon 30; Surfonic N-300, NB-5; Triton N-302); insecticides/pesticides (Conco NI-187; Ethylan N30; Macol NP-30, NP-30(70); Rexol 25/30, 25/307)

Food applications: indirect food additives (Surfonic N-300, NB-5)

Household detergents: (Conco NI-187; Teric N30); detergent base (Triton N-302); high-temperature detergents (Igepal CO-880; Renex 650; T-Det N-30, N-307); liquid detergents (Merpoxen NO300); paste detergents (Merpoxen NO300); powdered detergents (Merpoxen NO300; Teric N30)

Industrial applications: (Igepal CO-887); dyes and pigments (Serdox NNP30/70); floor finishes (Ablunol NP30, NP30 70%); glass processing (Igepal CO-880); latex applications (Ethylan N30; Igepal CO-880; Renex 650); lubricating/cutting oils (Surfonic N-300, NB-5); paint mfg. (Ablunol NP30, NP30 70%; Chemax NP-30, NP-30/70; Teric N30); paper mfg. (Ablunol NP30, NP30 70%; Chemax NP-30, NP-30/70; Makon 30); petroleum industry (Alkasurf NP-30 70%; Conco NI-187; Igepal CO-880; Makon 30; Surfonic N-300, NB-5); plastics (Ablunol NP30, NP30 70%; Ethylan N30; Polystep F-9); polishes and waxes (Conco NI-187); polymers/polymerization (Carsonon N-30, N-30 70%; Conco NI-187; Serdox NNP30/70; Sermul EN30/70, EN145; Triton N-302; Trycol NP-30, NP-307); rubber (Surfonic N-300, NB-5); textile/leather processing (Ablunol NP30, NP30 70%; Chemax NP-30, NP-30/70; Conco NI-187; Macol NP-30, NP-30(70); Makon 30; Surfonic N-300, NB-5)

Industrial cleaners: (Conco NI-187; Makon 30); bottle washing (Conco NI-187); lime soap dispersing (Serdox NNP30/70); metal processing surfactants (Chemax NP-30, NP-30/70; Makon 30); sanitizers/germicides (Teric N30); specialty detergents (Triton N-302); textile scouring (Alkasurf NP-30 70%; Igepal CO-880, CO-887; Makon 30; T-Det N-30, N-307)

PROPERTIES:

Form:

Liquid (Ablunol NP30 70%; Alkasurf NP-30 70%; Cedepal CO-887; Chemax NP-30/70; Iconol NP-30-70%; Igepal CO-887; Macol NP-30(70); Rexol 25/307; Serdox NNP30/70; Sermul EN145; T-Det N-307; Trycol NP-307)

Clear liquid (Surfonic NB-5)

Clear to hazy liquid (Polystep F-9)

Clear viscous liquid (Triton N-302)

POE (30) nonyl phenyl ether (cont'd.)

Gel (Polystep F-9)
Liquid to semisolid (Carsonon N-30 70%)
Solid (Ablunol NP30; Carsonon N-30; Chemax NP-30; Renex 650; Rexol 25/30; Sermul EN30/70; Synperonic NP30; Trycol NP-30)
Flake (Teric N30)
Wax (Cedepal CO-880; Conco NI-187; Iconol NP-30; Igepal CO-880; Merpoxen NO300)
Waxy solid (Ethylan N30; Macol NP-30; Makon 30; Surfonic N-300; T-Det N-30)
Color:
White (Ethylan N30; Iconol NP-30; Merpoxen NO300; Renex 650; Surfonic N-300)
Off-white (Makon 30)
Pale yellow (Carsonon N-30 70%; Igepal CO-880, CO-887; T-Det N-30)
APHA 100 (T-Det N-307)
APHA 100 max. (Iconol NP-30-70%)
APHA 250 (Triton N-302)
Gardner 1 (Trycol NP-30, NP-307)
Hazen 100 (Teric N30)
Hazen 150 max. (Synperonic NP30)
Odor:
Negligible (Ethylan N30)
Mild (Triton N-302)
Mild, pleasant (Carsonon N-30)
Mild, aromatic (T-Det N-30)
Aromatic (Igepal CO-880, CO-887)
Composition:
68.5–71.5% active in water (Iconol NP-30-70%)
69–71% active in water (Triton N-302)
70% active (Ablunol NP30 70%; Cedepal CO-887; Chemax NP-30/70; Igepal CO-887; Macol NP-30(70); Polystep F-9; Rexol 25/307; Serdox NNP30/70; Sermul EN30/70, EN145; Surfonic NB-5; Trycol NP-307)
70% active min. (Alkasurf NP-30 70%)
70% active in water (Carsonon N-30 70%)
70 ± 0.5% active (T-Det N-307)
99% active min. (Iconol NP-30; Synperonic NP30)
99.5% active min. (T-Det N-30)
100% active (Carsonon N-30; Conco NI-187; Ethylan N30; Igepal CO-880; Makon 30; Merpoxen NO300; Renex 650; Rexol 25/30; Surfonic N-300; Teric N30; Trycol NP-30)
100% conc. (Ablunol NP30; Arkopal N-300; Cedepal CO-880; Macol NP-30)
Solubility:
Sol. in acetone (Surfonic N-300)
Sol. in alcohols (Synperonic NP30; Triton N-302)

Sol. in benzene (Teric N30)
Sol. in butyl Cellosolve (Igepal CO-880; T-Det N-30)
Sol. in carbon tetrachloride (Surfonic N-300)
Sol. in ethanol (Igepal CO-880; T-Det N-30; Teric N30)
Sol. in ethyl acetate (Teric N30)
Sol. in ethyl Icinol (Teric N30)
Sol. in ethylene dichloride (Igepal CO-880; Triton N-302)
Sol. in ethylene glycol (Igepal CO-880; T-Det N-30)
Disp. in glycerol trioleate (@ 5%) (Trycol NP-307)
Sol. in glycol ethers (Synperonic NP30)
Sol. in isopropanol (@ 10%) (Renex 650)
Sol. in methanol (Surfonic N-300)
Sol. in perchloroethylene (Teric N30)
Sol. in propylene glycol (@ 10%) (Renex 650)
Sol. in water (Ablunol NP30, NP30 70%; Ethylan N30; Iconol NP-30, NP-30-70%;
    Igepal CO-880; Makon 30; Surfonic N-300; Synperonic NP30; T-Det N-30; Teric
    N30; Triton N-302); (@ 10%) (Alkasurf NP-30 70%; Renex 650); (@ 5%) (Trycol
    NP-30, NP-307)
Sol. in xylene (Igepal CO-880; Surfonic N-300; T-Det N-30); (@ 10%) (Renex 650)
Ionic Nature:
Nonionic (Ablunol NP30, NP30 70%; Arkopal N-300; Carsonon N-30; Cedepal CO-
    880, CO-887; Conco NI-187; Ethylan N30; Iconol NP-30, NP-30-70%; Igepal CO-
    880, CO-887; Macol NP-30, NP-30(70); Makon 30; Merpoxen NO300; Renex 650;
    Rexol 25/30, 25/307; Serdox NNP30/70; Sermul EN30/70, EN145; Surfonic N-
    300, NB-5; Synperonic NP30; T-Det N-30, N-307; Teric N30; Triton N-302; Trycol
    NP-30, NP-307)
M.W.:
1535 (Iconol NP-30); (active) (NP-30-70%)
1540 (Surfonic N-300)
1543 (Triton N-302)
Sp.gr.:
1.064 (60 C) (Ethylan N30)
1.066 (50 C) (Teric N30)
≈ 1.07 (132 F) (T-Det N-30)
1.08 (50/25 C) (Iconol NP-30); (50 C) (Igepal CO-880)
1.09 (Carsonon N-30 70%; Iconol NP-30-70%; Igepal CO-887)
1.092 (Triton N-302)
1.15 (Renex 650)
1.17 (Carsonon N-30)
Density:
1.074 g/ml (50 C) (Synperonic NP30)
1.09 g/ml (Alkasurf NP-30 70%)

POE (30) nonyl phenyl ether (cont'd.)

9.0 lb/gal (Makon 30; Trycol NP-307)
9.1 lb/gal (T-Det N-307; Triton N-302); (20 C) (Surfonic N-300)
9.75 lb/gal (Carsonon N-30)

Visc.:
30 cps (100 C) (Iconol NP-30)
150 cps (50 C) (Synperonic NP30; Teric N30)
649 cps (Brookfield, #2 spindle, 12 rpm) (Triton N-302)
806 cps (Brookfield) (T-Det N-307)
1100 cps (Iconol NP-30-70%)
90 cs (60 C) (Ethylan N30)
260 cSt (100 F) (Trycol NP-307)
160 SUS (210 F) (Surfonic N-300)

F.P.:
40 C (Synperonic NP30)
44 C (Surfonic N-300)

M.P.:
40 C (Trycol NP-30)
40 ± 2 C (Teric N30)
41 C (Iconol NP-30)

Pour Pt.:
0 C (Iconol NP-30-70%)
5 C (Trycol NP-307)
39 C (Ethylan N30)
40 C (Makon 30)
≈ 30 F (T-Det N-307)
34 F (Chemax NP-30/70)
34 ± 2 F (Igepal CO-887)
91 F (Renex 650)
109 F (Chemax NP-30)
109 ± 2 F (Igepal CO-880)
≈ 110 F (T-Det N-30)

Solidification Pt.:
40 C (Makon 30)
25 ± 2 F (Igepal CO-887)
104 ± 2 F (Carsonon N-30; Igepal CO-880)

Flash Pt.:
> 93 C (SFCC) (Triton N-302)
> 200 F (PMCC) (Igepal CO-880, CO-887)
> 212 F (TOC) (T-Det N-307)
> 300 F (Renex 650)
> 400 F (COC) (Ethylan N30)
> 500 F (T-Det N-30); (OC) (Surfonic N-300)

POE (30) nonyl phenyl ether (cont'd.)

Cloud Pt.:

   73–75 C (1 g in 100 ml 10% NaCl sol'n.) (Merpoxen NO300)

   74 C (1% in 10% NaCl) (Alkasurf NP-30 70%)

   74–76 C (1% in 10% sodium chloride) (Rexol 25/30, 25/307)

   92 C (5% saline) (Trycol NP-307)

   Clear @ 100 C (Makon 30); (1% sol'n.) (Igepal CO-887)

   > 100 C (Trycol NP-30); (1% aq.) (Iconol NP-30, NP-30-70%; Surfonic N-300; Synperonic NP30; Triton N-302); (1%) (T-Det N-307); (1% in hard water) (Teric N30)

   167–171 F (Carsonon N-30 70%)

   Clear @ 212 F (Renex 650); (1% sol'n.) (Carsonon N-30; Igepal CO-880; T-Det N-30)

   > 212 F (Conco NI-187)

HLB:

   17.0 (Ethylan N30; Serdox NNP30/70; T-Det N-30)

   17.1 (Ablunol NP30, NP30 70%; Alkasurf NP-30 70%; Renex 650; Surfonic N-300; Synperonic NP30; Triton N-302; Trycol NP-30, NP-307)

   17.2 (Cedepal CO-880, CO-887; Iconol NP-30, NP-30-70%; Igepal CO-880, CO-887; Macol NP-30, NP-30(70); Rexol 25/30, 25/307; Teric N30)

Hydroxyl No.:

   37 (Synperonic NP30)

Stability:

   Stable to hydrolysis by acids and alkalis (Trycol NP-30, NP-307)

   Stable to acids, bases, salts (Surfonic N-300; T-Det N-30)

   Stable in presence of moderate concs. of acids, bases, salts (Triton N-302)

   Stable to electrolytes, hard water, extremes of pH and temp. (Ethylan N30)

   Very good in acids, alkalis, and hard water (Renex 650)

   Stable to acids, alkalis, hard water, foam (Conco NI-187)

   Good in hard or saline waters and in reasonable concs. of acids and alkalis (Teric N30)

   Stable to acids, alkalis, dilute sol'ns. of many oxidizing and reducing agents (Igepal CO-880, CO-887)

Ref. Index:

   1.4690 (50 C) (Surfonic N-300)

pH:

   Neutral (1% sol'n.) (Carsonon N-30)

   5.0–7.0 (5% aq.) (Triton N-302)

   5.0–8.0 (5% DW) (Alkasurf NP-30 70%)

   6.0–7.0 (3%) (Carsonon N-30 70%)

   6.0–7.5 (5% aq.) (Iconol NP-30, NP-30-70%)

   6.0–8.0 (1% aq.) (Ethylan N30; Synperonic NP30; Teric N30)

   6.5–8.0 (1%) (T-Det N-30)

   8.5 (1% sol'n.) (Makon 30)

POE (30) nonyl phenyl ether (cont'd.)

Surface Tension:
   42 dynes/cm (0.1% aq.) (Iconol NP-30, NP-30-70%); (0.01%) (Renex 650)
   42.8 dynes/cm (0.1%, 20 C) (Synperonic NP30; Teric N30)
   43 dynes/cm (0.01% sol'n.) (Igepal CO-880)
Biodegradable: (Triton N-302)
TOXICITY/HANDLING:
   May cause slight to moderate eye and skin irritation (Triton N-302)
   May cause skin and eye irritation; spillages are slippery (Teric N30)
   Eye irritant; skin irritant on prolonged contact with conc. form (Synperonic NP30)
   Avoid prolonged contact with conc. form; spillages may be slippery (Ethylan N30)
STORAGE/HANDLING:
   Contact with conc. oxidizing or reducing agents may be explosive (Ethylan N30; Igepal CO-880, CO-887; T-Det N-30, N-307; Triton N-302)
   Store at ambient temps.; corrosive to copper and brass on long storage (Triton N-302)
STD. PKGS.:
   200-kg net iron drums (Merpoxen NO300)
   200-kg net mild steel drums or bulk (Ethylan N30)
   55-gal (450 lb net) steel drums (Iconol NP-30)
   55-gal (450 lb net) lined fiber drums (Iconol NP-30-70%)
   480-lb net steel removable head drums (T-Det N-30)
   480-lb net closed-head lined steel drums (T-Det N-307)

POE (40) nonyl phenyl ether

SYNONYMS:
   Nonoxynol-40 (CTFA)
   Nonyl phenol ethoxylate (40 moles EO)
   Nonyl phenol 40 polyglycol ether
   Nonylphenoxy polyethoxy ethanol (40 moles EO)
   Nonylphenoxy poly(ethyleneoxy) ethanol (40 moles EO)
   PEG-40 nonyl phenyl ether
   PEG 2000 nonyl phenyl ether
   POE (40) nonyl phenol
   Polyethoxylated nonylphenol (40 EO)
STRUCTURE:

   $C_9H_{19}C_6H_4(OCH_2CH_2)_nOH$
      where avg. $n = 40$
CAS No.:
   9016-45-9 (generic); 26027-38-3 (generic); 37205-87-1 (generic)
   RD No.: 977057-37-6

POE (40) nonyl phenyl ether (cont'd.)

TRADENAME EQUIVALENTS:
Ablunol NP40, NP40 70% [Taiwan Surfactant]
Alkasurf NP-40 [Alkaril]
Carsonon N-40 70% [Carson]
Cedepal CO-890 [Domtar]
Chemax NP-40, NP-40/70 [Chemax]
Chemcol NPE-400 [Chemform]
Conco NI-190, NI-197 [Continental Chem.]
Hetoxide NP-40 [Heterene]
Hyonic NP-407 [Diamond Shamrock]
Iconol NP-40, NP-40-70% [BASF Wyandotte]
Igepal CO-890, CO-897 [GAF]
Peganol NP40, NP40 (70%) [Borg-Warner]
Polystep F-10 [Stepan]
Siponic NP40 [Alcolac]
Surfonic N-400, NB-14 [Jefferson]
Synperonic NP40 [ICI Petrochem. Div.]
T-Det N-40, N-407 [Thompson-Hayward]
Tergitol NP-40, NP-40 (70% Aqueous), NP-44 [Union Carbide]
Teric N40 [ICI Australia Ltd.]
Triton N-401 [Rohm & Haas]
Trycol NP-40, NP-407 [Emery]

CATEGORY:
Stabilizer, dispersant, solubilizer, wetting agent, emulsifier, coemulsifier, detergent, penetrant, foaming agent, demulsifier, emollient, viscosity control agent, lubricant, dyeing assistant, intermediate, surfactant

APPLICATIONS:
Cosmetic industry preparations: (Conco NI-190, NI-197; Hetoxide NP-40; Surfonic N-400, NB-14); perfumery (Hetoxide NP-40; Teric N40)

Farm products: agricultural oils/sprays (Chemcol NPE-400; Hyonic NP-407; Surfonic N-400, NB-14); insecticides/pesticides (Conco NI-190, NI-197)

Food applications: indirect food additives (Hyonic NP-407; Surfonic N-400, NB-14)

Household detergents: (Chemcol NPE-400; Conco NI-190, NI-197; Hetoxide NP-40; Teric N40); high-temperature detergents (Hyonic NP-407; T-Det N-40, N-407; Trycol NP-40, NP-407)

Industrial applications: construction (Tergitol NP-40, NP-40 (70% Aqueous)); dyes and pigments (Alkasurf NP-40; Hetoxide NP-40; Igepal CO-890, CO-897; Peganol NP40, NP40 (70%); Tergitol NP-40, NP-40 (70% Aqueous), NP-44); latex applications (Chemax NP-40, NP-40/70; Conco NI-190, NI-197; Hyonic NP-407; Igepal CO-890, CO-897; Peganol NP40, NP40 (70%)); lubricating/cutting oils (Surfonic N-400, NB-14); paint mfg. (Ablunol NP40, NP40 70%; Alkasurf NP-40; Rexol 25/40; T-Det N-407; Teric N40); paper mfg. (Ablunol NP40, NP40 70%); petroleum

POE (40) nonyl phenyl ether (cont'd.)

industry (Chemcol NPE-400; Conco NI-190, NI-197; Surfonic N-400, NB-14); plastics (Ablunol NP40, NP40 70%; Hyonic NP-407; Igepal CO-890, CO-897; Peganol NP40, NP40 (70%); Polystep F-10; Siponic NP40); polishes and waxes (Ablunol NP40, NP40 70%; Conco NI-190, NI-197; Igepal CO-890, CO-897; Peganol NP40, NP40 (70%); Siponic NP40); polymers/polymerization (Alkasurf NP-40; Chemax NP-40, NP-40/70; Conco NI-190, NI-197; Igepal CO-890, CO-897; Peganol NP40, NP40 (70%); Surfonic N-400, NB-14; T-Det N-40, N-407; Tergitol NP-40, NP-40 (70% Aqueous); Triton N-401); printing inks (Alkasurf NP-40); rubber (Surfonic N-400, NB-14); textile/leather processing (Ablunol NP40, NP40 70%; Alkasurf NP-40; Chemcol NPE-400; Conco NI-190, NI-197; Hetoxide NP-40; Surfonic N-400, NB-14; Trycol NP-40, NP-407)

Industrial cleaners: (Conco NI-190, NI-197); bottle washing (Conco NI-190, NI-197); metal processing surfactants (Hetoxide NP-40); sanitizers/germicides (Teric N40); textile scouring (Hetoxide NP-40; T-Det N-40, N-407; Tergitol NP-44)

PROPERTIES:

Form:

Liquid (Ablunol NP40 70%; Alkasurf NP-40; Chemax NP-40/70; Conco NI-197; Iconol NP-40-70%; Igepal CO-897; Peganol NP40 (70%); Siponic NP40; T-Det N-407; Triton N-401; Trycol NP-407); (@ 30 C) (Hetoxide NP-40)

Clear liquid (Hyonic NP-407; Surfonic NB-14; Tergitol NP-40 (70% Aqueous))

Clear to hazy liquid (Polystep F-10)

Gel (Polystep F-10)

Semisolid (Carsonon N-40 70%)

Solid (Ablunol NP40; Chemax NP-40; Chemcol NPE-400; Rexol 25/40; Synperonic NP40; Tergitol NP-40, NP-44; Trycol NP-40)

Flakes (Teric N40)

Wax (Cedepal CO-890; Conco NI-190; Iconol NP-40; Igepal CO-890; Peganol NP40)

Waxy solid (Surfonic N-400; T-Det N-40)

Color:

Light (Alkasurf NP-40)

Slightly colored (Tergitol NP-40 (70% Aqueous))

White (Iconol NP-40; Surfonic N-400; Tergitol NP-40)

White to yellow (Carsonon N-40 70%)

Off-white (Chemcol NPE-400; Igepal CO-890)

Pale yellow (Igepal CO-897; T-Det N-40)

APHA 100 (T-Det N-407)

APHA 100 max. (Iconol NP-40-70%); (50 C) (Peganol NP40)

APHA 250 (Triton N-401)

Gardner 1 (Trycol NP-40, NP-407)

Gardner 5 (Tergitol NP-44)

Hazen 100 (Teric N40)

Odor:

Low (Alkasurf NP-40)

Aromatic (Igepal CO-890, CO-897)
Mild (Tergitol NP-40 (70% Aqueous))
Mild, characteristic (Tergitol NP-40)
Mild, aromatic (T-Det N-40)

Composition:

68.5–71.5% active in water (Iconol NP-40-70%)

70% active (Ablunol NP40 70%; Alkasurf NP-40; Conco NI-197; Hyonic NP-407; Igepal CO-897; Polystep F-10; Siponic NP40; Triton N-401; Trycol NP-407)

70% active in water (Carsonon N-40 70%; Peganol NP40 (70%); Surfonic NB-14; Tergitol NP-40 (70% Aqueous))

70 ± 0.5% active (T-Det N-407)

99% active (Iconol NP-40)

99.5% active (Peganol NP40)

99.5% active min. (T-Det N-40)

100% active (Ablunol NP40; Chemcol NPE-400; Conco NI-190; Igepal CO-890; Rexol 25/40; Surfonic N-400; Tergitol NP-40; Teric N40)

100% conc. (Cedepal CO-890; Synperonic NP40; Tergitol NP-44)

Solubility:

Sol. in acetone (Surfonic N-400)
Sol. in benzene (Teric N40)
Disp. in butyl stearate (@ 5%) (Trycol NP-407)
Sol. in carbon tetrachloride (Surfonic N-400)
Sol. in ethanol (Igepal CO-890; Teric N40)
Sol. in ethyl acetate (Teric N40)
Sol. in ethyl Icinol (Teric N40)
Sol. in ethylene dichloride (Igepal CO-890)
Sol. in glycerol trioleate (@ 5%) (Trycol NP-40)
Sol. in isopropanol (Hetoxide NP-40)
Sol. in methanol (Surfonic N-400)
Insol. in white min. oil (Peganol NP40)
Sol. in perchloroethylene (Teric N40); insol. (Peganol NP40)

Sol. in water (Ablunol NP40, NP40 70%; Alkasurf NP-40; Hetoxide NP-40; Hyonic NP-407; Iconol NP-40, NP-40-70%; Igepal CO-890; Peganol NP40; Siponic NP40; Surfonic N-400; T-Det N-40; Tergitol NP-40, NP-40 (70% Aqueous), NP-44; Teric N40); (@ 5%) (Trycol NP-40, NP-407)

Sol. in xylene (Surfonic N-400); (@ 5%) (Trycol NP-40); insol. (Peganol NP40)

Ionic Nature:

Nonionic (Ablunol NP40, NP40 70%; Alkasurf NP-40; Cedepal CO-890; Chemcol NPE-400; Conco NI-190, NI-197; Hyonic NP-407; Iconol NP-40; Igepal CO-890, CO-897; Peganol NP40, NP40 (70%); Rexol 25/40; Siponic NP40; Surfonic N-400; Synperonic NP40; T-Det N-40, N-407; Tergitol NP-40, NP-40 (70% Aqueous), NP-44; Teric N40; Triton N-401)

POE (40) nonyl phenyl ether *(cont'd.)*

M.W.:
  1975 (Iconol NP-40); (active) (Iconol NP-40-70%)
  1980 (Tergitol NP-40, NP-44); (avg.) (Tergitol NP-44; Triton N-401)
Sp.gr.:
  ≈ 1.07 (132 F) (T-Det N-40)
  1.080 (T-Det N-407); (50/20 C) (Tergitol NP-40); (50 C) (Teric N40)
  1.081 (Peganol NP40 (70%))
  1.082 (58/20 C) (Tergitol NP-44)
  1.087 (50 C) (Peganol NP40)
  1.09 (50 C) (Igepal CO-890); (50/25 C) (Iconol NP-40)
  1.10 (Conco NI-197; Iconol NP-40-70%; Igepal CO-897)
  1.104 (20/20 C) (Tergitol NP-40 (70% Aqueous))
Density:
  1.10 g/ml (Hyonic NP-407)
  8.97 lb/gal (55 C) (Tergitol NP-40, NP-44)
  9.0 lb/gal (T-Det N-407)
  9.1 lb/gal (20 C) (Surfonic N-400)
  9.19 lb/gal (20 C) (Tergitol NP-40 (70% Aqueous))
  9.2 lb/gal (Triton N-401; Trycol NP-407)
Visc.:
  40 cps (100 C) (Iconol NP-40)
  250 cps (50 C) (Teric N40)
  850 cps (Triton N-401)
  900 cps (Brookfield) (T-Det N-407)
  1400 cps (Iconol NP-40-70%)
  385 cSt (100 F) (Trycol NP-407)
  44 cs (100 C) (Peganol NP40)
  533 cs (20 C) (Tergitol NP-40 (70% Aqueous))
  210 SUS (210 F) (Surfonic N-400)
F.P.:
  45 C (Surfonic N-400)
  50–60 F (Alkasurf NP-40)
B.P.:
  > 250 C (dec.) (Tergitol NP-40)
M.P.:
  40 C (Trycol NP-40)
  45 ± 2 C (Teric N40)
  48 C (Iconol NP-40; Tergitol NP-40)
Pour Pt.:
  –6 C (Hyonic NP-407)
  7 C (Iconol NP-40-70%; Trycol NP-407)
  45 C (Peganol NP40)

≈ 48 C (Chemcol NPE-400)
18 F (Triton N-401)
≈ 30 F (T-Det N-407)
46 F (Chemax NP-40/70)
46 ± 2 F (Igepal CO-897)
112 F (Chemax NP-40)
112 ± 2 F (Igepal CO-890)
≈ 120 F (T-Det N-40)
Solidification Pt.:
–2 C (Tergitol NP-40 (70% Aqueous))
48 C (Tergitol NP-44)
41 ± 2 F (Igepal CO-897)
106 ± 2 F (Igepal CO-890)
Flash Pt.:
> 200 C (Peganol NP40)
> 200 F (PMCC) (Igepal CO-890, CO-897)
> 212 F (TOC) (T-Det N-407)
> 230 F (Setaflash CC) (Triton N-401)
> 450 F (TOC) (Chemcol NPE-400)
> 500 F (T-Det N-40); (OC) (Surfonic N-400)
525 F (COC) (Tergitol NP-40, NP-44)
560 F (Trycol NP-40)
None (Tergitol NP-40 (70% Aqueous))
Cloud Pt.:
74–76 C (1% in 10% sodium chloride) (Siponic NP40)
90 C (5% saline) (Trycol NP-40, NP-407)
100 C (Peganol NP40); (1% aq.) (Hyonic NP-407); (0.5% aq.) (Tergitol NP-40)
> 100 C (1%) (Triton N-401); (1% aq.) (Chemcol NPE-400; Iconol NP-40, NP-40-70%; Surfonic N-400; Tergitol NP-44); (1% in hard water) (Teric N40)
Clear @ 212 F (1% sol'n.) (Carsonon N-40 70%; Igepal CO-890, CO-897; T-Det N-40)
> 212 F (Conco NI-190, NI-197); (1%) (T-Det N-407)
HLB:
17.0 (Hetoxide NP-40)
17.6 (Hyonic NP-407)
17.7 (Chemcol NPE-400; Iconol NP-40, NP-40-70%; Peganol NP40 (70%)); Siponic NP40; T-Det N-40)
17.8 (Ablunol NP40, NP40 70%; Cedepal CO-890; Igepal CO-890; Peganol NP40; Rexol 25/40; Surfonic N-400; Synperonic NP40; Tergitol NP-40, NP-44; Triton N-401; Trycol NP-40, NP-407)
18.0 (Teric N40)
19.0 (T-Det N-407)

POE (40) nonyl phenyl ether (cont'd.)

Hydroxyl No.:
28 (Peganol NP40; Tergitol NP-44)
Stability:
Very stable against hydrolysis by acids and alkalis (Trycol NP-40, NP-407)
Stable to acids, bases, salts (Surfonic N-400; T-Det N-40, N-407)
Stable to acids, alkalis, hard water, and foam (Conco NI-190, NI-197)
Stable to acids, alkalis, dilute sol'ns. of many oxidizing and reducing agents (Igepal CO-890, CO-897)
Good in hard or saline waters and in reasonable concs. of acids and alkalis (Teric N40)
pH:
3.5–4.5 (10% sol'n.) (Chemcol NPE-400)
4.0–8.0 (10%) (Tergitol NP-44); (10% aq.) (Tergitol NP-40)
6.0–7.0 (3% sol'n.) (Carsonon N-40 70%)
6.0–7.5 (5% aq.) (Iconol NP-40, NP-40-70%; Peganol NP40)
6.0–8.0 (1% aq.) (Teric N40)
6.5–8.0 (1%) (T-Det N-40)
7.0 (1% aq.) (Hyonic NP-407)
Foam (Ross Miles):
120 mm initial, 100 mm after 5 min (0.05% sol'n.) (Hyonic NP-407)
Surface Tension:
41.0 dynes/cm (0.1%, 20 C) (Teric N40)
45 dynes/cm (0.1% aq.) (Tergitol NP-40)
Biodegradable: Fair (Tergitol NP-44)
TOXICITY/HANDLING:
Eye irritant (Tergitol NP-44)
Toxic to aquatic life; burning can produce carbon dioxide and/or carbon monoxide (Tergitol NP-40, NP-40 (70% Aqueous))
May cause skin and eye irritation; spillages are slippery (Teric N40)
STORAGE/HANDLING:
Contact with conc. oxidizing or reducing agents may be explosive (Igepal CO-890, CO-897; T-Det N-40)
STD. PKGS.:
55-gal drums, 1- and 5-gal containers (Tergitol NP-44)
55-gal (200 l) steel drums, bulk (Hyonic NP-407)
55-gal (450 lb net) steel drums (Iconol NP-40)
55-gal (450 lb net) lined fiber drums (Iconol NP-40-70%)
55-gal (470 lb net) drums, 5-gal (45 lb net) pails, 1-gal (9 lb net) jugs (Tergitol NP-40)
55-gal (490 lb net) drums, 5-gal (45 lb net) pails, 1-gal (9 lb net) jugs (Tergitol NP-40 (70% Aqueous))
480-lb net removable-head steel drums (T-Det N-40)
480-lb net closed-head lined steel drums (T-Det N-407)

272

SYNONYMS:

Nonoxynol-50 (CTFA)
Nonyl phenol ethoxylate (50 moles EO)
Nonyl phenol 50 polyglycol ether
Nonylphenoxy polyethoxy ethanol (50 moles EO)
Nonylphenoxy poly(ethyleneoxy) ethanol (50 moles EO)
PEG-50 nonyl phenyl ether
PEG (50) nonyl phenyl ether
POE (50) nonyl phenol

STRUCTURE:

$C_9H_{19}C_6H_4(OCH_2CH_2)_nOH$
where avg. $n = 50$

CAS No.:

9016-45-9 (generic); 26027-38-3 (generic); 37205-87-1 (generic)
RD No.: 977057-38-7

TRADENAME EQUIVALENTS:

Ablunol NP50, NP50 70% [Taiwan Surfactants]
Alkasurf NP-50 70% [Alkaril]
Carsonon N-50 70% [Carson]
Chemax NP-50, NP-50/70 [Chemax]
Conco NI-2000 [Continental Chem.]
Ethylan N50 [Lankro Chem. Ltd.]
Hyonic NP-500 [Diamond Shamrock]
Iconol NP-50 [BASF Wyandotte]
Igepal CO-970, CO-977 [GAF]
Peganol NP50, NP50 (70%) [Borg-Warner]
Rexol 25/50, 25/507 [Hart Chem. Ltd.]
Synperonic NP50 [ICI Petrochem. Div.]
T-Det N-50, N-507 [Thompson-Hayward]
Trycol NP-50, NP-507 [Emery]

CATEGORY:

Emulsifier, stabilizer, dispersant, solubilizer, wetting agent, foaming agent, penetrant, demulsifier, dyeing assistant, detergent, coemulsifier

APPLICATIONS:

Cosmetic industry preparations: (Conco NI-2000); perfumery (Ethylan N50)
Farm products: agricultural oils/sprays (Hyonic NP-500); insecticides/pesticides (Conco NI-2000)
Food applications: indirect food additives (Hyonic NP-500; Iconol NP-50)
Household detergents: (Conco NI-2000); high-temperature detergents (Chemax NP-50, NP-50/70; Hyonic NP-500; T-Det N-50, N-507; Trycol NP-50, NP-507)
Industrial applications: dyes and pigments (Igepal CO-970, CO-977; Peganol NP50, NP50 (70%)); latex applications (Alkasurf NP-50 70%; Chemax NP-50, NP-50/70;

Conco NI-2000; Ethylan N50; Igepal CO-970, CO-977; Peganol NP50, NP50 (70%)); paint mfg. (Ablunol NP50, NP50 70%; Alkasurf NP-50 70%; T-Det N-507); paper mfg. (Ablunol NP50, NP50 70%; Alkasurf NP-50 70%); petroleum industry (Conco NI-2000); plastics (Ablunol NP50, NP50 70%; Hyonic NP-500; Igepal CO-970, CO-977; Peganol NP50, NP50 (70%)); polishes and waxes (Ablunol NP50, NP50 70%; Conco NI-2000; Igepal CO-970, CO-977; Peganol NP50, NP50 (70%); Rexol 25/50, 25/507); polymers/polymerization (Alkasurf NP-50 70%; Chemax NP-50, NP-50/70; Conco NI-2000; Ethylan N50; Igepal CO-970, CO-977; Peganol NP50, NP50 (70%)); textile/leather processing (Ablunol NP50, NP50 70%; Alkasurf NP-50 70%; Conco NI-2000; Trycol NP-50, NP-507')

Industrial cleaners: (Conco NI-2000); bottle washing (Conco NI-2000); textile scouring (T-Det N-50, N-507)

PROPERTIES:

Form:

Liquid (Ablunol NP50 70%; Alkasurf NP-50 70%; Chemax NP-50/70; Igepal CO-977; Rexol 25/507; T-Det N-507; Trycol NP-507)

Semisolid (Carsonon N-50 70%)

Solid (Ablunol NP50; Chemax NP-50; Iconol NP-50; Rexol 25/50; Synperonic NP50; Trycol NP-50)

Clear solid (Hyonic NP-500)

Wax (Conco NI-2000; Igepal CO-970; Peganol NP50)

Waxy solid (Ethylan N50; T-Det N-50)

Color:

White (Ethylan N50)

White to yellow (Carsonon N-50 70%)

Off-white (Igepal CO-970)

Pale yellow (Igepal CO-977; T-Det N-50)

APHA 100 (T-Det N-507)

APHA 100 max. (50 C) (Peganol NP50)

Gardner 1 (Trycol NP-50, NP-507)

Odor:

Negligible (Ethylan N50)

Mild, aromatic (T-Det N-50)

Aromatic (Igepal CO-970, CO-977)

Composition:

70% active (Ablunol NP50 70%; Iconol NP-50; Rexol 25/507; Trycol NP-507)

70% active min. (Alkasurf NP-50 70%)

70% active in water (Carsonon N-50 70%; Peganol NP50 (70%))

70 ± 0.5% active (T-Det N-507)

> 99% active (Hyonic NP-500)

99.5% active min. (Peganol NP50; T-Det N-50)

100% active (Ablunol NP50; Conco NI-2000; Ethylan N50; Igepal CO-970; Rexol 25/50; Synperonic NP50)

Solubility:
Sol. in ethanol (Igepal CO-970)
Sol. in ethylene dichloride (Igepal CO-970)
Insol. in white min. oil (Peganol NP50)
Insol. in perchloroethylene (Peganol NP50)
Sol. in water (Ablunol NP50, NP50 70%; Ethylan N50; Hyonic NP-500; Igepal CO-970, CO-977; Peganol NP50); (@ 10%) (Alkasurf NP-50 70%); (@ 5%) (Trycol NP-50, NP-507)
Sol. in xylene (@ 5%) (Trycol NP-50); insol. (Peganol NP50)

Ionic Nature:
Nonionic (Ablunol NP50, NP50 70%; Ethylan N50; Hyonic NP-500; Iconol NP-50; Igepal CO-970; Peganol NP50; Rexol 25/50, 25/507; Synperonic NP50; T-Det N-50, N-507)

Sp.gr.:
1.067 (Peganol NP50 (70%))
1.07 (132 F) (T-Det N-50)
1.073 (60 C) (Ethylan N50)
1.09 (T-Det N-507)
1.095 (50 C) (Peganol NP50)
1.10 (Igepal CO-977); (50 C) (Igepal CO-970)

Density:
1.08 g/ml (Hyonic NP-500)
1.09 g/ml (Alkasurf NP-50 70%)
9.0 lb/gal (Trycol NP-507)
9.1 lb/gal (T-Det N-507)

Visc.:
55 cs (100 C) (Peganol NP50)
135 cs (60 C) (Ethylan N50)
760 cps (Brookfield) (T-Det N-507)
440 cSt (100 F) (Trycol NP-507)

M.P.:
54 C (Trycol NP-50)

Pour Pt.:
–4 C (Trycol NP-507)
24 C (Hyonic NP-500)
43 C (Ethylan N50)
46 C (Peganol NP50)
≈ 40 F (T-Det N-507)
52 F (Chemax NP-50/70)
52 ± 2 F (Igepal CO-977)
114 F (Chemax NP-50)
114 ± 2 F (Igepal CO-970)

POE (50) nonyl phenyl ether (cont'd.)

≈ 120 F (T-Det N-50)

Solidification Pt.:

46 ± 2 F (Igepal CO-977)

108 ± 2 F (Igepal CO-970)

Flash Pt.:

> 200 C (Peganol NP50)

> 200 F (PMCC) (Igepal CO-970, CO-977)

> 212 F (TOC) (T-Det N-507)

> 400 F (COC) (Ethylan N50)

> 500 F (T-Det N-50)

520 F (Trycol NP-50)

Cloud Pt.:

74–76 C (1% in 10% NaCl) (Rexol 25/50, 25/507)

76 C (1% in 10% NaCl) (Alkasurf NP-50 70%); (10% saline) (Trycol NP-50, NP-507)

100 C (1% aq.) (Hyonic NP-500)

> 100 C (Peganol NP50); (1%) (T-Det N-507); (1% aq.) (Ethylan N50)

163 F (in 10% NaCl) (Conco NI-2000)

Clear @ 212 F (1% sol'n.) (Carsonon N-50 70%; Igepal CO-970, CO-977; T-Det N-50)

HLB:

18.0 (Alkasurf NP-50 70%; Hyonic NP-500; Iconol NP-50; T-Det N-50, N-507)

18.2 (Ablunol NP50, NP50 70%; Ethylan N50; Igepal CO-970, CO-977; Peganol NP50; Rexol 25/50, 25/507; Synperonic NP50; Trycol NP-50, NP-507)

Hydroxyl No.:

23 (Peganol NP50)

Stability:

Very stable against hydrolysis by acids and alkalis (Trycol NP-50, NP-507)

Stable to acids, bases, and salts (T-Det N-50, N-507)

Stable to acids, alkalis, hard water, and foam (Conco NI-2000)

Stable to acids, alkalis, dilute sol'ns. of many oxidizing and reducing agents (Igepal CO-970, CO-977)

Stable to electrolytes, hard water, extremes of pH and temperature (Ethylan N50)

pH:

5.0–8.0 (5% DW) (Alkasurf NP-50 70%)

6.0–7.0 (3% sol'n.) (Carsonon N-50 70%)

6.0–7.5 (5% in distilled water) (Peganol NP50)

6.0–8.0 (1% aq.) (Ethylan N50)

6.5–8.0 (1%) (T-Det N-50)

7.0 (1% aq.) (Hyonic NP-500)

Foam (Ross Miles):

105 mm initial, 95 mm after 5 min (0.05% sol'n.) (Hyonic NP-500)

POE (50) nonyl phenyl ether (cont'd.)

TOXICITY/HANDLING:
Avoid prolonged contact with conc. form; spillages may be slippery (Ethylan N50)
STORAGE/HANDLING:
Contact with conc. oxidizing or reducing agents may be explosive (Ethylan N50; Igepal CO-970, CO-977; T-Det N-50, N-507)
STD. PKGS.:
200-kg net mild-steel drums or bulk (Ethylan N50)
55-gal (200 l) steel drums, bulk (Hyonic NP-500)
480-lb net removable-head steel drums (T-Det N-50)
480-lb net closed-head lined steel drums (T-Det N-507)

POE (100) nonyl phenyl ether

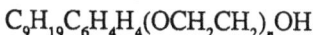

SYNONYMS:
Nonoxynol-100 (CTFA)
Nonyl phenol ethoxylate (100 moles EO)
Nonyl phenol 100 polyglycol ether
Nonylphenoxy polyethoxy ethanol (100 moles EO)
Nonylphenoxy poly(ethyleneoxy) ethanol (100 moles EO)
PEG-100 nonyl phenyl ether
PEG (100) nonyl phenyl ether
POE (100) nonyl phenol
STRUCTURE:
$C_9H_{19}C_6H_4H_4(OCH_2CH_2)_nOH$
where avg. $n = 100$
CAS No.:
9016-45-9 (generic); 26027-38-3 (generic); 37205-87-1 (generic)
RD No.: 977065-11-4
TRADENAME EQUIVALENTS:
Alkasurf NP-100 70% [Alkaril]
Carsonon N-100 70% [Carson]
Cedepal CO-990, CO-997 [Domtar]
Chemax NP-100, NP-100/70 [Chemax]
Iconol NP-100 [BASF Wyandotte]
Igepal CO-990, CO-997 [GAF]
Macol NP-100 [Mazer]
Peganol NP100, NP100 (70%) [Borg-Warner]
Rexol 25/100-70% [Hart Chem. Ltd.]
T-Det N-100, N-1007 [Thompson-Hayward]

POE (100) nonyl phenyl ether (cont'd.)

TRADENAME EQUIVALENTS *(cont'd.)*:
   Teric 100 [ICI Australia Ltd.]
   Triton N-998, N-998-70% [Rohm & Haas]
   Trycol NP-1007 [Emery]
CATEGORY:
   Emulsifier, stabilizer, wetting agent, surfactant, detergent, dyeing assistant, coemulsifier, solubilizer, solvent
APPLICATIONS:
   Cosmetic industry preparations: perfumery (Teric 100)
   Food applications: indirect food additives (Iconol NP-100)
   Household detergents: (Macol NP-100; Teric 100); high-temperature detergents (T-Det N-100, N-1007; Triton N-998, N-998-70%); flake detergents (Teric 100); powdered detergents (Teric 100)
   Industrial applications: construction (T-Det N-100, N-1007; Triton N-998, N-998-70%); dyes and pigments (Igepal CO-990, CO-997; Peganol NP100, NP100 (70%); Rexol 25/100-70%); latex applications (Alkasurf NP-100 70%; Chemax NP-100, NP-100/70; Igepal CO-990, CO-997; Peganol NP100, NP100 (70%); Trycol NP-1007); paint mfg. (Alkasurf NP-100 70%; Teric 100); paper mfg. (Alkasurf NP-100 70%); plastics (Igepal CO-990, CO-997; Peganol NP100, NP100 (70%)); polishes and waxes (Igepal CO-990, CO-997; Peganol NP100, NP100 (70%)); polymers/polymerization (Alkasurf NP-100 70%; Chemax NP-100, NP-100/70; Igepal CO-990, CO-997; Peganol NP100, NP100 (70%)); textile/leather processing (Alkasurf NP-100 70%)
   Industrial cleaners: sanitizers/germicides (Teric 100); textile scouring (T-Det N-100, N-1007; Triton N-998, N-998-70%)
PROPERTIES:
Form:
   Liquid (Cedepal CO-997; Chemax NP-100/70; Igepal CO-997; Trycol NP-1007)
   Clear liquid (Rexol 25/100-70%)
   Clear viscous liquid (Triton N-998-70%)
   Clear to opaque viscous liquid (T-Det N-1007)
   Paste (Alkasurf NP-100 70%)
   Semisolid (Carsonon N-100 70%)
   Solid (Chemax NP-100; Macol NP-100)
   Cast solid (Iconol NP-100)
   Flake (Iconol NP-100; Macol NP-100; Teric 100)
   Wax (Cedepal CO-990; Igepal CO-990; Peganol NP100)
   Waxy solid (T-Det N-100; Triton N-998)
Color:
   White to off-white (Carsonon N-100 70%)
   Off-white (Igepal CO-990)
   Pale beige (Triton N-998)

Pale yellow (Igepal CO-997; T-Det N-100)
Yellow (Triton N-998-70%)
APHA 100 max. (50 C) (Peganol NP100)
Gardner 1 (Trycol NP-1007)
Gardner 1 max. (Iconol NP-100)
Hazen 100 (Teric 100)

Odor:
Mild (Triton N-998-70%)
Mild, aromatic (T-Det N-100, N-1007)
Aromatic (Igepal CO-990, CO-997)

Composition:
69–71% active in water (Triton N-998-70%)
70% active (Cedepal CO-997; Trycol NP-1007)
70% active min. (Alkasurf NP-100 70%)
70% active in water (Carsonon N-100 70%; Igepal CO-997; Peganol NP100 (70%))
70 ± 0.5% active (T-Det N-1007)
85% active (Rexol 25/100-70%)
99.5% active (Peganol NP100)
> 99.5% active (T-Det N-100; Triton N-998)
100% active (Igepal CO-990; Teric 100)
100% conc. (Cedepal CO-990; Iconol NP-100; Macol NP-100)

Solubility:
Sol. in alcohols (Triton N-998, N-998-70%)
Sol. in benzene (Teric 100)
Sol. in ethanol (Igepal CO-990; T-Det N-100; Teric 100)
Sol. in ethyl acetate (Teric 100)
Sol. in ethyl Icinol (Teric 100)
Sol. in ethylene dichloride (Igepal CO-990; T-Det N-100; Triton N.998, N-998-70%)
Insol. in white min. oil (Peganol NP100)
Sol. in perchloroethylene (Teric 100); insol. (Peganol NP100)
Very high water-sol. (Carsonon N-100 70%); sol. in water (Cedepal CO-990; Iconol NP-100; Igepal CO-990; Peganol NP100, NP100 (70%); T-Det N-100; Teric 100; Triton N-998, N-998-70%); sol. in cold water (Rexol 25/100-70%); (@ 10%) (Alkasurf NP-100 70%); (@ 5%) (Trycol NP-1007)
Insol. in xylene (Peganol NP100)

Ionic Nature:
Nonionic (Cedepal CO-990, CO-997; Iconol NP-100; Igepal CO-990, CO-997; Peganol NP100, NP100 (70%); Rexol 25/100-70%; T-Det N-100, N-1007; Teric 100; Triton N-998, N-998-70%)

M.W.:
4315 (Iconol NP-100)
4630 (Triton N-998, N-998-70%)

POE (100) nonyl phenyl ether *(cont'd.)*

Sp.gr.:
    1.07 (65.6 C) (Triton N-998)
    ≈ 1.08 (135 F) (T-Det N-100)
    1.084 (Peganol NP100 (70%))
    1.10 (T-Det N-1007; Triton N-998-70%)
    1.11 (Igepal CO-997)
    1.113 (60 C) (Teric 100)
    1.12 (50/25 C) (Iconol NP-100); (50 C) (Igepal CO-990; Peganol NP100)

Density:
    1.09 g/ml (Alkasurf NP-100 70%)
    8.91 lb/gal (Triton N-998)
    9.16 lb/gal (Triton N-998-70%)
    9.2 lb/gal (Trycol NP-1007)

Visc.:
    120 cs (100 C) (Peganol NP100)
    150 cps (100 C) (Iconol NP-100)

M.P.:
    52 C (Iconol NP-100)
    52 ± 2 C (Teric 100)

Pour Pt.:
    18 C (Triton N-998-70%; Trycol NP-1007)
    46 C (Triton N-998)
    50 C (Peganol NP100)
    ≈ 62 F (T-Det N-1007)
    68 F (Chemax NP-100/70)
    68 ± 2 F (Igepal CO-997)
    122 F (Chemax NP-100)
    122 ± 2 F (Igepal CO-990)
    ≈ 127 F (T-Det N-100)

Solidification Pt.:
    63 ± 2 F (Igepal CO-997)
    116 ± 2 F (Igepal CO-990)

Flash Pt.:
    110 C (PMCC) (Triton N-998)
    > 200 C (Peganol NP100)
    > 200 F (PMCC) (Igepal CO-990, CO-997)
    > 500 F (T-Det N-100, N-1007); (TOC) (Triton N-998-70%)

Cloud Pt.:
    72 C (10% saline) (Trycol NP-1007)
    76 C (1% in 10% NaCl) (Alkasurf NP-100 70%)
    > 100 C (1% aq.) (Iconol NP-100; Peganol NP100; Triton N-998, N-998-70%); (1% in hard water) (Teric 100)

Clear @ 212 F (1% sol'n.) (Carsonon N-100 70%; Igepal CO-990, CO-997; T-Det N-100, N-1007)

HLB:
18.0 (Alkasurf NP-100 70%)
19.0 (Cedepal CO-990, CO-997; Iconol NP-100; Igepal CO-990, CO-997; Peganol NP100, NP100 (70%); Triton N-998, N-998-70%; Trycol NP-1007)
19.1 (Teric 100)

Hydroxyl No.:
12 (Peganol NP100)

Stability:
Stable to acids, bases, and salts (T-Det N-100)
Very stable against hydrolysis by acids and alkalis (Trycol NP-1007)
Stable in presence of moderate concs. of acids, bases, salts (Triton N-998, N-998-70%)
Stable to acids, alkalis, dilute sol'ns. of many oxidizing and reducing agents (Igepal CO-990, CO-997)
Good in hard or saline waters and in reasonable concs. of acids and alkalis (Teric 100)

pH:
5.0–8.0 (5% DW) (Alkasurf NP-100 70%)
6.0–7.0 (1% sol'n.) (T-Det N-1007); (3% sol'n.) (Carsonon N-100 70%); (5% aq.) (Triton N-998, N-998-70%)
6.0–7.5 (5% aq.) (Iconol NP-100); (5% in distilled water) (Peganol NP100)
6.0–8.0 (1% aq.) (Teric 100)

Surface Tension:
46.0 dynes/cm (0.1%, 20 C) (Teric 100)

Biodegradable: (Triton N-998, N-998-70%)

TOXICITY/HANDLING:
Severely irritating to eyes (possible permanent injury); irritating to skin on repeated/prolonged contact (Triton N-998)
Possibly irritating to eyes; irritating to skin on repeated/prolonged contact (Triton N-998-70%)
May cause skin and eye irritation; spillages are slippery (Teric 100)

STORAGE/HANDLING:
Avoid contact with strong oxidizing or reducing agents (Triton N-998; Triton N-998-70%)
Contact with conc. oxidizing or reducing agents may be explosive (Igepal CO-990, CO-997)
Store in SS or plastic-lined steel vessels (Triton N-998-70%)

STD. PKGS.:
55-gal (450 lb net) steel drums (cast solid); 55-gal (250 lb net) fiber drums (flake) (Iconol NP-100)
480-lb net removable-head steel drums (T-Det N-100)

POE (80) sorbitan monolaurate

SYNONYMS:
PEG-80 sorbitan laurate (CTFA)
PEG (80) sorbitan monolaurate

CAS No.:
9005-64-5 (generic)

TRADENAME EQUIVALENTS:
Hetsorb L80-72% [Heterene]
T-Maz 28 [Mazer]

CATEGORY:
Surfactant, emulsifier, solubilizer, wetting agent, visc. modifier, antistat, stabilizer, dispersant, counter-irritant

APPLICATIONS:
Cosmetic industry preparations: (T-Maz 28); perfumery (T-Maz 28); shampoos (Hetsorb L80-72%)
Food applications: (T-Maz 28)
Industrial applications: metalworking (T-Maz 28); textile/leather processing (T-Maz 28)
Pharmaceutical applications: (T-Maz 28)

PROPERTIES:
Form:
Liquid (T-Maz 28)
Clear liquid (Hetsorb L80-72%)
Color:
Yellow (T-Maz 28)
Gardner 5 max. (Hetsorb L80-72%)
Composition:
70% solids (T-Maz 28)
70–74% solids (Hetsorb L80-72%)
Solubility:
Sol. in acetone (T-Maz 28)
Sol. in ethanol (T-Maz 28)
Sol. in isopropanol (Hetsorb L80-72%)
Insol. in min. oil (Hetsorb L80-72%)
Sol. in water (Hetsorb L80-72%; T-Maz 28)
Sp.gr.:
1.0 (T-Maz 28)
Visc.:
1100 cps (T-Maz 28)
Acid No.:
2.0 max. (T-Maz 28)
3.0 max. (Hetsorb L80-72%)

POE (80) sorbitan monolaurate *(cont'd.)*

Saponification No.:
  5–15 (T-Maz 28)
  8–18 (Hetsorb L80-72%)
Hydroxyl No.:
  22–35 (Hetsorb L80-72%)

POE (20) sorbitol ether

SYNONYMS:
  PEG-20 sorbitol ether
  PEG 1000 sorbitol ether
  Sorbeth-20 (CTFA)
CAS No.:
  RD No. 977058-26-6
TRADENAME EQUIVALENTS:
  Ethosperse SL-20 [Lonza]
  Liponic SO-20 [Lipo]
  Trylox SS-20 [Emery]
CATEGORY:
  Humectant, plasticizer, intermediate
APPLICATIONS:
  Cosmetic industry preparations: (Ethosperse SL-20; Liponic SO-20); creams and
    lotions (Liponic SO 20); shaving preparations (Liponic SO-20); toiletries (Liponic
    SO-20)
  Industrial applications: (Ethosperse SL-20; Trylox SS-20)
PROPERTIES:
Form:
  Liquid (Trylox SS-20)
  Viscous liquid (Liponic SO-20)
Color:
  Yellow (Liponic SO-20)
  Gardner 1 (Trylox SS-20)
Odor:
  Bland, characteristic (Liponic SO-20)
Solubility:
  Sol. in alcohol (Liponic SO-20)
  Insol. (5%) in min. oil (Trylox SS-20)
  Insol. in oils (Liponic SO-20)
  Sol. in water (Liponic SO-20); sol. (5%) in water (Trylox SS-20)

283

POE (20) sorbitol ether (cont'd.)

Density:
  9.7 lb/gal (Trylox SS-20)
Visc.:
  200 cSt (Trylox SS-20)
Pour Pt.:
  7 C (Trylox SS-20)
Flash Pt.:
  435 F (Trylox SS-20)
Cloud Pt.:
  > 100 C (10% saline) (Trylox SS-20)
HLB:
  15.4 (Trylox SS-20)
Acid No.:
  1 max. (Liponic SO-20)
Hydroxyl No.:
  385–430 (Liponic SO-20)
Stability:
  Stable over pH range normally encountered in cosmetics (Liponic SO-20)

POP (12) POE (50) lanolin

SYNONYMS:
  POE (50) POP (12) lanolin
  PPG-12-PEG-50 lanolin (CTFA)
STRUCTURE:
  $R(OCHCH_2)_x(OCH_2CH_2)_yOH$

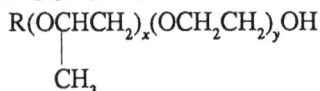

  CH_3
     where   R represents the lanolin radicals
            avg. x = 12, and
            avg. y = 50
CAS No.:
  68458-58-5 (generic)
  RD No. 977062-71-7
TRADENAME EQUIVALENTS:
  Lanexol AWS [Croda]
CATEGORY:
  Plasticizer, emollient, conditioner, superfatting agent, foam stabilizer, coupling agent, humectant, solubilizer, emulsifier, glossing agent

284

POP (12) POE (50) lanolin (cont'd.)

APPLICATIONS:
Cleansers: skin cleanser; soaps
Cosmetic industry preparations: hair sprays; shampoos
Pharmaceutical applications: antiperspirant/deodorant
PROPERTIES:
Form:
Liquid
Color:
Gardner 11 max.
Composition:
100% active
Solubility:
Sol. in alcohols
Sol. in oil
Sol. in water
Ionic Nature:
Nonionic
Acid No.:
2 max.
Iodine No.:
10 max.
Stability:
Unaffected by hard water, high electrolyte concentrations, or variations in pH
pH:
6.0–7.0 (1% aq. sol'n.)

Potassium alginate (CTFA)

SYNONYMS:
Alginic acid, potassium salt
CAS No.:
9005-36-1
TRADENAME EQUIVALENTS:
Kelmar, Kelmar Improved [Kelco]
CATEGORY:
Thickener, bodying agent, gelling agent, stabilizer, emulsifier, film former, slip agent
APPLICATIONS:
Cosmetic industry preparations: creams and lotions (Kelmar Improved); facial masks
(Kelmar Improved)

285

Potassium alginate (cont'd.)

Food applications: (Kelmar, Kelmar Improved)
Industrial applications: (Kelmar, Kelmar Improved); adhesives (Kelmar, Kelmar Improved); ceramics (Kelmar, Kelmar Improved); explosives (Kelmar, Kelmar Improved); latexes (Kelmar, Kelmar Improved); paper mfg. (Kelmar, Kelmar Improved; polishes and waxes (Kelmar, Kelmar Improved); resins (Kelmar); rubber (Kelmar); textile printing/dyeing (Kelmar, Kelmar Improved); toys (Kelmar, Kelmar Improved)
Industrial cleaners: (Kelmar, Kelmar Improved)
Pharmaceutical applications: dental preparations (Kelmar, Kelmar Improved)
PROPERTIES:
Form:
Fibrous particles (Kelmar Improved)
Granular (Kelmar)
Color:
Cream (Kelmar, Kelmar Improved)
Odor:
Odorless (Kelmar)
Solubility:
Sol. in cold to hot water (Kelmar, Kelmar Improved)
Sol'ns. will tolerate 30–40% ethyl and propyl alcohol (Kelmar, Kelmar Improved)
Visc.:
270 cps (1%, Brookfield LVF, 60 rpm) (Kelmar)
400 cps (1%, Brookfield LVF, 60 rpm) (Kelmar Improved)
pH:
Neutral (Kelmar, Kelmar Improved)

Potassium monoricinoleate

SYNONYMS:
12-Hydroxy-9-octadecenoic acid, monopotassium salt
9-Octadecenoic acid, 12-hydroxy-, monopotassium salt
Potassium ricinoleate (CTFA)
EMPIRICAL FORMULA:
$C_{18}H_{34}O_3 \cdot K$
STRUCTURE:

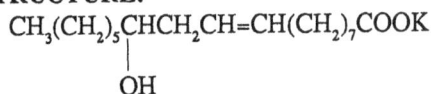

CH$_3$(CH$_2$)$_5$CHCH$_2$CH=CH(CH$_2$)$_7$COOK
|
OH

Potassium monoricinoleate (cont'd.)

CAS No.:
7492-30-0
TRADENAME EQUIVALENTS:
Seachem 55 [Seaboard]
Solricin 135 [NL Industries]
CATEGORY:
Stabilizer, emulsifier, germicide, lubricant, foam stabilizer
APPLICATIONS:
Cosmetic industry preparations: (Solricin 135)
Household products: (Solricin 135)
Industrial applications: latexes (Seachem 55); lubricating/cutting oils (Solricin 135); rubber (Solricin 135)
PROPERTIES:
Form:
Liquid (Seachem 55; Solricin 135)
Color:
Gardner 2 (Solricin 135)
Composition:
32% active (Solricin 135)
Ionic Nature:
Anionic (Seachem 55; Solricin 135)
Sp.gr.:
1.034 (Solricin 135)
Density:
8.6 lb/gal (Solricin 135)
Visc.:
0.9 stokes (Solricin 135)
STORAGE/HANDLING:
Keep from contact with oxidizing materials (Solricin 135)
STD. PKGS.:
55-gal drums (Solricin 135)

Propylene glycol alginate (CTFA)

SYNONYMS:
Alginic acid, ester with 1,2-propanediol
CAS No.:
9005-37-2
TRADENAME EQUIVALENTS:
Kelcoloid, D, DH, DO, DSF, HUF, LUF, O, S [Kelco]

Propylene glycol alginate *(cont'd.)*

CATEGORY:
Thickener, stabilizer, gelling agent, emulsifier

APPLICATIONS:
Food applications: (Kelcoloid, D, DH, DO, DSF, HUF, LUF, O, S)

Industrial applications: adhesives (Kelcoloid D, DH, DO, DSF, HUF, LUF, O, S); antifoams (Kelcoloid D, DH, DO, DSF, HUF, LUF, O, S); ceramics (Kelcoloid D, DH, DO, DSF, HUF, LUF, O, S); explosives (Kelcoloid D, DH, DO, DSF, HUF, LUF, O, S); latexes (Kelcoloid D, DH, DO, DSF, HUF, LUF, O, S); paper mfg. (Kelcoloid D, DH, DO, DSF, HUF, LUF, O, S); polishes and waxes (Kelcoloid D, DH, DO, DSF, HUF, LUF, O, S); textile printing/dyeing (Kelcoloid D, DH, DO, DSF, HUF, LUF, O, S); toys (Kelcoloid D, DH, DO, DSF, HUF, LUF, O, S)

Industrial cleaners: (Kelcoloid D, DH, DO, DSF, HUF, LUF, O, S)

PROPERTIES:

Form:
Fibrous particles (Kelcoloid D, HVF, LVF, O, S)
Agglomerated (Kelcoloid DH, DO, DSF)

Color:
Cream (Kelcoloid D, DH, DO, DSF, HUF, LUF, O, S)

Composition:
13% max. moisture (Kelcoloid D, DH, DO, DSF, HUF, LUF, O, S)

Solubility:
Solution will tolerate ethyl and propyl alcohol (Kelcoloid)
Insol. in nonaqueous solvents (Kelcoloid)
Sol. in cold and hot water (Kelcoloid)

Sp.gr.:
1.46 (Kelcoloid D, DH, DO, DSF, HUF, LUF, O, S)

Bulk Density:
33.71 lb/ft^3 (Kelcoloid D, DH, DO, DSF, HUF, LUF, O, S)

Visc.:
20 cps (1%, Brookfield LVF, 60 rpm) (Kelcoloid DSF, S)
25 cps (1%, Brookfield LVF, 60 rpm) (Kelcoloid DO, O)
120 cps (1%, Brookfield LVF, 60 rpm) (Kelcoloid LVF)
170 cps (1%, Brookfield LVF, 60 rpm) (Kelcoloid D)
400 cps (1%, Brookfield LVF, 60 rpm) (Kelcoloid DH, HVF)

Ref. Index:
1.3343 (20 C) (Kelcoloid D, DH, DO, DSF, HUF, LUF, O, S)

pH:
Acidic (2% sol'n.) (Kelcoloid)
4.0 (Kelcoloid DH, DSF, HVF, LVF, S)
4.3 (Kelcoloid DO, O)
4.4 (Kelcoloid D)

Surface Tension:
58 dynes/cm (Kelcoloid D, DH, DO, DSF, HUF, LUF, O, S)

Propylene glycol hydroxystearate (CTFA)

SYNONYMS:
Octadecanoic acid, 12-hydroxy-, monoester with 1,2-propanediol
EMPIRICAL FORMULA:
$C_{21}H_{42}O_4$
STRUCTURE:

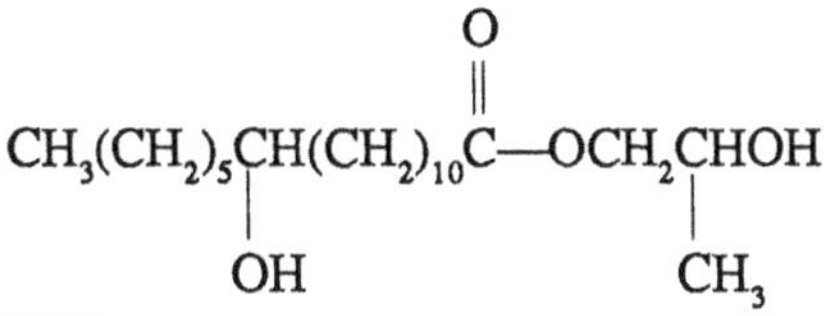

$$CH_3(CH_2)_5\underset{\underset{OH}{|}}{CH}(CH_2)_{10}\overset{\overset{O}{||}}{C}-OCH_2\underset{\underset{CH_3}{|}}{CH}OH$$

CAS No.:
33907-47-0
TRADENAME EQUIVALENTS:
Naturechem PGHS [CasChem]
Paricin 9 [CasChem]
CATEGORY:
Thickener, wax modifier, stabilizer
APPLICATIONS:
Cosmetic industry preparations: (Naturechem PGHS); creams and lotions (Naturechem PGHS); makeup (Naturechem PGHS)
PROPERTIES:
Form:
Flakes (Naturechem PGHS)
Color:
White (Naturechem PGHS)
Composition:
100% active (Naturechem PGHS; Paricin 9)
Ionic Nature:
Nonionic (Naturechem PGHS)
M.P.:
53 C (Naturechem PGHS)

Propylene glycol monoricinoleate

SYNONYMS:
12-Hydroxy-9-octadecenoic acid, monoester with 1,2-propanediol
9-Octadecenoic acid, 12-hydroxy-, monoester with 1,2-propanediol
Propylene glycol ricinoleate (CTFA)
EMPIRICAL FORMULA:
$C_{21}H_{40}O_4$

Propylene glycol monoricinoleate (cont'd.)

STRUCTURE:

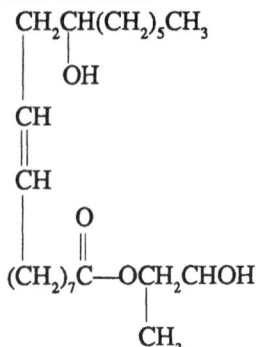

CH₂CH(CH₂)₅CH₃
│   │
│   OH
CH
‖
CH
│   O
│   ‖
(CH₂)₇C—OCH₂CHOH
│
CH₃

CAS No.:

26402-31-3

TRADENAME EQUIVALENTS:

Cithrol PGMR N/E [Croda]

Cithrol PGMR S/E [Croda] (self-emulsifying)

Flexricin 9 [NL Industries]

Naturechem PGR [CasChem]

CATEGORY:

Plasticizer, dispersant, emulsifier, coemulsifier, stabilizer, wetting agent, solvent, visc. reducer, coupling agent, emollient

APPLICATIONS:

Cosmetic industry preparations: (Cithrol PGMR N/E, PGMR S/E; Flexricin 9; Naturechem PGR); makeup (Naturechem PGR); nail polishes (Naturechem PGR); toiletries (Naturechem PGR)

Food applications: (Cithrol PGMR N/E, PGMR S/E)

Household products: (Flexricin 9; Naturechem PGR)

Industrial applications: (Cithrol PGMR N/E, PGMR S/E); dyes and pigments (Flexricin 9; Naturechem PGR); plastics (Cithrol PGMR N/E, PGMR S/E); polishes and waxes (Naturechem PGR); textile/leather processing (Flexricin 9)

Pharmaceutical applications: (Cithrol PGMR N/E, PGMR S/E)

PROPERTIES:

Form:

Liquid (Cithrol PGMR N/E, PGMR S/E; Flexricin 9)

Color:

Gardner 2+ (Flexricin 9)

Composition:

100% active (Flexricin 9)

100% conc. (Cithrol PGMR N/E, PGMR S/E)

Solubility:

Sol. in butyl acetate (Flexricin 9)

Propylene glycol monoricinoleate (cont'd.)

Sol. in ethanol (Flexricin 9)
Sol. in MEK (Flexricin 9)
Sol. in toluene (Flexricin 9)
Ionic Nature:
Nonionic (Cithrol PGMR N/E; Flexricin 9)
Anionic (Cithrol PGMR S/E)
Sp.gr.:
0.96 (Flexricin 9)
Visc.:
3 stokes (Flexricin 9)
STORAGE/HANDLING:
Keep containers closed; avoid contact with oxidizing materials (Flexricin 9)
STD. PKGS.:
55-gal drums (Flexricin 9)

Propyl gallate (CTFA)

SYNONYMS:
Benzoic acid, 3,4,5-trihydroxy-, propyl ester
n-Propyl 3,4,5-trihydroxybenzoate
3,4,5-Trihydroxybenzoic acid, propyl ester
EMPIRICAL FORMULA:
$C_{10}H_{12}O_5$
STRUCTURE:

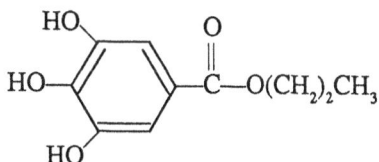

CAS No.:
121-79-9
TRADENAME EQUIVALENTS:
Sustane PG [UOP]
Tenox PG [Eastman]
CATEGORY:
Antioxidant, stabilizer, preservative
APPLICATIONS:
Food applications: (Sustane PG; Tenox PG)

Propyl gallate (cont'd.)

PROPERTIES:
Form:
  Crystalline powder (Sustane PG; Tenox PG)
Color:
  White (Sustane PG; Tenox PG)
Odor:
  Very slight (Sustane PG; Tenox PG)
Composition:
  100% active (Sustane PG)
Solubility:
  Sol. 121 g/100 g acetone (Sustane PG)
  Sol. 103 g/100 g ethanol (Sustane PG); sol. ≥ 50% (Tenox PG)
  Sol. 83 g/100 g ethyl ether (Sustane PG)
  Sol. 170 g/100 g methanol (Sustane PG)
  Sol. < 1% in min. oil (Tenox PG)
  Moderately sol. in edible oils (Tenox PG)
  Sol. in organic solvents (Tenox PG)
  Insol. in paraffin (Tenox PG)
  Sol. 67 g/100 g propylene glycol (Sustane PG); sol. ≥ 50% (Tenox PG)
  Moderately sol. in water (Tenox PG); sol. < 1% in water (Tenox PG)
M.W.:
  212 (Sustane PG; Tenox PG)
B.P.:
  Decomposes (Sustane PG)
  Decomposes above 148 C (Tenox PG)
M.P.:
  146–148 C (Sustane PG; Tenox PG)
TOXICITY/HANDLING:
  May cause an allergenic reaction or skin irritation; prolonged exposure may cause
    sensitization (Sustane PG)
  Wear protective gloves and glasses when handling to avoid irritant effects; use with
    adequate ventilation (Tenox PG)
STORAGE/HANDLING:
  Fine dust may create a dust explosion (Tenox PG)
STD. PKGS.:
  100 lb drums; 5 lb cartons (Sustane PG)
  2.27, 11.34, 22.68, or 45.36 kg fiber drums (Tenox PG)

Quaternium-18 bentonite (CTFA)

SYNONYMS:

Quaternary ammonium compounds, bis (hydrogenated tallow alkyl) dimethyl, chlorides, reaction products with bentonite

CAS No.:

68953-58-2

TRADENAME EQUIVALENTS:

Bentone 34 [NL Chemicals]

CATEGORY:

Thickener, gellant, thixotrope

APPLICATIONS:

Cosmetic industry preparations: eye care products; lip products

Industrial applications: solvent-based coatings

Pharmaceutical applications: antiperspirant creams and lotions

PROPERTIES:

Form:

Fine powder

Color:

Very light cream

Composition:

100% nonvolatiles

Density:

1.70 g/cm³

Quaternium-18 hectorite (CTFA)

SYNONYMS:

Quaternary ammonium compounds, bis (hydrogenated tallow alkyl) dimethyl, chlorides, reaction product with hectorite)

CAS No.:

RD No. 977062-10-4

TRADENAME EQUIVALENTS:

Bentone 38 [NL Chem.]

Bentone Gel MIO [NL Chem.] (in mineral oil)

Bentone Gel MIO A-40 [NL Chem.] (in mineral oil and SDA 40)

Quaternium-18 hectorite *(cont'd.)*

TRADENAME EQUIVALENTS *(cont'd.):*
Bentone Gel S-130, SS-71 [NL Chem.] (in mineral spirits)
Bentone Gel VS-5 [NL Chem.] (in cyclomethicone)
CATEGORY:
Thickener, gellant, thixotrope
APPLICATIONS:
Cosmetic industry preparations: anhydrous formulations (Bentone Gel MIO, Gel MIO A-40, S-130, SS-71, VS-5); eye care products (Bentone 38, Gel S-130, SS-71); lip products (Bentone 38)
Industrial applications: anhydrous formulations (Bentone 38, Gel MIO, Gel MIO A-40, Gel S-130, Gel SS-71, Gel VS-5); solvent-based systems (Bentone 38)
PROPERTIES:
Form:
Fine powder (Bentone 38)
Gel (Bentone Gel MIO A-40), Gel VS-5)
Color:
Creamy white (Bentone 38)
Composition:
100% nonvolatiles (Bentone 38)
Solubility:
Dispersible in many cosmetic oils (Bentone Gel MIO, Gel MIO A-40, Gel S-130, Gel SS-71, VS-5)
Density:
1.70 g/cm³ (Bentone 38)

Rosin (CTFA)

SYNONYMS:
Colophony
Gum rosin
Rosin gum
WW Wood rosin

CAS No.:
8050-09-7

TRADENAME EQUIVALENTS:
FF Wood Rosin [Harwick]
Foral AX [Hercules]
　　Generically sold by: [Harwick]

CATEGORY:
Plasticizer, tackifier, softener, processing aid

APPLICATIONS:
Industrial applications: adhesives (generic; Foral AX); coatings (Foral AX); latex
　　(generic); rubber (generic; FF Wood Rosin; Foral AX)

PROPERTIES:

Form:
Solid (generic; Foral AX)

Color:
Colorless to dark brown (generic)
Very pale (Foral AX)

Odor:
Low (Foral AX)
Typical (generic)

Solubility:
Sol. in alcohols (Foral AX)
Sol. in chlorinated solvents (Foral AX)
Sol. in esters (Foral AX)
Sol. in hydrocarbons (Foral AX)
Sol. in ketones (Foral AX)
Sol. in min. oils (Foral AX)
Insol. in water (Foral AX)

Sp.gr.:
1.044 (Foral AX)
1.08 (generic)
1.1 (FF Wood Rosin)

Rosin (cont'd.)

M.P.:
160 F (B&R) (FF Wood Rosin)
Softening Pt.:
75 C (Drop) (Foral AX)
Acid No.:
156 (FF Wood Rosin)
160 (Foral AX)
Stability:
Excellent heat stability and color retention (Foral AX)
Ref. Index:
1.4955 (100 C) (Foral AX)
STD. PKGS.:
227-kg net lightweight metal drums (Foral AX)

Sodium carboxymethyl cellulose

SYNONYMS:
- Carboxymethyl cellulose
- Cellulose, carboxymethyl ether
- Cellulose gum (CTFA)
- CMC
- Sodium CMC

CAS No.:
- 9004-32-4

TRADENAME EQUIVALENTS:
- Blanose Cellulose Gum, Refined CMC [Hercules BV]
- Cellulose Gum [Hercules]
- CMC-6CTL [Hercules] (crude grade)
- CMC-T [Hercules] (tech. grade)
- CMTC-T [Hercules]
- Hercules Cellulose [Hercules] (food grade)
- Hercules CMC [Hercules]
- Hercules CMC-6-DG-L [Hercules] (crude grade)
- Rycel 100, 105 [Ryco]

CATEGORY:
- Thickener, binder, stabilizer, suspending agent, film-former, rheology modifier, protective colloid, detergent

APPLICATIONS:
- Cosmetic industry preparations: (Cellulose Gum; Hercules Cellulose; Hercules CMC)
- Farm products: insecticides/pesticides (Cellulose Gum)
- Food applications: (Cellulose Gum; Hercules CMC)
- Household detergents: (Cellulose Gum; CMC-T); built detergents (Hercules CMC-6-DG-L; CMTC-T); laundry detergent (Cellulose Gum); laundry sizes (CMTC-T); powdered detergents (CMC-6CTL; Hercules CMC-6-DG-L)
- Industrial applications: (Hercules Cellulose); adhesives (Cellulose Gum; CMC-T; CMTC-T; Hercules CMC); ceramics (Hercules CMC); lithography (Cellulose Gum); paints and coatings mfg. (Cellulose Gum; CMC-T); paper mfg. (Cellulose Gum; CMC-T; Hercules CMC); petroleum industry (Cellulose Gum; Rycel 100, 105); textile/leather processing (Cellulose Gum; CMC-T; CMTC-T)
- Pharmaceutical applications: (Cellulose Gum; Hercules Cellulose; Hercules CMC)

PROPERTIES:
Form:
- Liquid (Hercules CMC)

Sodium carboxymethyl cellulose *(cont'd.)*

Granular (CMC-T)
Pellets (CMC-T)
Granular powder (CMC-6CTL; Hercules CMC-6-DG-L)
Powder (Blanose Cellulose Gum, Refined CMC; Cellulose Gum; CMC-T; CMTC-T;
    Hercules Cellulose)
Free-flowing powder (Rycel 100, 105)
Color:
White to light cream (Cellulose Gum)
Off-white to tan (CMC-T)
Odor:
Odorless (CMC-T)
Composition:
65% purity (CMC-6CTL)
90% purity (Hercules CMC-6-DG-L)
96% purity (CMC-T)
99.5% active (Hercules CMC)
99.5% min. purity (Cellulose Gum)
Solubility:
Insol. in organic solvents (films of CMC-6CTL)
Sol. in water (CMTC-T); sol. in hot or cold water (Cellulose Gum; CMC-6CTL; CMC-
    T; Hercules CMC); hydrates in hot or cold water (Rycel 100, 105)
Ionic Nature:
Anionic (Cellulose Gum; Hercules CMC)
Sp.gr.:
1.0068 (2% sol'n.) (Cellulose Gum; Hercules CMC)
Density:
0.65 g/ml (tamped) (CMC-6CTL)
0.75 g/ml (Cellulose Gum; CMC-T; Hercules CMC)
Fineness:
1% max. retained on 30-mesh; 5% max. retained 40-mesh (Rycel 100, 105)
Visc.:
< 20 cps (2% aq. sol'n.) to 5000 cps (1% aq. sol'n.) (Cellulose Gum)
20–600 cps (Brookfield, 2% sol'n.) (CMC-T)
25–150 cps (2% sol'n.) (CMC-6CTL; Hercules CMC-6-DG-L)
Stability:
Films of cellulose gum are unaffected by oils, greases, and organic solvents incl.
    alcohols, ketones, and hydrocarbons (Cellulose Gum)
Films of CMC-6CTL resist penetration by oils and greases (CMC-6CTL)
Films of CMC-T are resistant to oils, greases, and organic solvents (CMC-T)
Not subject to bacterial decomposition (Rycel 100, 105)
Ref. Index:
1.3355 (2% sol'n.) (Hercules CMC)

Sodium carboxymethyl cellulose *(cont'd.)*

pH:
7.5 (2% sol'n.) (Cellulose Gum; Hercules CMC)
Surface Tension:
71 dynes/cm (1% sol'n.) (Hercules CMC)
STORAGE/HANDLING:
Store in a cool, dry place (Cellulose Gum; CMC-6CTL; Hercules CMC-6-DG-L)
STD. PKGS.:
22.7-kg net multiwall bags (Rycel 100, 105)
39-lb net multiwall bags (Hercules CMC-6-DG-L)
50-lb multiwall bags (Cellulose Gum; CMC-6CTL)
50-lb bags (CMTC-T)

Sodium octyl sulfate *(CTFA)*

SYNONYMS:
Sodium 2-ethylhexyl sulfate
Sulfuric acid, mono (2-ethylhexyl) ester, sodium salt
EMPIRICAL FORMULA:
$C_8H_{18}O_4S \cdot Na$
STRUCTURE:

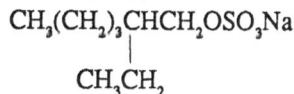

$$CH_3(CH_2)_3CHCH_2OSO_3Na$$
$$|$$
$$CH_3CH_2$$

CAS No.:
126-92-1
TRADENAME EQUIVALENTS:
Avirol SA-4106 [Henkel]
Carsonol SHS [Carson]
Duponol 80 [Du Pont] (tech.)
Emersal 6465 [Emery]
Merpinal EH 40, Q 147 [Kempen]
Niaproof 08 [Niacet]
Rewopol NEHS 40 [Rewo Chemische]
Serdet DSK 40 [Servo BV]
Sipex BOS, OLS [Alcolac]
Sulfotex OA [Henkel]
Sole-Terge TS-2-S [Hodag]
Witcolate D-510 [Witco]

Sodium octyl sulfate *(cont'd.)*

CATEGORY:
   Wetting agent, detergent, dispersant, penetrant, softener, emulsifier, stabilizer, mercerizing agent, hydrotrope, rinse aid, visc. control agent

APPLICATIONS:
   Degreasers: (Merpinal EH 40, Q 147)
   Food applications: fruit/vegetable washing (Carsonol SHS; Emersal 6465; Niaproof 08; Sipex BOS; Sulfotex OA); food packaging (Avirol SA-4106; Sulfotex OA)
   Household detergents: (Sulfotex OA); automatic dishwashing (Sipex BOS, OLS); bleaching powders (Witcolate D-510); hard surface cleaner (Merpinal EH 40, Q 147; Sipex BOS, OLS)
   Industrial applications: adhesives (Avirol SA-4106; Niaproof 08; Sulfotex OA); cellulosics (Witcolate D-510); dyes and pigments (Duponol 80; Sulfotex OA); electroplating (Niaproof 08; Rewopol NEHS 40); latex (Serdet DSK 40); metal processing (Duponol 80; Merpinal EH 40, Q 147; Sipex BOS); paint mfg. (Sipex BOS); paper/paperboard mfg. (Avirol SA-4106; Sulfotex OA); photography (Niaproof 08); plywood mfg. (Sipex BOS); polymers/polymerization (Niaproof 08; Avirol SA-4106); rubber (Sipex OLS); textile/leather processing (Duponol 80; Sipex BOS, OLS; Sole-Terge TS-2-S; Sulfotex OA)
   Industrial cleaners: (Niaproof 08; Sulfotex OA); alkaline cleaners (Rewopol NEHS 40; Witcolate D-510); lime soap dispersant (Witcolate D-510); metal processing surfactants (Duponol 80; Niaproof 08; Sipex OLS)

PROPERTIES:
Form:
   Liquid (Avirol SA-4106; Duponol 80; Emersal 6465; Merpinal EH 40; Niaproof 08; Rewopol NEHS 40; Serdet DSK 40; Sole-Terge TS-2-S)
   Clear liquid (Carsonol SHS; Merpinal Q 147; Sipex BOS, OLS; Witcolate D-510)
Color:
   Essentially colorless (Niaproof 08)
   Light amber (Sulfotex OA)
   Amber (Avirol SA-4106)
   Light yellow (Duponol 80)
   Yellowish (Merpinal Q 147)
   Gardner 3 (Carsonol SHS)
Composition:
   33% active (Sipex OLS)
   33–35% active (Duponol 80)
   35% conc. (Sole-Terge TS-2-S)
   38.5–40.5% active in water (Sulfotex OA)
   39–40% active (Sipex BOS)
   40% active (Carsonol SHS; Emersal 6465; Merpinal EH 40, Q 147; Niaproof 08; Rewopol NEHS 40; Serdet DSK 40)
   43–46% solids (Avirol SA-4106)

45% conc. (Witcolate D-510)

Solubility:

Sol. (5%) in isopropanol (Witcolate D-510)

Insol. (5%) in kerosene (Witcolate D-510)

Sol. in presence of many multivalent ions (Duponol 80)

Sol. (5%) in water (Witcolate D-510); miscible @ 20 C (Niaproof 08)

Insol. (5%) in xylene (Witcolate D-510)

Ionic Nature:

Anionic (Duponol 80; Emersal 6465; Merpinal EH 40, Q 147; Niaproof 08; Serdet DSK 40; Sipex BOS, OLS; Sole-Terge TS-2-S; Sulfotex OA)

M.W.:

232 (Duponol 80)

Sp.gr.:

1.10 (Avirol SA-4106); (25/4 C) (Witcolate D-510)

1.109 (20/20 C) (Niaproof 08)

1.15 (Carsonol SHS)

Density:

9.2 lb/gal (Avirol SA-4106; Sulfotex OA)

9.23 lb/gal (20 C) (Niaproof 08)

9.6 lb/gal (Carsonol SHS)

Visc.:

30 cps (27 C) (Duponol 80)

35 cps (Carsonol SHS)

50 cps (Sipex BOS)

100 cps (Sipex OLS)

200 cps max. (Brookfield) (Avirol SA-4106)

Flash Pt.:

> 93 C (Witcolate D-510)

Cloud Pt.:

5 C max. (Avirol SA-4106)

< 10 C (Sipex BOS)

HLB:

42 (Sipex BOS, OLS)

Stability:

Stable in 30% caustic sol'n.; excellent stability in presence of chlorine donors (Sulfotex OA)

Very stable to high electrolyte and in acid and basic solutions (Carsonol SHS)

Stable to high pH and temp., and in relatively high concs. of caustic soda or other electrolytes (Sipex BOS)

Stable under varying conditions of pH and temp.; stable to 20–25% caustic sol'ns. (Sipex OLS)

Alkali-stable (Serdet DSK 40)

Sodium octyl sulfate *(cont'd.)*

Storage Stability:
Stable under normal storage conditions; may be stored and used in a wide variety of containers or reaction vessels (e.g., SS, aluminum, monel) (Avirol SA-4106)

pH:
7.0–10.0 (10% sol'n.) (Avirol SA-4106)
7.3 (0.1% aq.) (Niaproof 08)
8.0 (10% sol'n.) (Sipex OLS)
8.0–10.0 (10% sol'n.) (Sulfotex OA)
9.5–10.5 (10% sol'n.) (Sipex BOS)
10.3 (10% sol'n.) (Carsonol SHS)
10.5 (10% aq.) (Witcolate D-510)

Surface Tension:
38 dynes/cm (@ CMC) (Avirol SA-4106)
51.7 dynes/cm (0.1% sol'n.) (Duponol 80)
63 dynes/cm (0.1% aq.) (Niaproof 08)

Biodegradable: (Avirol SA-4106; Merpinal Q147; Serdet DSK 40; Sipex BOS); fully biodegradable (Sulfotex OA)

TOXICITY/HANDLING:
Irritating to skin and eyes in conc. form; avoid ingestion (Sulfotex OA)
Moderate oral and skin penetration toxicity; avoid contact with skin or eyes; avoid breathing vapor (Niaproof 08)
Avoid contact with eyes or prolonged contact with skin (Sipex BOS)

STORAGE/HANDLING:
Store in closed containers above 7 C (Sulfotex OA)
Store above 60 F to avoid separation (Niaproof 08)
Containers should be kept closed when not in use to avoid evaporation of solvent (Avirol SA-4106)

STD. PKGS.:
440 lb net fiber drums, bulk, tank wagons, rail cars (Sulfotex OA)
480 lb net containers (Carsonol SHS)
Coated 200-kg net iron drums (Merpinal EH 40)
55-gal tight polyethylene and steel composite drums (Niaproof 08)

Sodium olefin sulfonate

SYNONYMS:
Sodium α-olefin sulfonate
Sodium C_{14-16} olefin sulfonate (CTFA)—CAS No. 68439-57-6
Sodium C_{16-18} olefin sulfonate (CTFA)—RD No. 977067-81-4

302

Sodium olefin sulfonate (cont'd.)

TRADENAME EQUIVALENTS:
Bio Soft LD-70, LD-80 [Stepan]
Bio-Terge AS-40, AS-90 Beads, AS-90F [Stepan] (C_{14-16})
Conco AOS-40, AOS-90F [Continental] (C_{14-16})
Elfan OS46 [Akzo Chemie] (C_{14-16})
Polystep A-18 [Stepan] (C_{14-16})
Siponate 301-10F [Alcolac] (C_{14-16})
Siponate A-167, A-168 [Alcolac] (C_{16-18})
Siponate A-246, A-246L, A246LX [Alcolac] (C_{14-16})
Stepantan 29N, 39N [Stepan] (linear)
Sterling AOS [Canada Packers] (C_{14-16})
Sulframin AOS, AOS90 [Witco/Organics]
Surco AOS [Onyx]
Ultrawet AOK [Arco] (C_{14-16})
Witconate AOS [Witco/Org.] (C_{14-16})

CATEGORY:
Detergent, emulsifier, foaming agent, stabilizer, surfactant, viscosity builder, wetting agent

APPLICATIONS:
Automobile cleaners: car shampoo (Bio Soft LD-70; Bio-Terge AS-40; Conco AOS-40, AOS-90F; Siponate A-246L)
Bath products: bubble bath (Bio Soft LD-70; Bio-Terge AS-90 Beads; Conco AOS-40, AOS-90F; Elfan OS46; Siponate 301-10F, A246L; Sterling AOS; Sulframin AOS, AOS90; Surco AOS; Ultrawet AOK)
Cleansers: body cleansers (Elfan OS46; Surco AOS); hand cleanser (Conco AOS-40, AOS-90F; Siponate 301-10F, A246L; Surco AOS; Ultrawet AOK)
Cosmetic industry preparations: (Bio-Terge AS-90 Beads; Conco AOS-40, AOS-90F; Siponate A-246L); shampoos (Bio Soft LD-70; Conco AOS-40, AOS-90F; Siponate 301-10F, A246L; Sterling AOS; Sulframin AOS, AOS90; Surco AOS)
Household detergents: all-purpose cleaner (Elfan OS46; Siponate A-246L; Sterling AOS); carpet & upholstery shampoos (Siponate 301-10F; Sterling AOS); dishwashing (Bio Soft LD-70, LD-80; Bio-Terge AS-40; Conco AOS-40, AOS-90F; Elfan OS46; Siponate A-246L); laundry detergent (Bio Soft LD-70; Siponate 301-10F, A-246); light-duty cleaners (Siponate 301-10F); liquid detergents (Sterling AOS; Sulframin AOS, AOS90)
Industrial applications: (Bio-Terge AS-40; Conco AOS-40; Siponate A-168, A-246; Stepantan 29N, 39N; Witconate AOS); petroleum industry (Stepantan 29N, 39N); textile/leather processing (Conco AOS-40, AOS-90F; Witconate AOS)
Pet shampoos: (Siponate A246L)

PROPERTIES:
Form:
Liquid (Bio-Terge AS-40; Conco AOS-40; Elfan OS46; Polystep A-18; Siponate A

Sodium olefin sulfonate *(cont'd.)*

167, A-168, A-246, A246L, A246LX; Sterling AOS; Sulframin AOS; Surco AOS; Witconate AOS)
Clear liquid (Bio Soft LD-70, LD-80)
Free-flowing beads (Bio-Terge AS-90 Beads)
Flake (Bio-Terge AS-90F; Conco AOS-90F; Siponate 301-10F; Sulframin AOS90; Ultrawet AOK)
Powder (Conco AOS-90F)
Color:
Pale (Polystep A-18)
Off-white (Bio-Terge AS-90 Beads)
Amber (Bio Soft LD-70, LD-80; Conco AOS 40, AOS-90F)
Yellow (Bio-Terge AS-40; Elfan OS46)
Klett 400 (5% active) (Stepantan 29N, 39N)
Odor:
Faint (Bio-Terge AS-90 Beads)
Composition:
20% active (Stepantan 29N)
30% active (Siponate A-167, A-168)
35% active (Bio Soft LD-70)
37% active (Elfan OS46)
38–40% active (Surco AOS)
39% active (Sterling AOS)
40% active (Bio-Terge AS-40; Conco AOS-40; Polystep A-18; Siponate A-246, A-246L; Stepantan 39N)
41% solids (Bio Soft LD-80)
90% active (Bio-Terge AS-90 Beads, AS-90F; Conco AOS-90F; Siponate 301-10F)
90% conc. (Ultrawet AOK)
Solubility:
Sol. in water (Bio Soft LD-80; Bio-Terge AS-40, AS-90 Beads, AS-90F; Elfan OS46; Witconate AOS
Ionic Nature:
Anionic (Bio-Terge AS-40; Conco AOS-40, AOS-90F; Elfan OS46; Polystep A-18; Siponate 301-10F, A-168, A-246, A-246L; Sterling AOS; Witconate AOS)
Nonionic (Bio Soft LD-70)
Sp.gr.:
1.05 (Stepantan 29N, 39N); (25/20 C) (Elfan OS46; Surco AOS)
Density:
0.30 g/ml (Bio-Terge AS-90 Beads)
8.7 lb/gal (Bio Soft LD-70)
8.8 lb/gal (Bio Soft LD-80)
8.9 lb/gal (Bio-Terge AS-40)
Viscosity:
225 cps (Bio Soft LD-80)

250 cps (20 C) (Elfan OS46)
300 cps (Bio Soft LD-70; Stepantan 39N)
F.P.:
< 0 C (Elfan OS46)
Flash Pt.:
> 200 F (Surco AOS)
Cloud Pt.:
< 5 C (Bio Soft LD-80)
> 40 F (Bio Soft LD-70)
Stability:
Stable to min. and organic acids, dilute alkalies, reducing agents; moderately stable to oxidizing agents (Bio-Terge AS-90 Beads)
pH:
7.5 (5% sol'n.) (Conco AOS-40, AOS-90F)
8.0 ± 1 (Elfan OS46)
8.0–9.0 (Surco AOS)
9.0 (Stepantan 29N, 39N)
9.8 (5% aq.) (Bio-Terge AS-90 Beads)
Foam (Ross Miles):
180 mm initial, 155 mm after 5 min (0.1% conc. in dist. water) (Bio-Terge AS-90 Beads)
Surface Tension:
46.2 dynes/cm (0.1%) (Bio-Terge AS-90 Beads)
Wetting (Draves):
17 s (0.1% conc. in dist. water) (Bio-Terge AS-90 Beads)
Biodegradable: (Bio Soft LD-70, LD-80; Bio-Terge AS-40, AS-90 Beads, AS-90F; Ultrawet AOK)
TOXICITY/HANDLING:
Avoid excessive skin contact (Bio-Terge AS-90 Beads; Conco AOS-40, AOS-90F)
Avoid inhalation of dust (Bio-Terge AS-90 Beads)
STORAGE/HANDLING:
Store in a cool, dry place to prevent agglomeration and deterioration in color (Bio-Terge AS-90 Beads)
Store between 10–40 C (Elfan OS46)
STD. PKGS.:
55-gal (150 lb net) fiber drums with polyethylene liner (Bio-Terge AS-90 Beads)
200-lb net fiber drums (Conco AOS-90F)
400-lb net open-head lined steel drums, bulk (Stepantan 39N)
Polyethylene-lined liquipaks, 304 or 316 stainless tanks, and certain fiber glass tanks (Conco AOS-40)

Sorbitan monolaurate

SYNONYMS:
1,4-Anhydro-D-glucitol, 6 dodecanoate
Anhydrosorbol monolaurate
D-Glucitol, 1,4-anhydro-, 6 dodecanoate
Sorbitan laurate (CTFA)
Sorbitan, monododecanoate

EMPIRICAL FORMULA:
$C_{18}H_{34}O_6$

STRUCTURE:

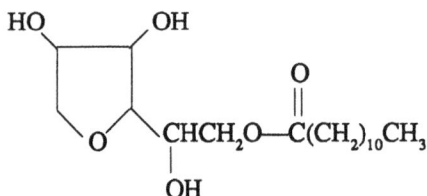

CAS No.
1338-39-2; 5959-89-7

TRADENAME EQUIVALENTS:
Ahco 759 [ICI United States]
Alkamuls SML [Alkaril]
Arlacel 20 [ICI]
Armotan ML [Armak]
Crill 1 [Croda]
Drewmulse SML [PVO Int'l.]
Durtan 20 [Durkee/SCM]
Emsorb 2515 [Emery]
Glycomul L, LC [Glyco]
Hodag SML [Hodag]
Ionet S-20 [Sanyo]
Kuplur SML [BASF Wyandotte]
Liposorb L [Lipo]
Montane 20 [Seppic]
Newcol 20 [Nippon Nyukazai]
Nissan Nonion LP-20R, LP-20RS [Nippon Oil & Fats]
Radiamuls 125, SORB 2125 [Oleofina]
Radiasurf 7125 [Oleofina]
S-Maz 20, 20R [Mazer]
Soprofor S/20 [Mario Geronazzo]
Sorbax SML [Chemax]
Sorbon S-20 [Toho]
Sorgen 90 [Dai-ichi Kogyo Seiyaku]
Span 20 [ICI]

CATEGORY:
Emulsifier, stabilizer, anticorrosive agent, antifoam, antistaling agent, antistat, descouring aid, detergent, dispersant, dryness improver, fog aid, lubricant, solubilizer, spray-drying aid, superfatting and bodying aid, thickener, whipping aid

APPLICATIONS:
Cosmetic industry preparations: (Armotan ML; Crill 1; Drewmulse SML; Durtan 20; Glycomul LC; Ionet S-20; Liposorb L; Newcol 20; Nissan Nonion LP-20R, LP-20RS; Radiasurf 7125; Sorbax SML; Sorgen 90); creams and lotions (Drewmulse SML); perfumery (Drewmulse SML); shaving preparations (Drewmulse SML)

Farm products: insecticides/pesticides (Crill 1; Glycomul LC; Ionet S-20; Liposorb L; Newcol 20; Nissan Nonion LP20R; Radiasurf 7125)

Food applications: (Crill 1; Glycomul LC; Liposorb L; Newcol 20; Nissan Nonion LP-20R, LP-20RS; Radiamuls 125, SORB 2125; Sorgen 90); food emulsifying (Radiamuls SORB 2125)

Household detergents: (Ionet S-20; Radiasurf 7125)

Industrial applications: dyes and pigments (Crill 1; Glycomul LC; Ionet S-20; Liposorb L; Newcol 20; Nissan Nonion LP20R; Radiasurf 7125); paper mfg. (Crill 1; Glycomul LC; Liposorb L; Newcol 20; Nissan Nonion LP20R; Radiasurf 7125); plastics (Crill 1; Glycomul LC; Liposorb L; Newcol 20; Nissan Nonion LP20R; Radiasurf 7125); polishes and waxes (Crill 1; Glycomul LC; Liposorb L; Newcol 20; Nissan Nonion LP20R; Radiasurf 7125); textile/leather processing (Ahco 759; Armotan ML; Crill 1; Glycomul LC; Ionet S-20; Liposorb L; Newcol 20; Nissan Nonion LP-20R, LP-20RS; Radiasurf 7125; Soprofor S/20; Sorbax SML; Sorgen 90)

Industrial cleaners: drycleaning compositions (Crill 1; Glycomul LC; Liposorb L; Newcol 20; Nissan Nonion LP20R; Radiasurf 7125)

Pharmaceutical applications: (Armotan ML; Crill 1; Drewmulse SML; Glycomul LC; Liposorb L; Newcol 20; Nissan Nonion LP-20R, LP-20RS; Radiasurf 7125; Sorbax SML; Sorgen 90); germicides (Drewmulse SML); ointments (Drewmulse SML); vitamins (Drewmulse SML)

PROPERTIES:
Form:
Liquid (Ahco 759; Alkamuls SML; Armotan ML; Drewmulse SML; Emsorb 2515; Hodag SML; Ionet S-20; Kuplur SML; Montane 20; Radiamuls 125, SORB 2125; Radiasurf 7125; Sorbax SML; Sorbon S-20; Sorgen 90)

Clear, viscous liquid (Crill 1; Glycomul LC; Liposorb L; Newcol 20)

Oily liquid (Nissan Nonion LP20R)

Color:
Amber (Alkamuls SML; Glycomul LC; Radiasurf 7125)

Yellow (Crill 1; Kuplur SML; Liposorb L; Newcol 20; Nissan Nonion LP20R)

Gardner 7 (Emsorb 2515)

Sorbitan monolaurate *(cont'd.)*

Composition:
   97% active (Alkamuls SML)
   100% active (Arlacel 20; Emsorb 2515; Glycomul L, LC; Liposorb L; Newcol 20; S-Maz 20; Span 20)
Solubility:
   Sol. in acetone (Nissan Nonion LP20R); miscible in certain proportions (Glycomul LC)
   Sol. in aromatic solvent (Alkamuls SML)
   Sol. cloudy in benzene (Radiasurf 7125)
   Sol. in butyl stearate (Emsorb 2515)
   Sol. in ethanol (Crill 1; Glycomul LC; Nissan Nonion LP20R)
   Miscible in certain proportions with ethyl acetate (Glycomul LC)
   Sol. in ethyl ether (Nissan Nonion LP20R)
   Sol. in glycerol trioleate (Emsorb 2515)
   Sol. in hexane (Radiasurf 7125)
   Sol. in hydrocarbons (Emsorb 2515)
   Sol. cloudy in isopropanol (Radiasurf 7125)
   Sol. in kerosene (Nissan Nonion LP20R)
   Sol. in methanol (Glycomul LC; Nissan Nonion LP20R)
   Sol. in min. oil (Alkamuls SML; Emsorb 2515; Hodag SML; Radiasurf 7125); miscible in certain proportions (Glycomul LC)
   Sol. in oleyl alcohol (Crill 1)
   Sol. in perchloroethylene (Emsorb 2515); insol. (Alkamuls SML)
   Sol. in Stoddard solvent (Emsorb 2515)
   Miscible in certain proportions with toluol (Glycomul LC)
   Sol. in trichlorethylene (Radiasurf 7125)
   Sol. cloudy in veg. oil (Radiasurf 7125); miscible in certain proportions (Glycomul LC)
   Disp. in water (Alkamuls SML; Emsorb 2515; Glycomul LC; Kuplur SML; Nissan Nonion LP20R; Radiasurf 7125)
   Sol. in xylene (Nissan Nonion LP20R)
Ionic Nature: Nonionic
M.W.:
   450 avg. (Radiasurf 7125)
Sp.gr.:
   1.0 (Glycomul LC)
   1.01 (Kuplur SML)
   1.025 (37.8 C) (Radiasurf 7125)
Density:
   1.0 g/ml (Alkamuls SML)
   8.8 lb/gal (Emsorb 2515)
Visc.:
   450 cps (Kuplur SML)

308

1193 cps (37.8 C) (Radiasurf 7125)
5000 cps (Glycomul LC)
4500 cs (Emsorb 2515)
Solidification Pt.:
10 C max. (Nissan Nonion LP20R)
Flash Pt.:
198 C (COC) (Radiasurf 7125)
Cloud Pt.:
23 C (Radiasurf 7125)
HLB:
6.8 (Radiamuls 125; Radiasurf 7125)
7.4 (Durtan 20)
7.6 (Emsorb 2515; Radiamuls SORB 2125)
8.6 (Alkamuls SML; Crill 1; Drewmulse SML; Glycomul LC; Kuplur SML; Nissan Nonion LP20R; Sorbax SML)
8.6 ± 1 (Liposorb L)
Acid No.:
7 max. (Kuplur SML; Nissan Nonionc LP20R; Radiamuls 125; Radiasurf 7125)
10 max. (Glycomul LC)
Iodine No.:
10 max. (Radiamuls 125; Radiasurf 7125)
Saponification No.:
158–170 (Crill 1; Durtan 20; Kuplur SML; Liposorb L)
159–171 (Drewmulse SML)
160–170 (Alkamuls SML)
154–163 (Glycomul LC)
155–175 (Radiasurf 7125)
Hydroxyl No.:
324–356 (Glycomul LC)
320–330 (Alkamuls SML)
330–358 (Durtan 20; Kuplur SML)
Ref. Index:
1.4718 (Radiasurf 7125)
STD. PKGS.:
190-kg bung drums or bulk (Radiamuls 125, SORB 2125; Radiasurf 7125)
55-gal (450 lb net) steel drums (Kuplur SML)

Sorbitan trioleate (CTFA)

SYNONYMS:
Anhydrosorbitol trioleate
Sorbitan, tri-9-octadecenoate
EMPIRICAL FORMULA:
$C_{60}H_{108}O_8$
STRUCTURE:

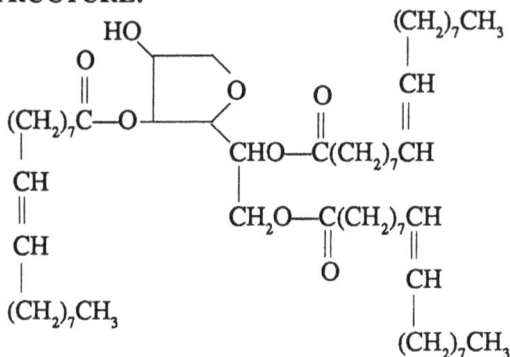

CAS No.:
26266-58-0
TRADENAME EQUIVALENTS:
Ahco FO-18 [ICI Americas]
Alkamuls STO [Alkaril]
Arlacel 85 [ICI United States]
Atlas G-4885 [ICI Specialty Chem.] (tech.)
Atlox 4885 [ICI Specialty Chem.]
Crill 45 [Croda]
Emasol O-30 [Kao]
Emsorb 2503 [Emery]
Glycomul TO [Glyco]
Hodag STO [Hodag]
Ionet S-85 [Sanyo]
Liposorb TO [Lipo]
Lonzest STO [Lonza]
Montane 85 [Seppic]
Nikkol SO-30 [Nikko]
Nissan Nonion OP-85•R [Nippon Oil & Fats]
Radiamuls SORB 2355 [Oleofina]
S-Maz 85 [Mazer]
Soprofor S/85 [Geronazzo SpA]
Sorbax STO [Chemax]
Span 85 [ICI United States]

CATEGORY:
Stabilizer, thickener, emulsifier, coemulsifier, surfactant, coupling agent, solubilizer, lubricant, antistat, corrosion inhibitor, softener, opacifier, wetting agent, dispersant, antifoam

APPLICATIONS:
Cosmetic industry preparations: (Alkamuls STO; Crill 45; Glycomul TO; Ionet S-85; Nissan Nonion OP-85•R; S-Maz 85; Sorbax STO; Span 85); creams and lotions (Ionet S-85); makeup (Crill 45); ointments (Ionet S-85)

Farm products: fungicides (Alkamuls STO); herbicides (Alkamuls STO; Atlox 4885); insecticides/pesticides (Alkamuls STO; Atlox 4885; Crill 45; Ionet S-85)

Food applications: (Alkamuls STO; Crill 45; Glycomul TO; Lonzest STO; Nissan Nonion OP-85•R; Radiamuls SORB 2355; S-Maz 85; Span 85)

Household products: (S-Maz 85); cleaners (Span 85)

Industrial applications: (Crill 45; Glycomul TO; Hodag STO; Ionet S-85; Lonzest STO; S-Maz 85); aerosols (Crill 45); dyes and pigments (Emasol O-30; Ionet S-85); lubricating/cutting oils (Crill 45); metalworking (Crill 45; Ionet S-85); petroleum industry (Crill 45); polishes and waxes (Crill 45); printing inks (Crill 45); surface coatings (Crill 45); textile/leather processing (Ahco FO-18; Alkamuls STO; Crill 45; Emsorb 2503; Ionet S-85; Lonzest STO; Nissan Nonion OP-85•R; S-Maz 85; Soprofor S/85; Sorbax STO; Span 85)

Industrial cleaners: drycleaning compositions (Crill 45)

Pharmaceutical applications: (Crill 45; Glycomul TO; Nissan Nonion OP-85•R; Span 85); medicaments (Crill 45); vitamins (Radiamuls SORB 2355)

PROPERTIES:
Form:
Liquid (Ahco FO-18; Alkamuls STO; Atlas G-4885; Atlox 4885; Emasol O-30; Emsorb 2503; Hodag STO; Ionet S-85; Liposorb TO; Lonzest STO; Montane 85; Nikkol SO-30; Radiamuls SORB 2355; S-Maz 85; Soprofor S/85; Sorbax STO; Span 85)

Oily liquid (Arlacel 85; Glycomul TO; Nissan Nonion OP-85•R)

Viscous liquid (Crill 45)

Color:
Amber (Alkamuls STO; Crill 45; Glycomul TO; Liposorb TO; S-Maz 85; Span 85)
Yellow (Arlacel 85)
Gardner 7 (Emsorb 2503)
Gardner 9 max. (Nissan Nonion OP-85•R)

Odor:
Typical (Alkamuls STO)

Composition:
100% active (Alkamuls STO; Arlacel 85; Emsorb 2503; Liposorb TO; Span 85)
100% conc. (Ahco FO-18; Atlas G-4885; Atlox 4885; Emasol O-30; Glycomul TO; Hodag STO; Ionet S-85; Lonzest STO; Montane 85; Nissan Nonion OP-85•R; Radiamuls SORB 2355; S-Maz 85; Soprofor S/85)

Sorbitan trioleate (cont'd.)

Solubility:
Miscible hot in certain proportions with acetone (S-Maz 85)
Sol. in butyl stearate (Emsorb 2503)
Sol. in corn oil (Arlacel 85)
Sol. in cottonseed oil (Arlacel 85); (@ 1%) (Span 85)
Miscible hot in certain proportions with ethanol (S-Maz 85)
Sol. in ethyl acetate (Glycomul TO)
Sol. in glycerol trioleate (Emsorb 2503)
Sol. in isopropanol (Arlacel 85); (@ 1%) (Span 85)
Sol. in isopropyl myristate (Crill 45)
Sol. in min. oil (Arlacel 85; Crill 45; Emsorb 2503; Glycomul TO; S-Maz 85); (@ 1%) (Span 85)
Sol. in naphtha (Glycomul TO; S-Maz 85)
Sol. in oils (Nissan Nonion OP-85•R)
Sol. in oleic acid (Crill 45)
Sol. in oleyl alcohol (Crill 45)
Sol. in olive oil (Crill 45)
Sol. in perchloroethylene (Emsorb 2503); (@ 1%) (Span 85)
Sol. in Stoddard solvent (Emsorb 2503)
Sol. in toluol (Glycomul TO; S-Maz 85)
Sol. in veg. oil (Glycomul TO; S-Maz 85)
Disp. in water (Glycomul TO; S-Maz 85); insol. in water (Alkamuls STO)
Sol. in xylene (@ 1%) (Span 85)
Ionic Nature:
Nonionic (Ahco FO-18; Alkamuls STO; Arlacel 85; Atlas G-4885; Emasol O-30; Emsorb 2503; Glycomul TO; Hodag STO; Ionet S-85; Liposorb TO; Lonzest STO; Montane 85; Nikkol SO-30; Nissan Nonion OP-85•R; Radiamuls SORB 2355; S-Maz 85; Soprofor S/85; Span 85)
Sp.gr.:
0.95 (Arlacel 85; Glycomul TO)
1.0 (S-Maz 85)
Density:
7.9 lb/gal (Emsorb 2503)
Visc.:
200 cps (Glycomul TO; S-Maz 85)
200 cs (Emsorb 2503)
210 cs (Span 85)
250 cps (Arlacel 85)
Solidification Pt.:
0 C max. (Nissan Nonion OP-85•R)
Flash Pt.:
> 300 F (Arlacel 85)

Fire Pt.:
  > 300 F (Arlacel 85)
HLB:
  1.8 (Ahco FO-18; Alkamuls STO; Arlacel 85; Atlas G-4885; Atlox 4885; Crill 45; Emasol O-30; Glycomul TO; Hodag STO; Ionet S-85; Montane 85; Nissan Nonion OP-85•R; S-Maz 85; Sorbax STO; Span 85)
  1.8 ± 1 (Liposorb TO)
  2.2 (Radiamuls SORB 2355)
  2.7 (Emsorb 2503)
  4.0 (Nikkol SO-30)
Acid No.:
  7.0 max. (Nissan Nonion OP-85•R)
  13.5 max. (Glycomul TO)
  14 max. (S-Maz 85)
Saponification No.:
  171–185 (Glycomul TO; Liposorb TO)
  172–186 (Crill 45; S-Maz 85)
Hydroxyl No.:
  56–68 (S-Maz 85)
  58–69 (Glycomul TO; Liposorb TO)

Sorbitan tristearate (CTFA)

SYNONYMS:
  Anhydrosorbitol tristearate
  Sorbitan, trioctadecanoate
EMPIRICAL FORMULA:
  $C_{60}H_{114}O_8$
STRUCTURE:

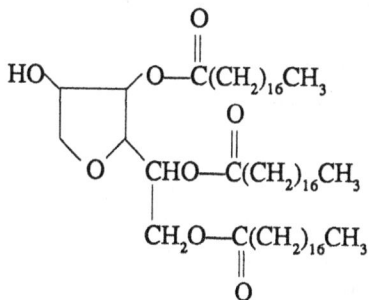

313

Sorbitan tristearate (cont'd.)

CAS No.:
26658-19-5
TRADENAME EQUIVALENTS:
Ahco FS-21 [ICI Americas]
Alkamuls STS [Alkaril]
Crill 35 [Croda]
Drewmulse STS [PVO Int'l.]
Emasol S-30 [Kao]
Emsorb 2507 [Emery]
Glycomul TS [Glyco]
Grindtek STS [Grinsted]
Hodag STS [Hodag]
Kuplur STS [BASF Wyandotte]
Liposorb TS [Lipo]
Lonzest STS [Lonza]
Montane 65 [Seppic]
Nikkol SS-30 [Nikko]
Radiamuls 345 [Oleofina S.A.]
Radiamuls SORB 2345 [Oleofina S.A.]
S-Maz 65 [Mazer]
Soprofor S-65 [Geronazzo SpA]
Sorbax STS [Chemax]
Span 65 [ICI United States]
CATEGORY:
Emulsifier, stabilizer, thickener, dispersant, lubricant, antistat, surfactant, softener, defoamer, opacifier, coemulsifier, solubilizer, wetting agent, detergent, viscosity control agent, corrosion inhibitor, spray drying aid
APPLICATIONS:
Cosmetic industry preparations: (Alkamuls STS; Crill 35; Drewmulse STS; Glycomul TS; S-Maz 65; Sorbax STS; Span 65); creams and lotions (Crill 35; Drewmulse STS); perfumery (Drewmulse STS); shampoos (Drewmulse STS); shaving preparations (Drewmulse STS)

Farm products: animal feed (Radiamuls SORB 2345); herbicides (Crill 35); insecticides/pesticides (Crill 35)

Food applications: (Alkamuls STS; Crill 35; Glycomul TS; Lonzest STS; Radiamuls 345, SORB 2345; S-Maz 65; Span 65); flavors (Drewmulse STS; Radiamuls SORB 2345); food emulsifying (Radiamuls 345)

Household products: (S-Maz 65); cleaners (Span 65)

Industrial applications: (Glycomul TS); dyes and pigments (Alkamuls STS); industrial processing (Hodag STS; Lonzest STS; S-Maz 65); polishes and waxes (Crill 35); textile/leather processing (Ahco FS-21; Crill 35; Emsorb 2507; Lonzest STS; S-Maz 65; Soprofor S-65; Sorbax STS; Span 65)

Industrial cleaners: metal processing surfactants (Crill 35)

Pharmaceutical applications: (Crill 35; Drewmulse STS; Glycomul TS; Span 65); germicides (Drewmulse STS); ointments (Drewmulse STS); vitamins (Drewmulse STS; Radiamuls SORB 2345)

PROPERTIES:
Form:
Solid (Ahco FS-21; Alkamuls STS; Drewmulse STS; Emasol S-30; Hodag STS; Lonzest STS; Montane 65; Soprofor S-65; Sorbax STS; Span 65)
Beads (Glycomul TS; Liposorb TS)
Flake (Liposorb TS; Nikkol SS-30; Radiamuls 345, SORB 2345; S-Maz 65)
Powder (Grindtek STS; Radiamuls 345, SORB 2345)
Wax (Kuplur STS)
Waxy solid (Emsorb 2507)
Hard waxy solid (Crill 35)
Color:
Cream (Liposorb TS; S-Maz 65; Span 65)
Light tan (Crill 35; Glycomul TS)
Tan (Alkamuls STS; Grindtek STS; Kuplur STS)
Gardner 4 (Emsorb 2507)
Odor:
Typical (Alkamuls STS)
Composition:
100% active (Alkamuls STS; Liposorb TS; Span 65)
100% conc. (Ahco FS-21; Emasol S-30; Emsorb 2507; Glycomul TS; Hodag STS; Lonzest STS; Montane 65; Nikkol SS-30; S-Maz 65; Soprofor S-65)
Solubility:
Disp. in acetone (Glycomul TS; S-Maz 65)
Sol. in butyl stearate (Emsorb 2507)
Partly sol. warm in ethanol (Grindtek STS)
Poorly sol. in ethyl acetate (Glycomul TS)
Sol. in glycerol trioleate (Emsorb 2507)
Sol. in isopropanol (@ 1%) (Span 65)
Partly sol. in isopropyl myristate (Crill 35)
Partly sol. in min. oil (Crill 35); disp. (Emsorb 2507; Glycomul TS; S-Maz 65)
Disp. in naphtha (Glycomul TS; S-Maz 65)
Partly sol. in oleic acid (Crill 35)
Parly sol. in oleyl alcohol (Crill 35)
Partly sol. in olive oil (Crill 35)
Partly sol. warm in paraffin oil (Grindtek STS)
Partly sol. warm in peanut oil (Grindtek STS)
Sol. in perchloroethylene (Emsorb 2507); (@ 1%) (Span 65)

Sorbitan tristearate *(cont'd.)*

Sol. in Stoddard solvent (Emsorb 2507)

Sol. warm in toluene (Grindtek STS); partly sol. (S-Maz 65); poorly sol. (Glycomul TS)

Disp. in veg. oil (Glycomul TS; S-Maz 65)

Insol. in water (Alkamuls STS; Emsorb 2507; Kuplur STS)

Sol. warm in white spirit (Grindtek STS)

Sol. in xylene (@ 1%) (Span 65)

Ionic Nature:

Nonionic (Ahco FS-21; Emasol S-30; Emsorb 2507; Glycomul TS; Hodag STS; Kuplur STS; Liposorb TS; Lonzest STS; Montane 65; Nikkol SS-30; Radiamuls 345; S-Maz 65; Soprofor S-65; Span 65)

Sp.gr.:

1.0 (S-Maz 65)

1.01 (Kuplur STS)

M.P.:

48 C (Crill 35)

50–60 C (Radiamuls 345, SORB 2345)

53–55 C (S-Maz 65)

54 C (Emsorb 2507)

Pour Pt.:

53 C (Span 65)

HLB:

2.1 (Ahco FS-21; Alkamuls STS; Crill 35; Drewmulse STS; Emasol S-30; Glycomul TS; Hodag STS; Kuplur STS; Nikkol SS-30; S-Maz 65; Sorbax STS; Span 65)

2.1 ± 1 (Liposorb TS)

2.3 (Grindtek STS; Radiamuls SORB 2345)

2.4 (Emsorb 2507)

2.7 (Montane 65; Radiamuls 345)

Acid No.:

15 max. (Glycomul TS; Kuplur STS; S-Maz 65)

17 max. (Radiamuls 345)

Iodine No.:

1.0 max. (Radiamuls 345)

Saponification No.:

170–190 (Drewmulse STS)

175–190 (Glycomul TS)

175–190 (Liposorb TS)

176–188 (Crill 35; Kuplur STS; S-Maz 65)

Hydroxyl No.:

65–80 (Glycomul TS; Liposorb TS)

66–80 (Kuplur STS; S-Maz 65)

STD. PKGS.:
25-kg net paper bags (Radiamuls SORB 2345)
25-kg net multi-ply paper bags (Radiamuls 345)
55-gal (450 lb net) steel drums (Kuplur STS)

Soya diethanolamide

SYNONYMS:
Amides, soya, N,N-bis (hydroxyethyl)-
N,N-Bis (hydroxyethyl) soya amides
Soya amides, N,N-bis (hydroxyethyl)-
Soyamide DEA (CTFA)
Soya DEA

STRUCTURE:

$$\begin{array}{c} O \\ \parallel \\ RC\!-\!N(CH_2CH_2OH)_2 \end{array}$$

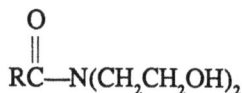

where RCO⁻ represents the soya acid radical

CAS No.:
68425-47-8

TRADENAME EQUIVALENTS:
Mackamide S [McIntyre]
Mazamide SS-10 [Mazer]

CATEGORY:
Thickener, conditioning agent, emulsifier, corrosion inhibitor, lubricant, buffing compound

APPLICATIONS:
Cosmetic industry preparations: conditioners (Mackamide S); hair preparations (Mackamide S)
Industrial applications: (Mazamide SS-10); soluble oils (Mazamide SS-10)

PROPERTIES:

Form:
Liquid (Mackamide S; Mazamide SS-10)

Composition:
100% conc. (Mackamide S)

Ionic Nature:
Nonionic (Mackamide S; Mazamide SS-10)

Sp.gr.:
0.98–1.00 (Mazamide SS-10)

Soya diethanolamide (cont'd.)

pH:
9.0–10.5 (Mazamide SS-10)
Biodegradable: (Mazamide SS-10)
TOXICITY/HANDLING:
Defatting to skin (Mazamide SS-10)

Stearalkonium hectorite (CTFA)

SYNONYMS:
Reaction product of hectorite and stearalkonium chloride
CAS No.:
RD No. 977067-62-1
TRADENAME EQUIVALENTS:
Bentone 27 [NL Chem.]
Bentone Gel CAO [NL Chem.] (in castor oil)
Bentone Gel IPM [NL Chem.] (in isopropyl myristate)
Bentone Gel LOI [NL Chem.] (in lanolin/isopropyl palmitate)
CATEGORY:
Thickener, gellant, thixotrope
APPLICATIONS:
Cosmetic industry preparations: anhydrous formulations (Bentone 27, Gel CAO, Gel IPM, Gel LOI); eye care products (Bentone 27); lip products (Bentone 27)
Industrial applications: anhydrous formulations (Bentone Gel IPM); solvent-based coatings (Bentone 27)
PROPERTIES:
Form:
Gel (Bentone Gel LOI)
Fine powder (Bentone 27)
Color:
Creamy white (Bentone 27)
Composition:
100% nonvolatiles (Bentone 27)
Solubility:
Disperses in many cosmetic oils (Bentone Gel CAO, Gel IPM, Gel LOI)
Density:
1.80 g/cm³ (Bentone 27)

SYNONYMS:
N,N-Bis (2-hydroxyethyl) octadecanamide
N,N-Bis (2-hydroxyethyl) stearamide
Diethanolmine stearic acid amide
Octadecanamide, N,N-bis (2-hydroxyethyl)-
Stearic acid diethanolamide
Stearoyl diethanolamide
Stearic DEA

EMPIRICAL FORMULA:
$C_{22}H_{45}NO_3$

STRUCTURE:

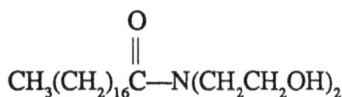

CAS No.:
93-82-3

TRADENAME EQUIVALENTS:
Alkamide HTDE [Alkaril]
Clindrol 200-S, 868 [Clintwood]
Clindrol Superamide 100S [Clintwood]
Cyclomide DS 280/S [Cyclo] (1:1)
Lipamide S [Lipo] (1:1)
Monamid 718 [Mona] (1:1)
Nopcogen 14-S [Diamond Shamrock]
Onyxol 42 [Onyx] (1:1)
Unamide S [Lonza] (1:1)

CATEGORY:
Thickener, gelling agent, visc. builder, emulsifier, softener, suspending agent, antisettling agent, emulsion stabilizer, lubricant, pearlescent, opacifier, dye carrier, foam
booster, foam stabilizer, detergent, wetting agent, corrosion inhibitor, dispersant

APPLICATIONS:
Bath products: bubble bath (Clindrol 200-S, Superamide 100S; Monamid 718)
Cleansers: cleansing creams (Clindrol 200-S); hand cleanser (Clindrol 868, Superamide 100S; Monamid 718); personal cleansers (Lipamide S); waterless hand
cleaner (Clindrol 200-S; Monamid 718)
Cosmetic industry preparations: (Clindrol 200-S, 868; Monamid 718; Onyxol 42);
absorption bases (Clindrol Superamide 100S); conditioners (Clindrol Superamide
100S; Cyclomide DS 280/S; Monamid 718); cold waves (Lipamide S); creams and
lotions (Clindrol Superamide 100S; Lipamide S; Monamid 718); hair preparations
(Clindrol 200-S, Superamide 100S; Lipamide S); personal care products (Clindrol
200-S; Unamide S); shampoos (Clindrol 200-S, 868, Superamide 100S; Lipamide

Stearic diethanolamide (cont'd.)

S; Monamid 718); shaving preparations (Clindrol 200-S); skin preparations (Clindrol Superamide 100S; Lipamide S); toiletries (Clindrol Superamide 100S)

Farm products: agricultural oils/sprays (Monamid 718)

Household applications: (Clindrol 868; Unamide S); carpet & upholstery shampoos (Monamid 718); dishwashing (Monamid 718)

Industrial applications: (Unamide S); aerosols (Clindrol 200-S); dyes and pigments (Monamid 718); lubricating/cutting oils (Monamid 718); metalworking (Clindrol 200-S); petroleum industry (Monamid 718); polishes and waxes (Clindrol 200-S; Monamid 718); textile/leather processing (Monamid 718; Nopcogen 14-S; Onyxol 42; Unamide S)

Industrial cleaners: metal processing surfactants (Monamid 718)

Pharmaceutical applications: antiperspirant/deodorant (Lipamide S)

PROPERTIES:

Form:
Liquid (Cyclomide DS 280/S)
Solid (Alkamide HTDE; Clindrol 868; Monamid 718; Nopcogen 14-S)
Wax (Lipamide S; Onyxol 42)
Waxy solid (Clindrol 200-S, Superamide 100S; Unamide S)

Color:
White (Clindrol 868, Superamide 100S; Unamide S)
Off-white (Lipamide S)
Cream (Alkamide HTDE)
GVCS-33: 4 max. (Monamid 718)

Odor:
Mild, waxy (Lipamide S)

Composition:
90% amide (Alkamide HTDE)
100% active (Clindrol 868; Lipamide S; Monamid 718; Unamide S)
100% conc. (Cyclomide DS 280/S; Nopcogen 14-S)

Solubility:
Sol. in alcohols (Clindrol 200-S, 868, Superamide 100S)
Sol. in aliphatic hydrocarbons (Clindrol 200-S); disp. (Clindrol Superamide 100S)
Sol. in aromatic hydrocarbons (Clindrol 200-S, 868, Superamide 100S); (@ 10%) (Alkamide HTDE; Monamid 718)
Sol. in chlorinated solvents (Clindrol 200-S, 868, Superamide 100S); disp. (@ 10%) (Monamid 718)
Sol. in esters (Clindrol 200-S, Superamide 100S)
Sol. in ethanol (@ 10%) (Monamid 718)
Sol. in ethers (Clindrol 200-S)
Sol. in glycols (Clindrol 200-S, Superamide 100S)
Sol. in ketones (Clindrol 200-S, 868, Superamide 100S)
Sol. in warm min. oil (Clindrol 200-S); disp. (@ 10%) (Alkamide HTDE)

Sol. in min. spirits (@ 10%) (Alkamide HTDE)
Sol. in naphtha (Clindrol 200-S)
Misc. hot with common oil phase ingredients (Lipamide S)
Misc. hot with organic solvents (Lipamide S)
Sol. in perchloroethylene (@ 10%) (Alkamide HTDE)
Sol. in toluol (Clindrol 200-S)
Disp. in water (Clindrol 200-S, Superamide 100S); disp. in hot water (Lipamide S); (@ 10%) (Alkamide HTDE); gels in water (@ 10%) (Monamid 718)
Ionic Nature:
Nonionic (Clindrol 200-S, 868; Cyclomide DS 280/S; Lipamide S; Monamid 718; Onyxol 42; Unamide S)
Sp.gr.:
0.96 (45 C) (Monamid 718)
0.97 (25/20 C) (Onyxol 42)
1.00 (Clindrol 200-S)
Density:
8.00 lb/gal (Monamid 718)
M.P.:
36 C (Clindrol 200-S)
41–46 C (Lipamide S)
48 C (Nopcogen 14-S)
50–55 C (Alkamide HTDE)
58–63 C (Clindrol 868, Superamide 100S)
Flash Pt.:
> 200 F (Onyxol 42)
Acid No.:
11–15 (Unamide S)
12–18 (Lipamide S)
18–24 (Onyxol 42)
21 ± 3 (Monamid 718)
Alkali No.:
30–70 (Onyxol 42)
45–65 (Monamid 718)
Amine No.:
44–52 (Unamide S)
Stability:
Stable over a broad pH range; excellent electrolyte tolerance (Lipamide S)
pH:
Alkaline (Clindrol 868)
8–11 (1% DW) (Alkamide HTDE)
9–10 (1% disp.) (Clindrol 200-S)
9.0–10.5 (1% aq.) (Lipamide S); (1% disp.) (Clindrol Superamide 100S)

Stearic diethanolamide (cont'd.)

9.3–10.3 (10% sol'n.) (Monamid 718)
Biodegradable: (Monamid 718)
TOXICITY/HANDLING:
Skin and eye irritant (Clindrol 868)
STD. PKGS.:
40-lb Cubitainers (Clindrol 868)
55-gal (450 lb net) open-head steel drums (Monamid 718)

Stearic monoethanolamide

SYNONYMS:
N-(2-Hydroxyethyl) octadecanamide
N-(2-Hydroxyethyl) stearamide
Monoethanolamine stearic acid amide
Octadecanamide, N-(2-hydroxyethyl)-
Stearamide MEA (CTFA)
Stearic acid monoethanolamide
Stearoyl monoethanolamide
Stearic MEA
EMPIRICAL FORMULA:
$C_{20}H_{41}NO_2$
STRUCTURE:

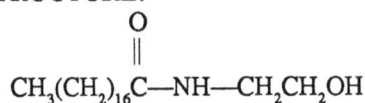

CAS No.:
111-57-9
TRADENAME EQUIVALENTS:
Alkamide HTME [Alkaril]
Ardet WB [Ardmore]
Clindrol 200MS, LT14-8 [Clintwood]
Comperlan HS [Henkel KGaA]
CPH-380N [C.P. Hall]
Cyclomide S 280 [Cyclo]
Incromide SM [Croda Surfactants]
Loramine S 280 [Dutton & Reinisch]
Monamid S [Mona] (1:1)
Ninol SNR [Stepan]
Schercomid SME [Scher]

Stearic monoethanolamide (cont'd.)

TRADENAME EQUIVALENTS *(cont'd.):*
  Schercomid SME-A, SME-M [Scher] (1:1)
  Schercopearl EA-100 [Scher]
  Witcamide 70 [Witco/Organics]
CATEGORY:
  Thickener, gelling agent, opacifier, pearlescent, clouding agent, softener, conditioner, dye carrier, detergent, superfatting agent, skin protectant, emulsion stabilizer, release agent, binder, foam booster, foam stabilizer, emulsifier, wetting agent, corrosion inhibitor, lubricant, dispersant
APPLICATIONS:
  Bath products: bubble bath (Alkamide HTME; Monamid S)
  Cleansers: hand cleanser (Clindrol 200MS; Monamid S); waterless hand cleaners (Monamid S)
  Cosmetic industry preparations: (Alkamide HTME; Clindrol 200MS; Incromide SM; Monamid S; Schercomid SME, SME-A, SME-M; Witcamide 70); conditioners (Monamid S; Schercomid SME, SME-A, SME-M); creams and lotions (Cyclomide S 280; Schercomid SME, SME-A, SME-M); personal care products (Clindrol LT14-8); shampoos (Alkamide HTME; Clindrol 200MS; Schercomid SME, SME-A, SME-M; Schercopearl EA-100); skin preparations (Clindrol 200MS); stick make-up (Comperlan HS); toiletries (Cyclomide S 280; Witcamide 70)
  Farm products: agricultural oils/sprays (Monamid S)
  Household products: (Clindrol 200MS); carpet & upholstery shampoos (Monamid S); dishwashing (Monamid S)
  Industrial applications: dyes and pigments (Monamid S); lubricating/cutting oils (Monamid S); petroleum industry (Monamid S); polishes and waxes (Monamid S); textile/leather processing (Incromide SM; Monamid S)
  Industrial cleaners: drycleaning compositions (Monamid S); metal processing surfactants (Monamid S)
  Pharmaceutical applications: stick deodorants (Schercomid SME-M)
PROPERTIES:
Form:
  Liquid (Ardet WB)
  Solid (Alkamide HTME; CPH-380-N; Ninol SNR; Schercopearl EA-100)
  Crystalline solid (Schercomid SME)
  Flake (Clindrol 200MS, LT14-8; Comperlan HS; Cyclomide S 280; Incromide SM; Loramine S 280)
  Granular (Monamid S)
  Wax (Schercomid SME-A, SME-M; Witcamide 70)
Color:
  Pale (Schercomid SME-A, SME-M)
  White (CPH-380-N; Loramine S 280)
  Off-white (Cyclomide S 280)

Cream (Alkamide HTME)

Tan (Clindrol 200MS, LT14-8; Monamid S; Schercomid SME)

Odor:

Ammoniacal (Clindrol 200MS)

Slight, ammoniacal (Schercomid SME, SME-A, SME-M)

Typical, slight (Loramine S 280)

Composition:

90% amide (Alkamide HTME)

100% active (Clindrol 200MS; Cyclomide S 280; Loramine S 280; Monamid S; Schercomid SME, SME-A, SME-M)

100% conc. (Ardet WB; Clindrol LT14-8; Comperlan HS; Incromide SM; Ninol SNR; Schercopearl EA-100)

Solubility:

Sol. in alcohols (Clindrol 200MS, LT14-8; Schercomid SME, SME-A, SME-M)

Sol. in aliphatic hydrocarbons (Schercomid SME, SME-A, SME-M)

Sol. in aromatic hydrocarbons (Clindrol 200MS; Schercomid SME, SME-A, SME-M); disp. (@ 10%) (Monamid S); insol. (@ 10%) (Alkamide HTME)

Sol. in chlorinated hydrocarbons (Schercomid SME, SME-A, SME-M); disp. (@ 10%) (Monamid S)

Sol. in esters (Clindrol 200MS)

Disp. in natural fats and oils (@ 10%) (Monamid S)

Partly sol. in glycerol (Schercomid SME, SME-A, SME-M)

Sol. in glycol ethers (Schercomid SME, SME-A, SME-M)

Sol. in glycols (Clindrol 200MS; Schercomid SME, SME-A, SME-M)

Disp. in kerosene (@ 10%) (Monamid S)

Sol. in ketones (Clindrol 200MS)

Disp. in white min. oil (@ 10%) (Monamid S); insol. in min. oil (@ 10%) (Alkamide HTME)

Disp. in min. spirits (@ 10%) (Monamid S); insol. (@ 10%) (Alkamide HTME)

Insol. in perchloroethylene (@ 10%) (Alkamide HTME)

Partly sol. in triols (Schercomid SME, SME-A, SME-M)

Disp. in water (Schercomid SME, SME-A, SME-M); disp. (@ 10%) (Monamid S); insol. (Clindrol LT14-8); insol. (@ 10%) (Alkamide HTME)

Ionic Nature:

Nonionic (Ardet WB; Clindrol LT14-8; Comperlan HS; Incromide SM; Loramine S 280; Monamid S; Ninol SNR; Schercomid SME, SME-A, SME-M; Schercopearl EA-100)

M.W.:

600 (CPH-380-N)

Sp.gr.:

0.998 (Clindrol 200MS)

Stearic monoethanolamide (cont'd.)

M.P.:
  91 ± 2 C (Schercomid SME-A)
  92 C (Loramine S 280)
  98 C (Clindrol 200MS)
  98 ± 2 C (Schercomid SME, SME-M)
Solidification Pt.:
  87 ± 2 C (Monamid S)
Flash Pt.:
  180 C (OC) (Schercomid SME, SME-M)
  > 180 C (OC) (Schercomid SME-A)
Acid No.:
  0–1 (Monamid S)
  2.0 max. (Schercomid SME)
  4.0 max. (Schercomid SME-M)
  5.0 max. (Schercomid SME-A)
Alkali No.:
  5–18 (Monamid S)
  14 max. (Schercomid SME-M)
  15 max. (Schercomid SME, SME-A)
pH:
  Alkaline (Clindrol 200MS; Loramine S 280)
  8–11 (1% DW) (Alkamide HTME)
  8.7 ± 0.5 (3% aq. disp.) (Schercomid SME-M)
  9.5–11.0 (10% sol'n.) (Monamid S)
Biodegradable: (Monamid S)
TOXICITY/HANDLING:
  Skin and eye irritant (Clindrol 200MS)
STD. PKGS.:
  47–gal (200 lb net) fiberboard drums (Monamid S)
  55-gal (200 lb net) open-head fiber drums (Clindrol 200MS)
  Paper sacks (Loramine S 280)

Stearic monoethanolamide stearate

SYNONYMS:
  Octadecanoic acid, 2-[(1-oxooctadecyl) amino] ethyl ester
  2-[(1-Oxooctadecyl) amino] octadecanoic acid, ethyl ester
  Stearamide MEA-stearate (CTFA)
EMPIRICAL FORMULA:
  $C_{38}H_{75}NO_3$

Stearic monoethanolamide stearate (cont'd.)

STRUCTURE:

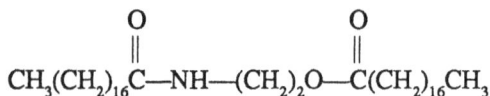

CAS No.:
14351-40-7

TRADENAME EQUIVALENTS:
Cerasynt D [Van Dyk]
Schercomid SME-S [Scher]
Witcamide MAS [Witco/Organics]

CATEGORY:
Emulsifier, thickener, pearling agent, opacifier, coating agent

APPLICATIONS:
Cosmetic industry preparations: (Schercomid SME-S); creams and lotions (Schercomid SME-S); shampoos (Cerasynt D; Schercomid SME-S; Witcamide MAS); shaving preparations (Cerasynt D); toiletries (Witcamide MAS)

Household detergents: liquid soaps (Witcamide MAS)

Industrial applications: aerosols (Cerasynt D); coatings/barriers (Witcamide MAS); mold release (Witcamide MAS); paper mfg. (Witcamide MAS); polishes and waxes (Witcamide MAS); textile/leather processing (Witcamide MAS); water-repellent compounds (Witcamide MAS)

Industrial cleaners: institutional cleaners (Witcamide MAS)

Pharmaceutical applications: (Schercomid SME-S)

PROPERTIES:

Form:
Flake (Cerasynt D)
Wax (Schercomid SME-S)
Waxy flakes (Witcamide MAS)

Color:
Light cream (Witcamide MAS)
Pale yellow (Schercomid SME-S)

Odor:
Mild, characteristic (Witcamide MAS)
Slight, typical (Schercomid SME-S)

Composition:
100% active (Cerasynt D; Schercomid SME-S)

Solubility:
Sol. in alcohols (Witcamide MAS)
Sol. in aliphatic hydrocarbons (Witcamide MAS)
Sol. in aromatic hydrocarbons (Witcamide MAS)
Sol. in chlorinated hydrocarbons (Witcamide MAS)
Sol. in esters (Witcamide MAS)

Stearic monoethanolamide stearate *(cont'd.)*

Insol. in 95% ethanol (Cerasynt D)
Gels in isopropyl myristate (Cerasynt D)
Sol. in ketones (Witcamide MAS)
Gels in min. oil (Cerasynt D)
Sol. warm in most hydrophobic organic solvents (Schercomid SME-S); disp. in most organic solvents (Schercomid SME-S)
Insol. in propylene glycol (Cerasynt D)
Disp. in water (Schercomid SME-S); insol. (Cerasynt D)
Ionic Nature:
Nonionic (Cerasynt D; Schercomid SME-S)
M.W.:
565 (Schercomid SME-S)
Sp.gr.:
0.844 (100/4 C) (Witcamide MAS)
M.P.:
77–82 C (Schercomid SME-S)
81 C (Witcamide MAS)
Flash Pt.:
> 180 C (OC) (Schercomid SME-S)
Acid No.:
5.0 (Witcamide MAS)
10 max. (Schercomid SME-S)
Alkali No.:
5 max. (Schercomid SME-S)
Iodine No.:
1.0 max. (Schercomid SME-S)
1.5 (Witcamide MAS)
TOXICITY/HANDLING:
Normal safety goggles and gloves should be worn during handling (Witcamide MAS)

Stearyl alcohol (CTFA)

SYNONYMS:
1-Octadecanol
Octadecyl alcohol
EMPIRICAL FORMULA:
$C_{18}H_{38}O$
STRUCTURE:
$CH_3(CH_2)_{16}CH_2OH$

Stearyl alcohol (cont'd.)

CAS No.:
112-92-5
TRADENAME EQUIVALENTS:
Adol 61NF, 62NF [Sherex] (NF grade)
Adol 63 [Sherex]
Adol 64 [Sherex] (tech. grade)
Alfol 18 Alcohol [Continental Oil]
Cyclogol Stearyl Alcohol [Cyclo]
Kalcohl 80 [Kao Corp.]
Lanol S [Seppic]
Stearal [Amerchol] (USP grade)
CATEGORY:
Emulsifier, detergent, intermediate, viscosity builder, texturizer, emollient
APPLICATIONS:
Cosmetic industry preparations: (Adol 61NF, 62NF; Cyclogol Stearyl Alcohol; Lanol
S); creams and lotions (Cyclogol Stearyl Alcohol)
Household detergents: detergent base (Adol 63)
Industrial applications: glass processing (Adol 62NF); wax formulations (Adol 61NF,
62NF)
PROPERTIES:
Form:
Solid (Kalcohl 80)
Crystalline solid (Adol 61NF)
Beads (Kalcohl 80)
Flake (Adol 61NF, 62NF; Cyclogol Stearyl Alcohol)
Wax (Lanol S)
Waxy solid (Alfol 18 Alcohol; Stearal)
Color:
White (Alfol 18 Alcohol; Stearal)
Bright white (Adol 61NF)
Off-white (Cyclogol Stearyl Alcohol)
Lovibond: 5Y/0.5R max. (5¹/‚˝) (Adol 62NF, 63, 64)
Odor:
Practically odorless (Adol 61NF, 62NF, 63, 64)
Mild, characteristic (Stearal)
Typical fatty alcohol (Alfol 18 Alcohol)
Composition:
68% C_{18} content (Adol 64)
93% C_{18} content (Adol 62NF)
97% C_{18} content (Adol 61NF)
98.7% active (Alfol 18 Alcohol)
100% conc. (Kalcohl 80; Lanol S)

Solubility:
  Sol. in acetone (g/100 g solvent): 12 g (Adol 63); 10 g (Adol 64); 7 g (Adol 61NF, 62NF)
  Sol. in ethyl ether (g/100 g solvent): 50 g (Adol 63); 40 g (Adol 64); 29 g (Adol 61NF, 62NF)
  Sol. in isopropanol (g/100 g solvent): 80 g (Adol 63); 66 g (Adol 64); 30 g (Adol 61NF, 62NF)
  Sol. in light min. oil (g/100 g solvent): 2 g (Adol 63); 1 g (Adol 61NF, 62NF, 64)
  Sol. in VMP naphtha (g/100 g solvent): 14 g (Adol 63); 8 g (Adol 64); 3 g (Adol 61NF, 62NF)
  Sol. in SDA-30 (g/100 g solvent): 47 g (Adol 63); 44 g (Adol 64); 27 g (Adol 61NF, 62NF)
Ionic Nature:
  Nonionic (Adol 61NF, 62NF; Cyclogol Stearyl Alcohol; Lanol S)
M.W.:
  268 (Adol 63)
  269 (Adol 64)
  272 (Adol 61NF, 62NF)
Sp.gr.:
  0.8075 (Alfol 18 Alcohol)
  0.816 (60/25 C) (Adol 63)
  0.817 (60/25 C) (Adol 61NF, 62NF, 64)
Density:
  6.71 lb/gal (Alfol 18 Alcohol)
Visc.:
  42 SSU (210 F) (Adol 62NF)
  44 SSU (210 F) (Adol 63, 64)
B.P.:
  312–344 C (760 mm, 90%) (Adol 63)
  337–360 C (760 mm, 90%) (Adol 62NF)
M.P.:
  55–60 C (Stearal)
  56–60 C (Cyclogol Stearyl Alcohol); (closed tube) (Adol 61NF, 62NF)
Solidification Pt.:
  48–53 C (Adol 63)
  51–54 C (Adol 64)
Acid No.:
  Nil (Cyclogol Stearyl Alcohol)
  0.5 max. (Adol 61NF)
  1.0 max. (Adol 62NF, 63, 64; Stearal)
Iodine No.:
  2.0 max. (Adol 61NF, 62NF, 63, 64; Stearal)

Stearyl alcohol *(cont'd.)*

Saponification No.:
   1.0 max. (Adol 61NF)
   2.0 max. (Stearal)
   3.0 max. (Adol 62NF, 63, 64)
Hydroxyl No.:
   200–212 (Adol 61NF, 62NF)
   200–220 (Stearal)
   204–216 (Adol 64)
   206–218 (Adol 63)
TOXICITY/HANDLING:
   Not considered hazardous (Adol 61NF, 62NF, 63, 64)
   Avoid inhaling vapors from the hot liquid; wash body parts contacting the alcohol with
     water (Alfol 18 Alcohol)
STORAGE/HANDLING:
   Should be kept as dry as possible; blanket the storage tank with inert gas (Adol 61NF,
     62NF, 63, 64)
   Combustible; treat with same precautions as high-boiling hydrocarbons; store in
     aluminum, SS, epoxy or heresite-lined steel, dry carbon steel (Alfol 18 Alcohol)
STD. PKGS.:
   50-lb multiwall paper bags, fiber drums (Adol 62NF, 63, 64)

Tall oil acid diethanolamide

SYNONYMS:
Amides, tall oil fatty, N,N-bis (2-hydroxyethyl)-
N,N-Bis (2-hydroxyethyl) tall oil acid amide
N,N-Bis (2-hydroxyethyl) tall oil fatty amides
Diethanolamine tall oil acid amide
Tallamide DEA (CTFA)

STRUCTURE:

O
‖
RC—N(CH₂CH₂OH)₂

where RCO⁻ is derived from tall oil

CAS No.:
68155-20-4

TRADENAME EQUIVALENTS:
Monamine T-100 [Mona] (1:2)
Schercomid SO-T, TO-2 [Scher]

CATEGORY:
Foam booster, foam stabilizer, emulsifier, thickener, visc. builder, detergent, wetting agent, corrosion inhibitor, lubricant, dispersant, conditioner

APPLICATIONS:
Bath products: bubble bath (Monamine T-100)
Cleansers: hand cleanser (Monamine T-100)
Cosmetic industry preparations: (Monamine T-100); conditioners (Monamine T-100; Schercomid TO-2); shampoos (Monamine T-100; Schercomid TO-2)
Farm products: agricultural oils/sprays (Monamine T-100)
Household detergents: carpet & upholstery shampoos (Monamine T-100); dishwashing (Monamine T-100)
Industrial applications: dyes and pigments (Monamine T-100; Schercomid TO-2); lubricating/cutting oils (Monamine T-100); metal polishes (Monamine T-100); petroleum industry (Monamine T-100); textile/leather processing (Monamine T-100)
Industrial cleaners: drycleaning compositions (Monamine T-100); metal processing surfactants (Monamine T-100); textile scouring (Monamine T-100)

PROPERTIES:
Form:
Liquid (Monamine T-100; Schercomid SO-T)
Clear liquid (Schercomid TO-2)

Tall oil acid diethanolamide *(cont'd.)*

Color:
Dark amber (Schercomid TO-2)
Gvcs-33 10 max. (Monamine T-100)

Odor:
Mild, characteristic (Schercomid TO-2)

Composition:
100% active (Monamine T-100; Schercomid SO-T, TO-2)

Solubility:
Sol. in alcohols (Schercomid TO-2)
Sol. in aliphatic hydrocarbons (Schercomid TO-2)
Sol. @ 10% in aromatic hydrocarbons (Monamine T-100); partly sol. (Schercomid TO-2)
Sol. in chlorinated hydrocarbons (Schercomid TO-2); sol. @ 10% (Monamine T-100)
Sol. @ 10% in ethanol (Monamine T-100)
Sol. in glycol ethers (Schercomid TO-2)
Sol. in glycols (Schercomid TO-2)
Sol. @ 10% in kerosene (Monamine T-100)
Sol. @ 10% in min. spirits (Monamine T-100)
Sol. in most organic solvents (Schercomid TO-2)
Disp. in water (Schercomid TO-2); disp. @10% (Monamine T-100)

Ionic Nature:
Nonionic/anionic (Monamine T-100; Schercomid SO-T, TO-2)

Sp.gr.:
0.97 (20 C) (Monamine T-100)
0.990 (Schercomid TO-2)

Density:
8.10 lb/gal (Monamine T-100)
8.25 lb/gal (Schercomid TO-2)

Congeal Pt.:
< −5 C (Schercomid TO-2)

Flash Pt.:
> 170 C (OC) (Schercomid TO-2)

Acid No.:
10–16 (Monamine T-100)
13–19 (Schercomid TO-2)

Alkali No.:
100–120 (Monamine T-100)
120–150 (Schercomid TO-2)

Stability:
Very stable in neutral or moderately alkaline or acid systems; subject to hydrolysis in high conc. of min. acids and caustic soda or potash sol'ns. (Monamine T-100)

Tall oil acid diethanolamide *(cont'd.)*

Storage Stability:
1 yr min. shelf life; alkali value decreases slightly with aging which does not affect performance (Schercomid TO-2)

pH:
10.0–11.0 (10% sol'n.) (Monamine T-100)

Biodegradable: (Monamine T-100)

STD. PKGS.:
55-gal (450 lb net) tighthead steel drums (Monamine T-100)
55-gal (450 lb net) bung-head steel drums (Schercomid TO-2)

Tall oil hydroxyethyl imidazoline *(CTFA)*

SYNONYMS:
4,5-Dihydro-7-nortall oil-1H-imidazole-1-ethanol
1H-Imidazole-1-ethanol, 4,5-dihydro-, 2-nortall oil
2-Nortall oil-1H-imidazole-1-ethanol, 4,5-dihydro-

STRUCTURE:

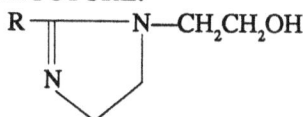

where R is derived from tall oil

CAS No.:
61791-39-7

TRADENAME EQUIVALENTS:
Miramine TOC [Miranol]
Monazoline T [Mona]

CATEGORY:
Thickener, wetting agent, emulsifier, detergent, dispersant, corrosion inhibitor, antistat, softener, bactericide, rinse aid, inhibitor, lubricant

APPLICATIONS:
Automobile cleaners: automatic car wash (Monazoline T)
Farm products: agricultural oils/sprays (Miramine TOC; Monazoline T); fungicides/herbicides (Monazoline T)
Industrial applications: adhesives/sealants (Monazoline T); asphalt (Miramine TOC); dyes, clays, pigments (Monazoline T); flotation (Monazoline T); lubricating/cutting oils (Miramine TOC); metalworking lubricants (Miramine TOC); paint mfg. (Monazoline T); petroleum industry (Miramine TOC; Monazoline T); plastics

Tall oil hydroxyethyl imidazoline *(cont'd.)*

(Monazoline T); printing inks (Monazoline T); protective metal coatings (Monazoline T); road construction (Monazoline T); textile/leather processing (Miramine TOC; Monazoline T)

PROPERTIES:

Form:
Liquid; may crystallize on aging (Monazoline T)
Solid (Miramine TOC)

Color:
Amber (Monazoline T)

Composition:
90% imidazoline min. (Monazoline T)
100% conc. (Miramine TOC)

Solubility:
Sol. in chlorinated hydrocarbons (@ 10%) (Monazoline T)
Sol. in ethanol (@ 10%) (Monazoline T)
Sol. in most hydrocarbon solvents (Monazoline T)
Sol. in kerosene (@ 10%) (Monazoline T)
Sol. in min. oil (@ 10%) (Monazoline T)
Sol. in min. spirits (@ 10%) (Monazoline T)
Sol. in most polar solvents (Monazoline T)
Sol. in toluene (@ 10%) (Monazoline T)
Sol. in veg. oil (@ 10%) (Monazoline T)
Disp. in water (@ 10%) (Monazoline T)

Ionic Nature:
Cationic (Miramine TOC; Monazoline T)

M.W.:
350 (Monazoline T)

Sp.gr.:
0.93 (Monazoline T)

Density:
7.75 lb/gal (Monazoline T)

Acid No.:
1.0 max. (Monazoline T)

pH:
10.0–11.5 (10% disp.) (Monazoline T)

TOXICITY/HANDLING:
Fairly strong organic bases—handle with caution; wear protective goggles and gloves (Monazoline T)

STD. PKGS.:
55-gal closed-head steel drums (Monazoline T)

SYNONYMS:
2H-1-Benzopyran-6-ol, 3,4-dihydro-2,5,7,8-tetramethyl-2-(4,8,12-trimethyltridecyl)-
3,4-Dihydro-2,5,7,8-tetramethyl-2-(4,8,12-trimethyltridecyl)-2H-1-benzopyran-6-ol
Mixed tocopherols
D-alpha tocopherol
DL-alpha tocopherol
Vitamin E

EMPIRICAL FORMULA:
$C_{29}H_{50}O_2$

STRUCTURE:

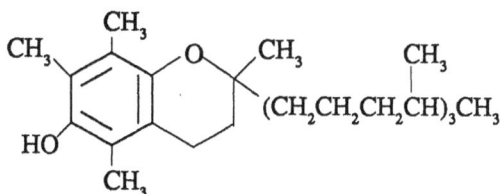

CAS No.:
59-02-9 (*d*-alpha)
10191-41-0 (*dl*-alpha)

TRADENAME EQUIVALENTS:
Covi-Ox T-50, T-70 [Henkel/Henkel Canada]
Covitol F-350M, F-600, F-1000 [Henkel] (pharmaceutical grade)
Generically sold by:
[BASF; Henkel; Henkel Canada; Hoffmann-La Roche]

CATEGORY:
Antioxidant, blocking agent, protectant

APPLICATIONS:
Cosmetic industry preparations: (Covi-Ox T-50); creams and lotions (Covi-Ox T-50); makeup (Covi-Ox T-50); perfumery (Covi-Ox T-50); shampoos (Covi-Ox T-50)
Food applications: (Covi-Ox T-70)
Pharmaceutical applications: (Covitol F-350M, F-600, F-1000); lip balms (Covi-Ox T-50)

PROPERTIES:
Form:
Viscous oil (Covi-Ox T-50)
Clear viscous oil (Covitol F-1000)
Oil (Covitol F-600, F-1000)
Powder (Covitol F-350M)
Color:
Cream (Covitol F-350M)

Tocopherol *(cont'd.)*

Clear yellow (Covi-Ox T-70)
Brownish-red (Covi-Ox T-50; Covitol F-1000)
Odor:
Characteristic (Covi-Ox T-50)
Mild (Covitol F-1000)
Taste:
Characteristic (Covi-Ox T-50)
Mild (Covitol F-1000)
Composition:
294 mg/g total tocopherols (Covitol F-350M)
403 mg/g total tocopherols (Covitol F-600)
500 mg/g total tocopherols (Covi-Ox T-50)
671 mg/g total tocopherols (Covitol F-1000)
700 mg/g total tocopherols (Covi-Ox T-70)
Solubility:
Sol. in oil (Covi-Ox T-50, T-70)
B.P.:
190–220 C (0.1 mm Hg) (Covi-Ox T-50)
Stability:
Unstable to air and light (Covitol F-350M)
Unstable to air and light esp. when in alkaline media (Covitol F-1000)
STORAGE/HANDLING:
Store in a cool, dry place protected from moisture (Covi-Ox T-50)
Store in a cool, dark environment for a minimum of time; prevent freezing (Covitol F-350M)
STD. PKGS.:
10- and 190-kilo metal drums (Covi-Ox T-50)
10- and 20-kilo fiber drums (Covitol F-350M)
20- and 190-kilo metal drums (Covitol F-600, F-1000)

Tribasic lead sulfate

STRUCTURE:
$3PbO \cdot PbSO_4 \cdot H_2O$
TRADENAME EQUIVALENTS:
Epistatic 100 [Eagle Picher]
Epistatic 103 [Eagle Picher] (surface treated)
Halbase 10 [Halstab]
Halbase 10-EP [Halstab] (coated)

Tribasic lead sulfate *(cont'd.)*

TRADENAME EQUIVALENTS *(cont'd.):*
   Tribase [Associated Lead]
   Tribase XL [Associated Lead] (coated)
   Tribase E Special, EXL Special [Associated Lead] (modified)
CATEGORY:
   Heat stabilizer, activator
APPLICATIONS:
   Industrial applications: electrical applications (Halbase 10, 10-EP; Tribase, E Special, EXL Special, XL); foam applications (Epistatic 100; Halbase 10, 10-EP; Tribase XL); phonograph records (Epistatic 100); pipe (Tribase XL); plastics/plastisols (Epistatic 100, 103; Halbase 10, 10-EP; Tribase, XL); profiles (Halbase 10, 10-EP; Tribase XL)
PROPERTIES:
Form:
   Fine powder (Tribase, XL)
Color:
   White (Tribase, XL)
Composition:
   67% PbO (Tribase E Special, EXL Special)
   74.5% lead (Epistatic 103)
   80.9% Pbo (Tribase XL)
   89.1% PbO (Tribase)
Sp.gr.:
   4.0 (Tribase EXL Special)
   4.4 (Tribase E Special)
   4.8 (Epistatic 103)
Fineness:
   99.9% −325 mesh (Epistatic 103)
Surface Area:
   2.7 m2/g (Epistatic 103)

Tributyl citrate *(CTFA)*

SYNONYMS:
   2-Hydroxy-1,2,3-propanetricarboxylic acid, tributyl ester
   1,2,3-Propanetricarboxylic acid, 2-hydroxy-, tributyl ester
EMPIRICAL FORMULA:
   $C_{18}H_{32}O_7$

Tributyl citrate (cont'd.)

STRUCTURE:

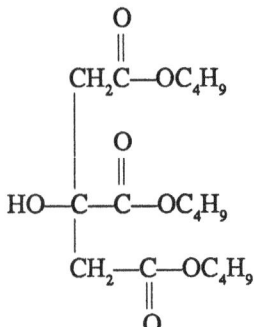

CAS No.:
77-94-1
TRADENAME EQUIVALENTS:
Citroflex 4 [Morflex]
TBC [Croda]
CATEGORY:
Plasticizer, defoaming agent
APPLICATIONS:
Industrial applications: cellulosics (Citroflex 4; TBC); plastics (Citroflex 4; TBC)
PROPERTIES:
Form:
Liquid (TBC)
TOXICITY/HANDLING:
Very low toxicity (Citroflex 4)

Tricresyl phosphate (CTFA)

SYNONYMS:
Phosphoric acid, tris (methylphenyl) ester
TCP
EMPIRICAL FORMULA:
$C_{21}H_{21}O_4P$
STRUCTURE:

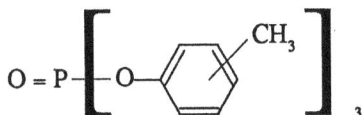

CAS No.:
1330-78-5
TRADENAME EQUIVALENTS:
Disflamoll TKP [Mobay]
Drimix [Kenrich Petrochemicals]
Kronitex TCP [FMC Corp.]
Lindol [Harwick]
Phosflex 179C [Harwick]
CATEGORY:
Flame retardant, plasticizer, processing aid, softener
APPLICATIONS:
Industrial applications: cellulosics (Phosflex 179C); coatings/lacquers (Disflamoll
TKP; Kronitex TCP); injection molded goods (Disflamoll TKP); plastics/plastisols
(Disflamoll TKP; Kronitex TCP; Lindol; Phosflex 179C); rubber (Kronitex TCP);
sheeting (Kronitex TCP); thermoplastic resins (Lindol)
PROPERTIES:
Form:
Liquid (Disflamoll TKP)
Clear liquid (Kronitex TCP)
Clear transparent liquid (Lindol)
Free-flowing powder (Drimix)
Color:
Colorless to slightly yellowish (Disflamoll TKP)
APHA 50 max. (Lindol)
Pt-Co 75 max. (Kronitex TCP)
Odor:
Very slight (Kronitex TCP)
Characteristic (Lindol)
Composition:
8.4% phosporus (Kronitex TCP)
75% active (Drimix)
Solubility:
Limited sol. in aliphatic hydrocarbons (Disflamoll TKP)
Limited sol. in min. oil (Disflamoll TKP)
Sol. in organic solvents (Disflamoll TKP)
Insol. in water (Disflamoll TKP)
M.W.:
368 (Lindol)
370 (Kronitex TCP)
Sp.gr.:
1.160–1.175 (20/20 C) (Kronitex TCP)
1.165 ± 0.005 (Phosflex 179C)

1.167 ± 0.005 (Lindol)
1.34 (Drimix)
Density:
1.175–1.185 g/cm³ (Disflamoll TKP)
1162 kg/m³ (20 C) (Kronitex TCP)
Visc.:
65–75 mPa•s (20 C) (Disflamoll TKP)
80 cp (20 C) (Kronitex TCP)
140 SUS (100 F) (Lindol)
B.P.:
240–250 C (Disflamoll TKP)
241–255 C (4 mm Hg) (Kronitex TCP)
509 F (10 mm) (Lindol)
Pour Pt.:
–28 C (Kronitex TCP)
Flash Pt.:
≥ 230 C (OC) (Disflamoll TKP)
252 C (COC) (Kronitex TCP)
465 F (Lindol)
Fire Pt.:
338 C (COC) (Kronitex TCP)
Acid No.:
≤ 0.2 (Disflamoll TKP)
Ref. Index:
1.554 (Kronitex TCP)
1.558–1.560 (20 C) (Disflamoll TKP)
TOXICITY/HANDLING:
Avoid contact with skin, eyes, and amucous membranes; use with adequate ventilation (Disflamoll TKP)
Avoid prolonged exposure; good ventilation recommended (Lindol)
STORAGE/HANDLING:
Storage life of at least 2 yrs. when kept in a cool place in sealed containers (Disflamoll TKP)
STD. PKGS.:
Drum (T/L, C/L, LTL) and bulk (T/T, T/C) (Lindol)
Drum, tankcar, TL, LTL, CL (Phosflex 179C)

Triethyl citrate (CTFA)

SYNONYMS:
2-Hydroxy-1,2,3-propanetricarboxylic acid, triethyl ester
1,2,3-Propanetricarboxylic acid, 2-hydroxy-, triethyl ester
EMPIRICAL FORMULA:
$C_{12}H_{20}O_7$
STRUCTURE:

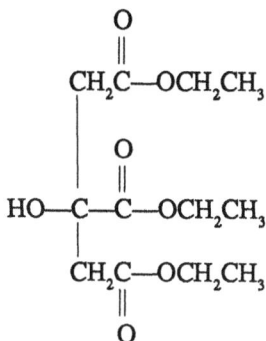

CAS No.:
77-93-0
TRADENAME EQUIVALENTS:
Citroflex 2 [Morflex]
CATEGORY:
Plasticizer, solvent
APPLICATIONS:
Industrial applications: cellulosics; lacquers; natural resins; plastics
PROPERTIES:
Solubility:
Limited oil solubility

Triisooctyl trimellitate

SYNONYMS:
TIOTM
STRUCTURE:
$C_6H_3(COOC_8H_{17})_3$
TRADENAME EQUIVALENTS:
Jayflex 80-TM [Exxon]
Plasthall TIOTM [C.P. Hall]
Staflex TIOTM [Reichhold]
    Generically sold by: [C.P. Hall]

Triisooctyl trimellitate (cont'd.)

CATEGORY:
Plasticizer, softener
APPLICATIONS:
Automotive applications: (Jayflex 80-TM)
Industrial applications: electrical insulation (generic; Jayflex 80-TM; Staflex TIOTM); film and sheeting (Jayflex 80-TM); plastics (generic; Jayflex 80-TM; Staflex TIOTM)
PROPERTIES:
Form:
Liquid (generic)
Clear liquid (Plasthall TIOTM)
Color:
APHA 100 (Staflex TIOTM)
Gardner 1–2 max. (generic)
Odor:
Mild (generic)
M.W.:
546 (Staflex TIOTM)
547 (Jayflex 80-TM)
550 (generic; Plasthall TIOTM)
Sp.gr.:
0.986 (23 C) (Staflex TIOTM)
Density:
8.2 lb/gal (Staflex TIOTM)
Visc.:
270 cps (generic)
1400 cSt (kinematic, 0 C) (Staflex TIOTM)
B.P.:
290 C (5 mm) (generic)
Pour Pt.:
–50 C (generic)
–40 F (Staflex TIOTM)
Flash Pt.:
490 F (Staflex TIOTM)
Cloud Pt.:
–40 F (Staflex TIOTM)
Ref. Index:
1.4830 (23 C) (Staflex TIOTM)

SYNONYMS:
Tri-2-ethylhexyl trimellitate
TOTM

STRUCTURE:
$C_6H_3(COOC_8H_{17})_3$

TRADENAME EQUIVALENTS:
Hatcol TOTM [Hatco]
Hexaflex TOTM [Hexagon]
Kodaflex TOTM [Eastman]
Palatinol TOTM [Badische]
Palatinol TOTM-E [Badische] (with 0.5% Bisphenol A)
Plasthall TOTM [C.P. Hall]
Polycizer TOTM [Harwick]
PX-338 [USS Chem.]
Staflex TOTM [Reichhold]
Uniflex TOTM [Union Camp]
   Generically sold by: [Ashland]

CATEGORY:
Plasticizer, softener

APPLICATIONS:
Automotive applications: (Hexaflex TOTM; Palatinol TOTM)
Industrial applications: coated fabrics (Hexaflex TOTM; Kodaflex TOTM); electrical insulation (Hatcol TOTM; Hexaflex TOTM; Kodaflex TOTM; Palatinol TOTM, TOTM-E; PX-338; Staflex TOTM); film and sheeting (Hatcol TOTM; Hexaflex TOTM; Kodaflex TOTM); lubricants (Uniflex TOTM); plastics (Hatcol TOTM; Hexaflex TOTM; Palatinol TOTM; Plasthall TOTM; Polycizer TOTM; PX-338; Staflex TOTM; Uniflex TOTM); rubber (Kodaflex TOTM; Plasthall TOTM; Polycizer TOTM; Uniflex TOTM)

PROPERTIES:
Form:
Liquid (Hatcol TOTM; Kodaflex TOTM; Polycizer TOTM)
Clear liquid (Plasthall TOTM)
Color:
Colorless (Polycizer TOTM)
APHA 75 max. (Hatcol TOTM)
APHA 100 (Staflex TOTM; Uniflex TOTM)
APHA 150 max. (Palatinol TOTM)
Pt-Co 100 ppm (Kodaflex TOTM)
Odor:
Mild (Hatcol TOTM)
Mild, characteristic (Palatinol TOTM)

343

Trioctyl trimellitate *(cont'd.)*

Composition:
99.0% min. ester content (Kodaflex TOTM; Palatinol TOTM)
Solubility:
Sol. 0.006 g/l @ 20 C in water (Kodaflex TOTM); < 0.01% @ 20 C (Palatinol TOTM)
M.W.:
546 (Hatcol TOTM; PX-338; Staflex TOTM)
547 (Hexaflex TOTM; Kodaflex TOTM)
550 (Plasthall TOTM)
Sp.gr.:
0.986–0.994 (20/20 C) (Palatinol TOTM)
0.987 (generic—Ashland); (23 C) (Staflex TOTM); 0.987 ± 0.003 (Hatcol TOTM)
0.989 (Polycizer TOTM); (20/20 C) (Kodaflex TOTM)
0.99 (Plasthall TOTM); (20/20 C) (Uniflex TOTM)
Density:
0.984 kg/l (20 C) (Kodaflex TOTM)
8.2 lb/gal (Staflex TOTM)
Visc.:
194 cP (Kodaflex TOTM)
260 cps (20 C) (Palatinol TOTM)
300 cps (Hatcol TOTM)
215 cSt (Uniflex TOTM)
1800 cSt (kinematic, 0 C) (Staflex TOTM)
F.P.:
–38 C (Kodaflex TOTM)
B.P.:
283 C (3 mm Hg) (Palatinol TOTM)
287 C (5 mm) (Hatcol TOTM)
414 C (760 mm) (Kodaflex TOTM)
Pour Pt.:
–46 C (Palatinol TOTM)
–45 C (Uniflex TOTM)
–50 F (Hatcol TOTM)
–30 F (Staflex TOTM)
Flash Pt.:
250 C (COC) (Palatinol TOTM)
260 C (COC) (Uniflex TOTM)
263 C (COC) (Kodaflex TOTM)
> 500 F (Staflex TOTM)
505 F (Polycizer TOTM)
Fire Pt.:
285 C (COC) (Uniflex TOTM)
297 C (COC) (Kodaflex TOTM)

Cloud Pt.:
  −30 F (Staflex TOTM)
Acid No.:
  < 0.1 (Uniflex TOTM)
  0.2 max. (Palatinol TOTM)
Storage Stability:
  Indefinite storage life (Palatinol TOTM)
Ref. Index:
  1.4832 (Kodaflex TOTM; Polycizer TOTM)
  1.4838 (23 C) (Staflex TOTM)
  1.485 (Uniflex TOTM)
Dissipation Factor:
  0.7×10^{-2} (1 kHz) (Kodaflex TOTM)
Dielectric Constant:
  4.9 (1 kHz) (Kodaflex TOTM)
Volume Resistivity:
  5×10^{11} ohm-cm (Kodaflex TOTM)
TOXICITY/HANDLING:
  Appropriate ventilation may be necessary at operations with elevated temps. or where mists or aerosols are encountered (Kodaflex TOTM)
  Avoid repeated/prolonged skin and eye contact; avoid breathing vapors; use with adequate ventilation (Palatinol TOTM)
STORAGE/HANDLING:
  The atmosphere above TOTM in the tank should be moisture-free (Palatinol TOTM)
STD. PKGS.:
  Available bulk (Palatinol TOTM)

Triphenyl phosphite

STRUCTURE:
  $(C_6H_5O)_3P$
CAS No.:
  101-02-0
TRADENAME EQUIVALENTS:
  Mark TPP [Argus]
  Weston TPP [Borg Warner]
CATEGORY:
  Stabilizer

Triphenyl phosphite (cont'd.)

APPLICATIONS:
Industrial applications: adhesives (Weston TPP); fiber applications (Weston TPP); film applications (Weston TPP); plastics (Weston TPP); rubber (Weston TPP)
PROPERTIES:
Form:
Liquid (Mark TPP)
Clear liquid (Weston TPP)
Color:
Water-white (Mark TPP)
APHA 50 max. (Weston TPP)
Composition:
9.5% phosporus (Mark TPP)
10.0% phosphorus (Weston TPP)
M.W.:
310 (Mark TPP; Weston TPP)
Sp.gr.:
1.172–1.186 (Mark TPP)
1.180–1.186 (25/15.5 C) (Weston TPP)
Density:
9.86 lb/gal (Mark TPP)
1.18 g/ml (Weston TPP)
Visc.:
12.0 cps (38 C) (Weston TPP)
M.P.:
25 C (Mark TPP)
Flash Pt.:
146 C (PM) (Weston TPP)
155 C (COC) (Mark TPP)
Acid No.:
0.50 max. (Weston TPP)
Ref. Index:
1.5825–1.5900 (Mark TPP)
1.5880–1.5900 (Weston TPP)
TOXICITY/HANDLING:
May cause irritation; avoid contact with eyes, skin, and clothing (Mark TPP)
Avoid contact with skin and exposure to vapors; use with protective gloves and glasses in adequately ventilated areas (Weston TPP)
STORAGE/HANDLING:
Store in tightly closed drums out of contact with moist air and oxidizing agents; tends to supercool and may solidify at R.T.—preheat to moderate elevated temps. to facilitate handling (Mark TPP

Tris (nonylphenyl) phosphite (CTFA)

SYNONYMS:
Nonylphenyl phosphite (3:1)
Phenol, nonyl-, phosphite (3:1)
EMPIRICAL FORMULA:
$C_{45}H_{69}O_3P$
STRUCTURE:
$[C_9H_{19}C_6H_4]_3$—OPO_2H
CAS No.:
26523-78-4
TRADENAME EQUIVALENTS:
Doverbros #4 [Dover]
Mark 1178, 1178B, TNPP [Argus]
Polygard [Uniroyal]
Stave TNPP [Stave]
Weston 399, 399B, TNPP [Borg Warner]
Wytox 312 [Olin]
CATEGORY:
Antioxidant, stabilizer, chelating agent
APPLICATIONS:
Food applications: food packaging (Mark 1178, 1178B; Polygard; Wytox 312)
Industrial applications: adhesives (Polygard; Weston 399, 399B, TNPP); cellulosics (Mark 1178, 1178B); latexes (Polygard); plastics (Doverbros #4; Mark 1178, 1178B; Weston 399, 399B, TNPP; Wytox 312); rubber (Polygard; Stave TNPP; Weston 399, 399B, TNPP)
PROPERTIES:
Form:
Liquid (Mark TNPP; Wytox 312)
Clear liquid (Polygard; Stave TNPP; Weston 399, 399B, TNPP)
Slightly viscous liquid (Mark 1178, 1178B)
Color:
Low (Wytox 312)
Pale yellow (Mark 1178, 1178B, TNPP)
Light amber (Polygard)
Light straw (Stave TNPP)
Gardner 2 max. (Weston TNPP)
Gardner 3 max. (Weston 399, 399B)
Odor:
Low (Wytox 312)
Composition:
0.75% triisopropanolamine (Weston 399)
1.0% triisopropanolamine (Weston 399B)
4.3% phosphorus (Mark TNPP; Weston 399, 399B, TNPP)

Tris (nonylphenyl) phosphite (cont'd.)

Solubility:
Sol. in amyl alcohol (Wytox 312)
Sol. in kerosene (Wytox 312)
Sol. in MEK (Wytox 312)

M.W.:
688 (Weston 399, 399B, TNPP)
715 (Mark TNPP)

Sp.gr.:
0.972 (Mark 1178B)
0.980–0.992 (25/15.5 C) (Weston 399, 399B, TNPP)
0.980–1.000 (Mark TNPP)
0.985 (Mark 1178)
0.985 ± 0.005 (Stave TNPP)
0.99 (Polygard)

Density:
8.25 lb/gal (Mark TNPP)
0.98 g/ml (Weston 399, 399B, TNPP)

Visc.:
250 cps (60 C) (Weston 399, 399B, TNPP)

Flash Pt.:
185 C (COC) (Mark TNPP)
186 C (COC) (Mark 1178B)
196 C (PM) (Weston 399, 399B)
207 C (PM) (Weston TNPP)
210 C (COC) (Mark 1178)
360 F (Stave TNPP)

Acid No.:
0.10 max. (Weston TNPP)
0.50 max. (Weston 399, 399B)

Storage Stability:
Indefinite storage life when kept in closed containers (Mark 1178, 1178B)
Good storage stability (Polygard)

Ref. Index:
1.5150 (Mark 1178B)
1.522–1.1526 (Mark TNPP)
1.5250 (Mark 1178)
1.5250–1.5280 (Weston 399, 399B)
1.5255–1.5280 (Weston TNPP)
1.526–1.528 (Stave TNPP)

TOXICITY/HANDLING:
May cause irritation—avoid contact with eyes, skin, and clothing (Mark 1178, 1178B, TNPP)

Tris (nonylphenyl) phosphite (cont'd.)

Avoid contact with skin; use protective gloves and glasses in adequately ventilated areas (Weston 399, 399B, TNPP)

STORAGE/HANDLING:

Store in closed drums (Mark 1178, 1178B)

Store in tightly closed drums, out of contact with moist air and oxidizing agents (Mark TNPP)

Avoid exposure to vapors (Weston 399, 399B, TNPP)

Keep container closed to prevent contamination by moisture (Polygard)

STD. PKGS.:

425 lb drum, bulk (Weston 399, TNPP)

Xanthan gum (CTFA)

SYNONYMS:
Corn sugar gum
Xanthan
CAS No.:
11138-66-2
TRADENAME EQUIVALENTS:
Biozan [Hercules]
Biozan SPX 5423 [Hercules] (industrial grade)
Keltrol [Kelco] (food grade)
Keltrol F [Kelco] (food grade)
Kelzan, D, M, XC Polymer [Kelco] (industrial grade)
Kelzan D 35, S [Kelco]
Rhodigel 23 [R.T. Vanderbilt]
Rhodopol [R.T. Vanderbilt]
Xanflood [Kelco] (industrial grade)
CATEGORY:
Thickener, emulsifier, suspending agent, stabilizer, gelling agent, rheology modifier, lubricant, flocculant
APPLICATIONS:
Cosmetic industry preparations: (Biozan; Rhodigel 23); creams and lotions (Keltrol)
Farm products: fertilizers (Kelzan, D, M, XC Polymer); herbicides (Kelzan, D, M, XC Polymer)
Food applications: food additives (Biozan; Keltrol, F)
Industrial applications: (Biozan SPX 5423; Rhodopol); abrasives (Kelzan, D, M, XC Polymer); adhesives (Kelzan, D, M, XC Polymer); ceramics (Kelzan, D, M, XC Polymer); dyes and pigments (Kelzan, D, M, XC Polymer); explosives (Biozan); foundry (Biozan) ; latex applications (Kelzan); mining (Kelzan, D, M, XC Polymer); paint mfg. (Kelzan, D, M, XC Polymer); paper mfg. (Kelzan, D, M, XC Polymer); petroleum industry (Biozan; Kelzan, D, D 35, M, XC Polymer; Xanflood); polishes (Kelzan, D, M, XC Polymer); printing inks (Kelzan, D, M, XC Polymer); textile/leather processing (Kelzan, D, M, XC Polymer)
Industrial cleaners: (Kelzan, D, M, XC Polymer); acid and caustic cleaners (Biozan)
Pharmaceutical applications: (Biozan; Rhodigel 23); dental preparations (Keltrol)
PROPERTIES:
Form:
Powder (Keltrol, F; Kelzan, D, M, XC Polymer; Xanflood)
Free-flowing powder (Biozan SPX 5423)

350

Color:
   Cream (Keltrol, F; Kelzan, D, M, XC Polymer; Xanflood)
   Tan (Biozan SPX 5423)
Solubility:
   Full disperses in neutral to acidic pH sol'ns.; tolerates up to 50% alcohols (Kelzan D 35)
   Fully disperses in water sol'ns. where pH is neutral to acidic; tolerates up to 50% alcohols (Kelzan S)
   Swells in glycerin (Keltrol)
   Insol. in most organic solvents (Keltrol)
   Swells in propylene glycol (Keltrol)
   Sol. in water (Rhodigel 23; Rhodopol); sol. in cold and hot water (Keltrol; Kelzan; Xanflood)
Sp.gr.:
   1.5 (Keltrol, F)
   1.6 (Kelzan, D, M, XC Polymer; Xanflood)
Bulk Density:
   52.2 lb/ft³ (Keltrol, F)
   52.4 lb/ft³ (Kelzan, D, M, XC Polymer; Xanflood)
Fineness:
   40 mesh size (Kelzan, D, M, XC Polymer; Xanflood)
   80 mesh size (Keltrol)
   200 mesh size (Keltrol F)
Visc.:
   850 cps (Brookfield LVF, 60 rpm) (Kelzan, D, M, XC Polymer; Xanflood)
   1300–1600 cps (Brookfield #3 spindle, 30 rpm, 1% sol'n.) (Biozan SPX 5423)
   1400 cps (Brookfield, LVF, 60 rpm, 1% visc. with 1% electrolyte added) (Keltrol, F)
F.P.:
   0 C (1% sol'n. in dist. water) (Keltrol, F; Kelzan, D, M, XC Polymer; Xanflood)
Stability:
   Tolerant for extremes of pH, temp., and dissolved salt content (Biozan)
   Excellent pH and temp. stability; stable in presence of a wide range of acids, bases, salts, surfactants, and other thickeners (Biozan SPX 5423)
   Stable to temps., pH and salt; sol'ns. will tolerate 30–40% ethyl and propyl alcohols (Keltrol)
   Freeze-thaw stable (Kelzan)
   Stable to pH, temp., salts, acids, alkalies, and shear (Kelzan, D, M, XC Polymer; Xanflood)
   Stable to temps. and high concs. of salt (Kelzan S)
Ref. Index:
   1.3332 (20 C) (Kelzan, D, M, XC Polymer; Xanflood)
   1.3338 (20 C) (Keltrol, F)

351

Xanthan gum *(cont'd.)*

pH:
  7.0 (Keltrol, F; Kelzan, D, M, XC Polymer; Xanflood)
Surface Tension:
  75 dynes/cm (Keltrol, F; Kelzan, D, M, XC Polymer; Xanflood)
STD. PKGS.:
  50-kg cartons with polyethylene liners (Biozan SPX 5423)

Zinc oxide

SYNONYMS:
C1 77947
Germany C-Weiss 8
EMPIRICAL FORMULA:
OZn
STRUCTURE:
ZnO
CAS No.:
1314-13-2
TRADENAME EQUIVALENTS:
Actox-14 [NJ Zinc] (Amer. process)
AZO-55, -66, -77 [Amer. Smelting & Refining]
AZO-55TT, -66TT, -77TT [Amer. Smelting & Refining] (surface-treated with zinc
  propionate)
Azodox-55, -55TT [Amer. Smelting & Refining] (deaerated)
Florence Green Seal-8 [N.J. Zinc] (French process)
Horse Head USP-20, XX-78, XX-85 [N.J. Zinc] (French process)
Horse Head XX-4, XX-32, XX-503, XX-504, XX-600, XX-601, XX-631 [N.J. Zinc]
  (Amer. process)
Kadox-15, -25, -72, -215, -272, -515 [N.J. Zinc] (French process)
Ottalume 2100 [Ferro] (fluorescent)
Photox-7, -8, -77, -80, -85, -90, -801 [N.J. Zinc] (French process)
Protox-169 [N.J. Zinc] (surface treated, with propionic acid)
Zinc Oxide 35 [Akron] (precipitated)
Zinc Oxide 500, 911 [C.P. Hall] (French process)
Zinc Oxide 500-36, 911-36 [C.P. Hall] (French process, surface treated)
Zinox-Grade 350 [Amer. Chemet] (precipitated)
Zoco 100, 103, 104 [Zochey] (French process)
Zoco 172 [Zochey] (French process, surface treated)
  Generically sold by:
    [Aceto; Asarco; R.E. Carroll; G.H. Chem.; H.M. Royal; N.J. Zinc; Royce; Sipi
    Metals; Smith Chem.; Summit Chem.]
CATEGORY:
Activator, accelerator, pigment, reinforcing agent, stabilizer, UV stabilizer, opacifier,
  UV absorber, crosslinking agent
APPLICATIONS:
Automotive applications: lubricants (Horse Head XX-4)

Zinc oxide *(cont'd.)*

Cosmetic industry preparations: (Horse Head USP-20); makeup (Horse Head USP-20)

Food applications: food can mfg. (Kadox-25; Zoco 100)

Industrial applications: (AZO-55); abrasive sheets (Horse Head XX-4, XX-32); adhesives (Horse Head XX-4, XX-32, XX-32; Kadox-25; Zinc Oxide 35); batteries (Horse Head USP-20); ceramics (AZO-66; Florence Green Seal-8; Horse Head XX-32, XX-32, XX-503, XX-504; Kadox-72, -272); coatings (Kadox-25); elastomers (Zinc Oxide 35); flooring (Horse Head XX-32); footwear (Kadox-72, -272); glass processing (Horse Head XX-32); latexes (AZO-66, -77; Horse Head XX-32; Kadox-15, Kadox-72, -215, -272; Protox-169; Zinc Oxide 35; Zoco 100, 103, 104, 172); lubricating/cutting oils (Horse Head XX-4, XX-32; Kadox-15); mechanical goods (Kadox-15, -72, -272; Zinc Oxide 35); metalworking (Horse Head XX-32); paint mfg. (AZO-55, -77; Azodox-55; Florence Green Seal-8; Horse Head XX-503, Horse Head XX-600, Horse Head XX-601, Horse Head XX-631; Kadox-515; Ottalume 2100); photography (Photox-7, -8, -77, -80, -85, -90, -801); plastics (AZO-77; Ottalume 2100); printing inks (Florence Green Seal-8); rubber (Actox-14; AZO-55, -55TT, -66, -66TT, -77, -77TT; Azodox-55, -55TT; Horse Head XX-4, XX-32, XX-32, XX-85; Kadox-15; Protox-169; Zinc Oxide 35, 911-36; Zinox-Grade 350; Zoco 100, 103, 104, 172); textile/leather processing (Horse Head XX-32; Kadox-72, -272); tires (Kadox-72, -272); wire and cable (AZO-66; Horse Head XX-32; Kadox-15, -72, -272)

Pharmaceutical applications: dental preparations (Horse Head USP-20); dermatological preparations (Horse Head USP-20); ointments (Horse Head USP-20); sunscreens (Horse Head USP-20); surgical tapes (AZO-66; Horse Head XX-32)

PROPERTIES:

Form:

Powder (Zinc Oxide 35)

Fine powder (Kadoz-15; Zoco 100, 103, 104)

Micron-sized powder (Ottalume 2100)

Nodular (Zinox-Grade 350)

Small nodular (AZO-55, -55TT, -66)

Very small nodular (AZO-77)

0.11 μ mean particle size (Kadox-25)

0.11 μ avg. particle size (Zinc Oxide 911, 911-36)

0.12 μ mean particle size (Kadox-15; Protox-169)

0.12 μ mean particle size pellets (Kadox-215)

0.15 μ mean particle size (Kadox-515)

0.18 μ mean particle size (Kadox-72)

0.18 μ mean particle size pellets (Kadox-272)

0.20 μ mean particle size (Photox-7)

0.24 μ avg. particle size (Zinc Oxide 500, 500-36)

0.25 μ mean particle size (Photox-77)

0.27 μ mean particle size (Horse Head XX-4, XX-601)

0.27 μ mean particle size pellets (Horse Head XX-32)
0.30 μ mean particle size (Photox-90)
0.31 μ mean particle size (Florence Green Seal-8; Horse Head USP-20, XX-32, XX-85; Photox-8)
0.36 μ mean particle size (Actox-14)
0.37 μ mean particle size (Photox-80, -85, -801)
0.6 μ mean particle size (Horse Head XX-600, XX-631)
0.8 μ mean particle size (Horse Head XX-503)
1.0 μ mean particle size (Horse Head XX-504)

Color:
White (Kadox-15; Zoco 100, 103)
Bright blue-white (Horse Head USP-20)
White daylight; fluorescent light blue (Ottalume 2100)
White to yellowish green (Zinc Oxide 35)
Yellow-buff (Photox-85; Horse Head XX-85)

Odor:
Odorless (Kadox-15; Zoco 100, 103, 104)

Composition:
93–96% ZnO (Zinc Oxide 35)
98.9% ZnO (Actox-14)
99.0% ZnO (Horse Head XX-4, XX-32, XX-600, XX-601, XX-631)
99.1% ZnO (Protox-169)
99.2% ZnO (AZO-55, -55TT; Horse Head XX-503)
99.3% ZnO (Photox-85; Horse Head XX-85, XX-504; Zinc Oxide 500-36)
99.4% ZnO (Zinc Oxide 911-36)
99.6% ZnO (Kadox-15, -25, -215, -515)
99.7% ZnO (AZO-77; Kadox-72, -272; Photox-7; Zinc Oxide 500)
99.8% ZnO (AZO-66; Florence Green Seal-8; Horse Head USP-20, XX-32; Photox–8, -77, -80, 90, -801; Zinc Oxide 911)

Sp.gr.:
≈ 5.4 (20 C) (Zinc Oxide 35)
5.57 (Kadox-15)
5.6 (Actox-14; AZO-55, -55TT, -66, -77; Florence Green Seal-8; Horse Head USP-20, XX-4, XX-32, XX-32, XX-85, XX-503, XX-504, XX-600, XX-601, XX-631; Kadox-15, -72, -215, -272, -515; Protox-169; Zinox-Grade 350)

Sp. Volume:
0.18 (Actox-14; Horse Head XX-4, XX-32, XX-85; Kadox-15, -72, -215, -272; Protox-169)

Density:
46.7 lb/solid gal (Florence Green Seal-8; Horse Head USP-20, XX-32, XX-503, XX-504, XX-600, XX-631; Kadox-515)

Zinc oxide (cont'd.)

Apparent Density:
28 lb/ft³ (AZO-77)
30 lb/ft³ (AZO-66; Horse Head XX-4, XX-601; Kadox-15, -25, -515)
32 lb/ft³ (AZO-55, -77TT)
35 lb/ft³ (Actox-14; Horse Head XX-600, XX-631; Kadox-72; Photox-7, -77, -90)
36 lb/ft³ (AZO-66TT)
40 lb/ft³ (Florence Green Seal-8; Horse Head USP-20, XX-32; Photox-8; Protox-169)
42 lb/ft³ (AZO-55TT)
45 lb/ft³ (Photox-801)
48 lb/ft³ (Azodox-55)
50 lb/ft³ (Photox-80)
55 lb/ft³ (Photox-85; Horse Head XX-85)
58 lb/ft³ (Azodox-55TT)
60 lb/ft³ (Horse Head XX-503; Kadox-215, -272)
65 lb/ft³ (Horse Head XX-32, XX-504)

Fineness:
99% thru 325 mesh (Zinox-Grade 350)
99.9% thru 325 mesh (Zinc Oxide 500, 500-36, 911, 911-36)
99.95% thru 325 mesh (AZO-55, -55TT; Horse Head XX-503, XX-504, XX-600)
99.96% thru 325 mesh (AZO-66, -77)
99.97% thru 325 mesh (Actox-14; Horse Head XX-4, XX-601, XX-631)
99.99% thru 325 mesh (Florence Green Seal-8; Horse Head USP-20, XX-32, XX-85;
    Kadox-15, -25, -72, -272, -515; Protox-169)

Surface Area:
1.0 m²/g (Horse Head XX-504)
1.3 m²/g (Horse Head XX-503)
1.7 m²/g (Horse Head XX-600, XX-631)
2.8 m²/g (Horse Head XX-85)
3.5 m²/g (Horse Head USP-20, XX-32)
4.0 m²/g (Horse Head XX-4, XX-32, XX-601)
6 m²/g (Kadox-72, -272)
7 m²/g (Kadox-515)
9 m²/g (Kadox-15, -215; Protox-169)
10 m²/g (Kadox-25)
17 m²/g (Zinox-Grade 350)
35 m²/g (Zinc Oxide 35)

Oil Absorption:
8 lb oil/100 lb ZnO (Horse Head XX-504)
10 lb oil/100 lb ZnO (Horse Head XX-503)
11 lb oil/100 lb ZnO (Florence Green Seal-8; Photox-80, -801)
12 lb oil/100 lb ZnO (Photox-90; Horse Head USP-20, XX-32, XX-600, XX-631;
    Kadox-25; Photox-8, -77; Protox-169)

13 lb oil/100 lb ZnO (Actox-14; Kadox-72, -272, -515; Photox-7)
14 lb oil/100 lb ZnO (Kadox-215)
16 lb oil/100 lb ZnO (Horse Head XX-32, XX-601)
18 lb oil/100 lb ZnO (Horse Head XX-4)
19 lb oil/100 lb ZnO (Photox-85)

Stability:

Good heat stability; maintains fluorescence up to 900 C; excellent light resistance (Ottalume 2100)

TRADENAME PRODUCTS AND
GENERIC EQUIVALENTS

Abitol [Hercules]—Dihydroabietyl alcohol
Ablunol NP30, NP30 70% [Taiwan Surfactant]—POE (30) nonyl phenyl ether
Ablunol NP40, NP40 70% [Taiwan Surfactant]—POE (40) nonyl phenyl ether
Ablunol NP50, NP50 70% [Taiwan Surfactants]—POE (50) nonyl phenyl ether
Accomid C [Armstrong]—Coconut diethanolamide
Acetol 1706 [Emery]—Acetylated lanolin alcohol
Acetorb A [Aceto]—2-Hydroxy-4-methoxybenzophenone
Acetulan [Amerchol; Amerchol Europe]—Acetylated lanolin alcohol
Actox-14 [NJ Zinc]—Zinc oxide (Amer. process)
Actrasol C50, C75, C85, PSR [Southland]—Castor oil, sulfated
Adimoll DN [Bayer AG]—Diisononyl adipate
Adimoll DO [Bayer AG]—Dioctyl adipate
Admex 710, 711 [Sherex]—Epoxidized soybean oil
Admex ELO [Sherex]—Epoxidized linseed oil
Adol 61NF, 62NF [Sherex]—Stearyl alcohol (NF grade)
Adol 63 [Sherex]—Stearyl alcohol
Adol 64 [Sherex]—Stearyl alcohol (tech. grade)
Advastab TM-180 [Cincinnati Milacron]—Butyltin mercaptide
Ahco 759 [ICI United States]—Sorbitan monolaurate
Ahco AJ-110 [ICI]—Castor oil, sulfated
Ahco FO-18 [ICI Americas]—Sorbitan trioleate
Ahco FS-21 [ICI Americas]—Sorbitan tristearate
Akrochem Hydrated Alumina [Akron]—Aluminum hydroxide
Aldo MC [Glyco]—Glyceryl monococoate
Aldo MR [Glyco]—Glyceryl monoricinoleate
Alfol 18 Alcohol [Continental Oil]—Stearyl alcohol
Alkamide 1002, 2104, 2106, 2204 [Alkaril]—Coconut diethanolamide
Alkamide CDE, CDO [Alkaril]—Coconut diethanolamide
Alkamide CME, CMO [Alkaril]—Coconut monoethanolamide
Alkamide HTDE [Alkaril]—Stearic diethanolamide
Alkamide HTME [Alkaril]—Stearic monoethanolamide
Alkamide L9DE [Alkaril]—Lauric diethanolamide
Alkamide L9ME [Alkaril]—Lauric monoethanolamide
Alkamide LIPA [Alkaril]—Lauric monoisopropanolamide
Alkamide SDO [Alkaril]—Oleic diethanolamide
Alkamidox CME-5 [Alkaril]—POE (6) coconut amide (5 EO)
Alkamox CAPO [Alkaril]—Cocamidopropylamine oxide
Alkamuls 6000-DS [Alkaril]—POE (150) distearate
Alkamuls GMR [Alkaril]—Glyceryl monoricinoleate

Alkamuls SML [Alkaril]—Sorbitan monolaurate
Alkamuls STO [Alkaril]—Sorbitan trioleate
Alkamuls STS [Alkaril]—Sorbitan tristearate
Alkasurf NP-30 70% [Alkaril]—POE (30) nonyl phenyl ether
Alkasurf NP-40 [Alkaril]—POE (40) nonyl phenyl ether
Alkasurf NP-50 70% [Alkaril]—POE (50) nonyl phenyl ether
Alkasurf NP-100 70% [Alkaril]—POE (100) nonyl phenyl ether
Alrosol B [Ciba-Geigy]—Coconut diethanolamide
Alrosol O [Ciba-Geigy]—Oleic diethanolamide
Alumina Hydrate 753, 983 [Harwick]—Aluminum hydroxide
Amerlate LFA, WFA [Amerchol]—Lanolin acid
Amerscreen P, P80/20 [Amerchol]—Ethyl dihydroxypropyl PABA
Amides of Cocout Oil [Murphy-Phoenix]—Coconut oil amide
Amidox C-5 [Stepan]—POE (6) coconut amide
Amine O [Ciba-Geigy]—Oleyl hydroxyethyl imidazoline
Aminol CM [Finetex]—Coconut monoethanolamide
Aminol COR-4C, LM-30-C Special [Finetex]—Lauric diethanolamide
Aminol HCA [Finetex]—Coconut diethanolamide
Aminol LNO [Finetex]—Linoleoyl diethanolamide
Aminol OF [Finetex]—Oleic diethanolamide
Aminoxid WS35 [Goldschmidt AG]—Cocamidopropylamine oxide
Ammonyx CDO [Onyx]—Cocamidopropylamine oxide
Ammonyx CO [Onyx]—Cetyl dimethyl amine oxide
Ammonyx MCO, MO [Onyx]—Myristyl dimethyl amine oxide
AMS-33 [Hefti]—Ethylene glycol monostearate
Anti UVA [Aceto]—2-Hydroxy-4-methoxybenzophenone
Ardet CB, DC, DCAE, DCX, DMA, DMC, DMF, WO [Ardmore]—Coconut dietha-
    nolamide
Ardet CEA [Ardmore]—Coconut monoethanolamide
Ardet DLA, LDA, LMC [Ardmore]—Lauric diethanolamide
Ardet LEA [Ardmore]—Lauric monoethanolamide
Ardet LIPA [Ardmore]—Lauric monoisopropanolamide
Ardet WB [Ardmore]—Stearic monoethanolamide
Argus DLTDP [Witco/Argus]—Dilauryl thiodipropionate
Argus DMTDP [Witco/Argus]—Dimyristyl thiodipropionate
Argus DSTDP [Argus]—Distearyl thiodipropionate
Argus DTDTDP [Witco/Argus]—Ditridecyl thiodipropionate
Argus DXTDP [Witco/Argus]—Distearyl thiodipropionate
Arkopal N-300 [Hoechst AG]—POE (30) nonyl phenyl ether
Arlacel 20 [ICI]—Sorbitan monolaurate
Arlacel 85 [ICI United States]—Sorbitan trioleate
Armid C [Armak]—Coconut oil amide

Armid HT [Armak]—Hydrogenated tallow amide

Armotan ML [Armak]—Sorbitan monolaurate

Aromox DM14D-W [Akzo Chemie]—Myristyl dimethyl amine oxide

Aromox DM16 [Armak]—Cetyl dimethyl amine oxide

ATBC [Croda Chem. Ltd.]—Acetyl tributyl citrate

Atlas G-4885 [ICI Specialty Chem.]—Sorbitan trioleate (tech.)

Atlox 4885 [ICI Specialty Chem.]—Sorbitan trioleate

Attacote [Engelhard Min. & Chem.]—Attapulgite

Attaflow [Engelhard Min. & Chem.]—Attapulgite

Attagel 40, 50 [Engelhard Min. & Chem.]—Attapulgite (colloidal)

Attagel 150, 350 [Engelhard Min. & Chem.]—Attapulgite

Avicel PH-101, PH-102, PH-103, PH-105 [FMC]—Microcrystalline cellulose (NF grades)

Avicel RC-591 [FMC]—Microcrystalline cellulose

Avirol SA-4106 [Henkel]—Sodium octyl sulfate

AZO-55, -66, -77 [Amer. Smelting & Refining]—Zinc oxide

AZO-55TT, -66TT, -77TT [Amer. Smelting & Refining]—Zinc oxide (surface-treated with zinc propionate)

Azodox-55, -55TT [Amer. Smelting & Refining]—Zinc oxide (deaerated)

Barlox 14 [Lonza]—Myristyl dimethyl amine oxide

Barlox C [Lonza]—Cocamidopropylamine oxide

Bentone 27 [NL Chem.]—Stearalkonium hectorite

Bentone 34 [NL Chemicals]—Quaternium-18 bentonite

Bentone 38 [NL Chem.]—Quaternium-18 hectorite

Bentone Gel CAO [NL Chem.]—Stearalkonium hectorite (in castor oil)

Bentone Gel IPM [NL Chem.]—Stearalkonium hectorite (in isopropyl myristate)

Bentone Gel LOI [NL Chem.]—Stearalkonium hectorite (in lanolin/isopropyl palmitate)

Bentone Gel MIO [NL Chem.]—Quaternium-18 hectorite (in mineral oil)

Bentone Gel MIO A-40 [NL Chem.]—Quaternium-18 hectorite (in mineral oil and SDA 40)

Bentone Gel S-130, SS-71 [NL Chem.]—Quaternium-18 hectorite (in mineral spirits)

Bentone Gel VS-5 [NL Chem.]—Quaternium-18 hectorite (in cyclomethicone)

Benzoflex 2-45 [Velsicol]—Diethylene glycol dibenzoate

Benzoflex 9-88, 9-88 SG [Velsicol]—Dipropylene glycol dibenzoate

Bio Soft LD-70, LD-80 [Stepan]—Sodium olefin sulfonate

Bio-Terge AS-40, AS-90 Beads, AS-90F [Stepan]—Sodium olefin sulfonate (C_{14-16})

Biozan [Hercules]—Xanthan gum

Biozan SPX 5423 [Hercules]—Xanthan gum (industrial grade)

Blanose Cellulose Gum, Refined CMC [Hercules BV]—Sodium carboxymethyl cellulose

Butyl Oleate C-914 [C.P. Hall]—Butyl oleate

Butyl Oleate DLC [Harwick]—Butyl oleate

C-30, -30BF, -31, -230, -330, -331, -333, -430, -433 [Alcoa]—Aluminum hydroxide

Calamide C, CW-100 [Pilot]—Coconut diethanolamide

Calamide O [Pilot]—Oleic diethanolamide

CAO-1 [PMC Specialties; Sherex; Summit]—Butylated hydroxytoluene

CAO-3 [PMC Specialties; Sherex; Summit]—Butylated hydroxytoluene (food grade)

Caprol 10G40 [Capital City]—Decaglycerol tetraoleate

Carsamide 7644, C-3, CA, LE, SAC [Carson]—Coconut diethanolamide

Carsamide SAL-7, SAL-9, SAL-82 [Carson]—Lauric diethanolamide

Carsonol SHS [Carson]—Sodium octyl sulfate

Carsonon N-30, N-30 70% [Carson]—POE (30) nonyl phenyl ether

Carsonon N-40 70% [Carson]—POE (40) nonyl phenyl ether

Carsonon N-50 70% [Carson]—POE (50) nonyl phenyl ether

Carsonon N-100 70% [Carson]—POE (100) nonyl phenyl ether

Carstab 700, 705 [Cincinnati Milacron]—2-Hydroxy-4-(octyloxy) benzophenone

Carstab DLTDP [Cincinnati Milacron; Morton Thiokol/Carstab]—Dilauryl thiodipropionate

Carstab DMTDP [Cincinnati Milacron]—Dimyristyl thiodipropionate

Carstab DSTDP [Cincinnati Milacron; Morton Thiokol]—Distearyl thiodipropionate

Cationic Guar C-261 [Henkel/Henkel Canada]—Guar hydroxypropyltrimonium chloride

Cedemide MX [Domtar/CDC]—Lauric monoethanolamide

Cedepal CO-880, CO-887 [Domtar]—POE (30) nonyl phenyl ether

Cedepal CO-890 [Domtar]—POE (40) nonyl phenyl ether

Cedepal CO-990, CO-997 [Domtar]—POE (100) nonyl phenyl ether

Cellosize Hydroxyethyl Cellulose [Union Carbide]—Hydroxyethylcellulose

Cellosize Hydroxyethyl Cellulose QP 40, QP 100M, QP 300, QP 4400, QP 15,000, QP 30,000, QP 52,000 [Union Carbide]—Hydroxyethylcellulose (quick dispersion grades)

Cellosize Hydroxyethyl Cellulose WP 3, WP 09 [Union Carbide]—Hydroxyethylcellulose

Cellulose Gum [Hercules]—Sodium carboxymethyl cellulose

Ceraphyl 41 [Van Dyk]—C12-15 alcohols lactate

Ceraphyl 230 [VanDyk]—Diisopropyl adipate

Cerasynt 303 [Van Dyk]—Diethylaminoethyl stearate

Cerasynt D [Van Dyk]—Stearic monoethanolamide stearate

Cerasynt M [Van Dyk]—Ethylene glycol monostearate

Cerasynt MN [Van Dyk]—Ethylene glycol monostearate (self-emulsifying)

Cereclor 42, 42P, 50LV, 51-L, 52P, 70L, AP45, AP52, LP4446, LP4985, S45, S-52 [ICI Americas]—Chlorinated paraffin

Chemadox M [Richardson]—Myristyl dimethyl amine oxide

Chemax E-1000ML [Chemax]—POE (20) monolaurate

Chemax NP-30, NP-30/70 [Chemax]—POE (30) nonyl phenyl ether

Chemax NP-40, NP-40/70 [Chemax]—POE (40) nonyl phenyl ether

Chemax NP-50, NP-50/70 [Chemax]—POE (50) nonyl phenyl ether
Chemax NP-100, NP-100/70 [Chemax]—POE (100) nonyl phenyl ether
Chemax SCO [Chemax]—Castor oil, sulfated
Chemcol NPE-400 [Chemform]—POE (40) nonyl phenyl ether
Chimipal MC [La Tassilchimka]—Coconut monoethanolamide
Chimipal OLD [La Tessilchimica]—Oleic diethanolamide
Cithrol 10DL [Croda]—POE (20) dilaurate
Cithrol 10ML [Croda]—POE (20) monolaurate
Cithrol EGMS N/E [Croda]—Ethylene glycol monostearate
Cithrol EGMS S/E [Croda]—Ethylene glycol monostearate (self-emulsifying)
Cithrol GDL S/E [Croda Chem. Ltd.]—Glyceryl dilaurate (self-emulsifying)
Cithrol PGMR N/E [Croda]—Propylene glycol monoricinoleate
Cithrol PGMR S/E [Croda]—Propylene glycol monoricinoleate (self-emulsifying)
Citroflex 2 [Morflex]—Triethyl citrate
Citroflex 4 [Morflex]—Tributyl citrate
Citroflex A-2 [Morflex]—Acetyl triethyl citrate
Citroflex A-4 [Morflex]—Acetyl tributyl citrate
Clindrol 100MC, 100MCG [Clintwood]—Coconut monoethanolamide
Clindrol 101CG, 200CG, 200CGN, 200HC, 200RC, 202CGN, 203CG [Clintwood]—
    Coconut diethanolamide
Clindrol 101LI [Clintwood]—Lauric monoisopropanolamide
Clindrol 200-L [Clintwood]—Lauric diethanolamide
Clindrol 200-O [Clintwood]—Oleic diethanolamide
Clindrol 200-S, 868 [Clintwood]—Stearic diethanolamide
Clindrol 200MS, LT14-8 [Clintwood]—Stearic monoethanolamide
Clindrol 206CGN, 207CGN, 209CGN [Clintwood]—Coconut diethanolamide
Clindrol LT15-73-1 [Clintwood]—Linoleoyl diethanolamide (1:1)
Clindrol SEG [Clintwood]—Ethylene glycol monostearate
Clindrol Superamide 100C, 100CG [Clintwood]—Coconut diethanolamide
Clindrol Superamide 100L, Superamide 100LM [Clintwood]—Lauric diethanolamide
Clindrol Superamide 100S [Clintwood]—Stearic diethanolamide
CMC-6CTL [Hercules]—Sodium carboxymethyl cellulose (crude grade)
CMC-T [Hercules]—Sodium carboxymethyl cellulose (tech. grade)
CMTC-T [Hercules]—Sodium carboxymethyl cellulose
Comperlan F [Henkel KGaA]—Linoleoyl diethanolamide
Comperlan HS [Henkel KGaA]—Stearic monoethanolamide
Comperlan KD, KDO [Henkel Canada]—Coconut diethanolamide
Comperlan LD, LDO, LDS [Henkel Canada]—Lauric diethanolamide
Comperlan LM [Henkel/Canada]—Lauric monoethanolamide
Comperlan LMD [Henkel KgaA]—Lauric diethanolamide
Comperlan LP [Henkel]—Lauric monoisopropanolamide
Comperlan OD [Henkel KGaA]—Oleic diethanolamide

Comperlan VOD [Henkel (Canada)]—Linoleoyl diethanolamide (1:1)
Conco AOS-40, AOS-90F [Continental]—Sodium olefin sulfonate (C_{14-16})
Conco Emulsifier K [Continental]—Coconut diethanolamide
Conco NI-187 [Continental Chem.]—POE (30) nonyl phenyl ether
Conco NI-190, NI-197 [Continental Chem.]—POE (40) nonyl phenyl ether
Conco NI-2000 [Continental Chem.]—POE (50) nonyl phenyl ether
Conco XA-C [Continental]—Cetyl dimethyl amine oxide
Conco XA-M [Continental]—Myristyl dimethyl amine oxide
Condensate CO, NP [Continental]—Coconut diethanolamide
Condensate PA, PN, PO, PS [Continental]—Coconut diethanolamide
Condensate PE, PL [Continental]—Lauric diethanolamide
Consos Castor Oil [Consos]—Castor oil, sulfated
Cordon NU 890/50, 890/75 [Finetex]—Castor oil, sulfated
Cosmedia Guar C-261 [Henkel/Henkel Canada/Henkel KGaA]—Guar hydroxypro-
    pyltrimonium chloride
Covi-Ox T-50, T-70 [Henkel/Henkel Canada]—Tocopherol
Covitol F-350M, F-600, F-1000 [Henkel]—Tocopherol (pharmaceutical grade)
CPF-0001, -0003, -0008, -0019, -0022 [Pearsall/Witco]—Chlorinated paraffin
CPH-380N [C.P. Hall]—Stearic monoethanolamide
Crill 1 [Croda]—Sorbitan monolaurate
Crill 35 [Croda]—Sorbitan tristearate
Crill 45 [Croda]—Sorbitan trioleate
Crillon ODE [Croda Chem. Ltd.]—Oleic diethanolamide
Crodalan LA [Croda Inc.]—Acetylated lanolin alcohol
Crodamol DA [Croda]—Diisopropyl adipate
Croderol G7000 [Croda]—Glycerin
Crystal Crown [CasChem Inc.]—Castor oil (deodorized, refined)
Crystal O [CasChem Inc.]—Castor oil (deodorized, refined)
Cyanox 711 [Amer. Cyanamid]—Ditridecyl thiodipropionate
Cyanox LTDP [Amer. Cyanamid]—Dilauryl thiodipropionate
Cyanox MTDP [Amer. Cyanamid]—Dimyristyl thiodipropionate
Cyanox STDP [Amer. Cyanamid]—Distearyl thiodipropionate
Cyasorb UV 9 [Amer. Cyanamid]—2-Hydroxy-4-methoxybenzophenone
Cyasorb UV 24 [Amer. Cyanamid]—2,2'-Dihydroxy-4-methoxybenzophenone
Cyasorb UV 531 [Amer. Cyanamid]—2-Hydroxy-4-(octyloxy) benzophenone
Cyastab 908 [Amer. Cyanamid]—Dibasic lead phthalate
Cyastab 988 [Amer. Cyanamid]—Dibasic lead phthalate (coated)
Cyclochem EGMS [Cyclo]—Ethylene glycol monostearate
Cyclogol Stearyl Alcohol [Cyclo]—Stearyl alcohol
Cyclomide C212 [Cyclo]—Coconut monoethanolamide
Cyclomide DC212 [Cyclo]—Coconut diethanolamide
Cyclomide DC212/S, DC212/SE, KD [Cyclo]—Coconut diethanolamide

Cyclomide DC212M [Cyclo]—Coconut diethanolamide
Cyclomide DIN 295/S [Cyclo]—Linoleoyl diethanolamide (1:1)
Cyclomide DL203, DL203/S, DL207/S [Cyclo]—Lauric diethanolamide
Cyclomide DO280 [Cyclo]—Oleic diethanolamide (2:1)
Cyclomide DO280/S [Cyclo]—Oleic diethanolamide (1:1)
Cyclomide DOTS [Cyclo]—Linoleoyl diethanolamide (1:1)
Cyclomide DP 240/S [Cyclo]—Palm kernel oil acid diethanolamide (1:1)
Cyclomide DS 280/S [Cyclo]—Stearic diethanolamide (1:1)
Cyclomide L203 [Cyclo]—Lauric monoethanolamide
Cyclomide LIPA [Cyclo]—Lauric monoisopropanolamide
Cyclomide S 280 [Cyclo]—Stearic monoethanolamide
Cyclomox CO [Cyclo]—Cocamidopropylamine oxide
Cycloteric BET O-30 [Cyclo]—Oleamidopropyl betaine
Cycloteric BET-CS [Cyclo]—Cocamidopropyl hydroxysultaine
Dariloid QH [Kelco/Div. of Merck & Co.]—Algin
Diamond Shamrock Butyl Oleate, light [Diamond Shamrock]—Butyl oleate
Dibasic Lead Phthalate Conc. 9-73-A, 13-191, 14-81, 17-239 [Santech]—Dibasic lead
    phthalate
Diluex [Floridin]—Attapulgite
Disflamoll TKP [Mobay]—Tricresyl phosphate
DMP [Dynamit-Nobel]—Dimethyl phthalate
DOP Dry Liquid Conc. [Polymerics]—Dioctyl phthalate
Doverbros #4 [Dover]—Tris (nonylphenyl) phosphite
Drapex 4.4 [Argus]—Epoxidized octyl tallate
Drapex 6.8 [Argus]—Epoxidized soybean oil
Drapex 10.4 [Argus]—Epoxidized linseed oil
Drewmulse 10-4-O [PVO Int'l.]—Decaglycerol tetraoleate
Drewmulse 75 [PVO]—Glyceryl monococoate
Drewmulse CNO [PVO]—Glyceryl monococoate
Drewmulse EGMS [Drew Produtos]—Ethylene glycol monostearate
Drewmulse GMRO [Drew Produtos]—Glyceryl monoricinoleate
Drewmulse SML [PVO Int'l.]—Sorbitan monolaurate
Drewmulse STS [PVO Int'l.]—Sorbitan tristearate
Drewpol 10-4-O [PVO Int'l.]—Decaglycerol tetraoleate
Drimix [Kenrich Petrochemicals]—Tricresyl phosphate
Duponol 80 [Du Pont]—Sodium octyl sulfate (tech.)
Durtan 20 [Durkee/SCM]—Sorbitan monolaurate
Dyphos [Associated Lead]—Dibasic lead phosphite
Dyphos XL [Associated Lead]—Dibasic lead phosphite (coated)
Dythal [Associated Lead]—Dibasic lead phthalate
Dythal XL [Associated Lead]—Dibasic lead phthalate (coated)
Ecconol 628 [Essential]—Coconut oil amide

364

Elfacos C26 [Akzo Chemie BV]—Hydroxyoctacosanyl hydroxystearate
Elfan OS46 [Akzo Chemie]—Sodium olefin sulfonate (C_{14-16})
Emasol O-30 [Kao]—Sorbitan trioleate
Emasol S-30 [Kao]—Sorbitan tristearate
Emcol M [Witco/Organics]—Myristyl dimethyl amine oxide
Emerest 2301 [Emery]—Methyl oleate
Emerest 2328 [Emery]—Butyl oleate
Emerest 2350 [Emery]—Ethylene glycol monostearate
Emersal 6465 [Emery]—Sodium octyl sulfate
Emersol 210, 220, 233LL[Emery]—Oleic acid
Emersol 6321 [Emery]—Oleic acid (food grade)
Emery 652 [Emery]—Lauric acid
Emery Methyl Oleate [Emery/Oleochem Group]—Methyl oleate
Emid 6500 [Emery Ind.]—Coconut monoethanolamide
Emid 6510, 6511 [Emery]—Lauric diethanolamide
Emid 6513 [Emery]—Lauric diethanolamide
Emid 6514, 6515 [Emery]—Coconut diethanolamide
Emid 6519 [Emery]—Lauric diethanolamide
Emid 6531, 6533, 6534 [Emery]—Coconut diethanolamide
Emid 6540 [Emery]—Linoleoyl diethanolamide
Emid 6541 [Emery]—Lauric diethanolamide
Emid 6545 [Emery]—Oleic diethanolamide
Empigen OH [Albright & Wilson/Detergents]—Myristyl dimethyl amine oxide
Empilan 2502 [Albright & Wilson/Detergents]—Coconut diethanolamide
Empilan CDE [Albright & Wilson/Marchon]—Coconut diethanolamide
Empilan CDEY [Albright & Wilson/Australia]—Coconut diethanolamide
Empilan CDX [Albright&Wilson/Detergents]—Coconut diethanolamide
Empilan CIS [Albright & Wilson/Detergents]—Coconut monoisopropanolamide
Empilan CM [Albright & Wilson/Australia]—Coconut monoethanolamide
Empilan CME [Albright &Wilson/Marchon]—Coconut monoethanolamide
Empilan EGMS [Albright&Wilson/Marchon]—Ethylene glycol monostearate
Empilan FD, FD20, FE [Albright & Wilson/Australia]—Coconut diethanolamide
Empilan LDE, LDX [Albright & Wilson/Marchon]—Lauric diethanolamide
Empilan LIS [Albright&Wilson/Marchon]—Lauric monoisopropanolamide
Empilan LME [Albright & Wilson/Detergents]—Lauric monoethanolamide
Empilan MAA [Albright & Wilson]—POE (6) coconut amide
Emsorb 2503 [Emery]—Sorbitan trioleate
Emsorb 2507 [Emery]—Sorbitan tristearate
Emsorb 2515 [Emery]—Sorbitan monolaurate
Emulsynt GDL [Van Dyk]—Glyceryl dilaurate
Emulvis [C.P. Hall Co.]—POE (150) distearate
Epistatic 100 [Eagle Picher]—Tribasic lead sulfate

Epistatic 103 [Eagle Picher]—Tribasic lead sulfate (surface treated)

Epithal 120 [Eagle-Picher]—Dibasic lead phthalate

Epoxol 5-2E [Swift]—Epoxidized octyl tallate

Epoxol 7-4 [Swift]—Epoxidized soybean oil

Epoxol 9-5 [Swift]—Epoxidized linseed oil

ESI-Terge 10, B-15, C-5, S-10 [Emulsion Systems]—Coconut diethanolamide

Estabex 138-A [Interstab]—Epoxidized soybean oil

Ethosperse SL-20 [Lonza]—POE (20) sorbitol ether

Ethylan A15, LD, LDG, LDS [Lankro]—Coconut diethanolamide

Ethylan L10 [Lankro]—POE (20) monolaurate

Ethylan MLD [Lankro]—Lauric diethanolamide

Ethylan N30 [Lankro]—POE (30) nonyl phenyl ether

Ethylan N50 [Lankro Chem. Ltd.]—POE (50) nonyl phenyl ether

Eureka 102 [Atlas Refinery]—Castor oil, sulfated

Evanstab 12 [Evans Chemetics]—Dilauryl thiodipropionate

Evanstab 13 [Evans Chemetics]—Ditridecyl thiodipropionate

Evanstab 14 [Evans Chemetics]—Dimyristyl thiodipropionate

Evanstab 16 [Evans Chemetics]—Dicetyl thiodipropionate

Evanstab 18 [Evans Chemetics]—Distearyl thiodipropionate

Fancol ALA [Fanning]—Acetylated lanolin alcohol

Fancor LFA [Fanning Corp.]—Lanolin acid

Ferro 1288 [Ferro]—Barium-cadmium-zinc compound

FF Wood Rosin [Harwick]—Rosin

Flexchlor 0001, 0002, 0008, 0009, 0010, 0012, 0018, 0023 [Pearsall/Witco]—Chlorinated paraffin

Flexol Plasticizer EP-8 [Union Carbide]—Epoxidized octyl tallate

Flexol Plasticizer EPO [Union Carbide]—Epoxidized soybean oil

Flexol Plasticizer LOE [Union Carbide]—Epoxidized linseed oil

Flexon 580, 641, 680 [Exxon]—Naphthenic oil

Flexricin 9 [NL Industries]—Propylene glycol monoricinoleate

Flexricin 13 [NL Industries]—Glyceryl monoricinoleate

Flexricin P-8 [CasChem Inc.]—Glyceryl triacetyl ricinoleate

Florco X [Floridin]—Attapulgite

Florence Green Seal-8 [N.J. Zinc]—Zinc oxide (French process)

Florex [Floridin]—Attapulgite

Florex Granular Grades [Floridin]—Attapulgite

Foamole A [Van Dyk]—Linoleoyl diethanolamide

Foral AX [Hercules]—Rosin

Gafamide CDD-518 [GAF]—Coconut diethanolamide

Galactasol Series [Henkel]—Guar gum

Gelcharg HP4 [Hercules]—Hydroxypropyl guar

Genu 04CB or 04CG, 100–120 Grade [Hercules]—Pectin (low methoxyl)

Genu 12CB or 12CG, 85–100 Grade [Hercules]—Pectin (low methoxyl)
Genu 15AB or 104AS, 100–120 Grade [Hercules]—Pectin (low methoxyl)
Genu 18CB or 18CG, 70–90 Grade [Hercules]—Pectin (low methoxyl)
Genu 20AB or 20AS, 100 Grade [Hercules]—Pectin (low methoxyl)
Genu 21AB or 21AS [Hercules]—Pectin (low methoxyl)
Genu 102AS Buffered, 100 Grade [Hercules]—Pectin (low methoxyl)
Genu AA Medium-Rapid Set, 150 Grade [Hercules]—Pectin (high methoxyl)
Genu BA-King, 150 Grade [Hercules]—Pectin (high methoxyl)
Genu BB Rapid Set, 150 Grade [Hercules]—Pectin (high methoxyl)
Genu DD Extra-Slow Set, 150 Grade [Hercules]—Pectin (high methoxyl)
Genu DD Extra-Slow Set-C, 150 Grade [Hercules]—Pectin (high methoxyl)
Genu DD Slow Set, 150 Grade [Hercules]—Pectin (high methoxyl)
Genu JM [Hercules]—Pectin (high methoxyl)
Genugel Series [Hercules]—Carrageenan
Genulacta Series [Hercules]—Carrageenan
Genuvisco Series [Hercules]—Carrageenan
Glucamate DOE-120 [Amerchol]—POE (120) methyl glucose dioleate
Glycomul L, LC [Glyco]—Sorbitan monolaurate
Glycomul TO [Glyco]—Sorbitan trioleate
Glycomul TS [Glyco]—Sorbitan tristearate
Graden Butyl Oleate [Graden]—Butyl oleate
Grindtek STS [Grinsted]—Sorbitan tristearate
Groco 2 Oleic Acid, 4 Oleic Acid [A. Gross]—Oleic acid
Grocor 4000 [A. Gross]—Butyl oleate
Grocor 8002, 8008 [A. Gross]—Methyl oleate
H-30, H-36 [Kaiser]—Aluminum hydroxide
Halbase 10 [Halstab]—Tribasic lead sulfate
Halbase 10-EP [Halstab]—Tribasic lead sulfate (coated)
Hallco DBS [C.P. Hall]—Dibutyl sebacate
Halthal [Halstab]—Dibasic lead phthalate
Halthal-EP [Halstab]—Dibasic lead phthalate (coated)
Hamposyl C [W.R. Grace]—Cocoyl sarcosine
Hamposyl L [W.R. Grace]—Lauroyl sarcosine
Hamposyl O [W.R. Grace]—Oleyl sarcosine
Haroil SCO-50 [Graden]—Castor oil, sulfated
Haroil SCO-65 [Graden]—Castor oil, sulfated
Haroil SCO-75 [Graden]—Castor oil, sulfated
Haroil SCO-7525 [Graden]—Castor oil, sulfated
Hartamide 9137 [Hart Chem. Ltd.]—Oleic diethanolamide
Hartamide LDA70, LDA90 [Hart]—Lauric diethanolamide
Hartamide LMEA-90 [Hart]—Lauric monoethanolamide
Hartamide OD [Hart]—Coconut diethanolamide

Hartenol V-63 [Hart]—Castor oil, sulfated
Hartex V63, V64 [Hart]—Castor oil, sulfated
Hatcol BOP [Hatco]—Butyl octyl phthalate
Hatcol DBP [Hatco]—Dibutyl phthalate
Hatcol DIBA [Hatco]—Diisobutyl adipate
Hatcol DIDA [Hatco]—Diisodecyl adipate
Hatcol DIDP [Hatco]—Diisodecyl phthalate
Hatcol DIDP-EG [Hatco]—Diisodecyl phthalate (electrical grade)
Hatcol DOA [Hatco]—Dioctyl adipate
Hatcol DOP [Hatco]—Dioctyl phthalate
Hatcol DOS [Hatco]—Dioctyl sebacate
Hatcol TOTM [Hatco]—Trioctyl trimellitate
Hercules Cellulose [Hercules]—Sodium carboxymethyl cellulose (food grade)
Hercules CMC [Hercules]—Sodium carboxymethyl cellulose
Hercules CMC-6-DG-L [Hercules]—Sodium carboxymethyl cellulose (crude grade)
Hetamide LA, ML [Heterene]—Lauric acid
Hetamide MC, RC [Heterene]—Coconut diethanolamide
Hetamide ML [Heterene]—Lauric diethanolamide
Hetlan AC [Heterene]—Acetylated lanolin alcohol
Hetoxide NP-40 [Heterene]—POE (40) nonyl phenyl ether
Hetsorb L80-72% [Heterene]—POE (80) sorbitan monolaurate
Hexaflex DBP [Hexagon]—Dibutyl phthalate
Hexaflex DIDA [Hexagon]—Diisodecyl adipate
Hexaflex DIDP [Hexagon]—Diisodecyl phthalate
Hexaflex DIOP [Hexagon]—Diisooctyl phthalate
Hexaflex DOP [Hexagon]—Dioctyl phthalate
Hexaflex TOTM [Hexagon]—Trioctyl trimellitate
Hipochem Dispersol SCO [High Point]—Castor oil, sulfated
Hodag EGMS [Hodag]—Ethylene glycol monostearate
Hodag GMR, GMR-D [Hodag]—Glyceryl monoricinoleate
Hodag SML [Hodag]—Sorbitan monolaurate
Hodag STO [Hodag]—Sorbitan trioleate
Hodag STS [Hodag]—Sorbitan tristearate
Hodag SVO-1047 [Hodag]—Decaglycerol tetraoleate
Horse Head USP-20, XX-78, XX-85 [N.J. Zinc]—Zinc oxide (French process)
Horse Head XX-4, XX-32, XX-503, XX-504, XX-600, XX-601, XX-631 [N.J. Zinc]—
    Zinc oxide (Amer. process)
Hydral 705, 710 [Alcoa]—Aluminum hydroxide
Hymolon CWC, K-90 [Hart]—Coconut diethanolamide
Hyonic NP-407 [Diamond Shamrock]—POE (40) nonyl phenyl ether
Hyonic NP-500 [Diamond Shamrock]—POE (50) nonyl phenyl ether
Hystrene 9512 [Humko Sheffield]—Lauric acid

Iconol NP-30, NP-30-70% [BASF Wyandotte]—POE (30) nonyl phenyl ether
Iconol NP-40, NP-40-70% [BASF Wyandotte]—POE (40) nonyl phenyl ether
Iconol NP-50 [BASF Wyandotte]—POE (50) nonyl phenyl ether
Iconol NP-100 [BASF Wyandotte]—POE (100) nonyl phenyl ether
Igepal CO-880, CO-887 [GAF]—POE (30) nonyl phenyl ether
Igepal CO-890, CO-897 [GAF]—POE (40) nonyl phenyl ether
Igepal CO-970, CO-977 [GAF]—POE (50) nonyl phenyl ether
Igepal CO-990, CO-997 [GAF]—POE (100) nonyl phenyl ether
Imwitor 940 [Dynamit-Nobel]—Palm oil glycerides
Incromide CA [Croda Surfactants]—Coconut diethanolamide
Incromide CM, CME [Croda Surfactants]—Coconut monoethanolamide
Incromide L-90, LCL, LL, LM-70, LR [Croda Surfactants]—Lauric diethanolamide
Incromide LA [Croda Surfactants]—Linoleoyl diethanolamide
Incromide LI [Croda Surfactants]—Lauric monoisopropanolamide
Incromide OPD [Croda Surfactants]—Oleic diethanolamide
Incromide SM [Croda Surfactants]—Stearic monoethanolamide
Incromine Oxide M [Croda Surfactants]—Myristyl dimethyl amine oxide
Incronam OP-30 [Croda Surfactants]—Oleamidopropyl betaine
Incronam P-30 [Croda Surfactants]—Palmitamidopropyl betaine
Indalca CD 30 [Hercules]—Guar gum
Industrene 105 [Humko]—Oleic acid
Interstab 761-28, BC-109, BC-110 [Interstab]—Barium-cadmium-zinc compound (high
    zinc)
Interstab 761-28A, BC-103, BC-103L, R-4109, R-4114, R-4137 [Interstab]—Barium-
    cadmium-zinc compound (low zinc)
Interstab BC-100S [Interstab]—Barium-cadmium compound
Interstab BC-103A, BC-103C, R-4101 [Interstab]—Barium-cadmium-zinc compound
    (medium zinc)
Interstab BC-4362 [Interstab]—Barium-cadmium-zinc compound
Ionet S-20 [Sanyo]—Sorbitan monolaurate
Ionet S-85 [Sanyo]—Sorbitan trioleate
Ionol, Ionol CP [Shell]—Butylated hydroxytoluene
Jayflex 80-TM [Exxon]—Triisooctyl trimellitate
Jayflex DIDP, DIDP-E [Exxon]—Diisodecyl phthalate
Jayflex DINA [Exxon]—Diisononyl adipate
Jayflex DINP [Exxon]—Diisononyl phthalate
Jayflex DINP-E [Exxon]—Diisononyl phthalate (electrical grade, with Bisphenol-A)
Jayflex DTDP [Exxon]—Ditridecyl phthalate (with Bisphenol-A)
Jordamide 201 [Jordan]—Oleic diethanolamide (2:1)
Jordamide CMEA, CMEA Extra [Jordan]—Coconut monoethanolamide
Jordamide JT128, WC Conc. [Jordan]—Coconut diethanolamide
Jordamide LLD [Jordan]—Linoleoyl diethanolamide

Jordamox CAPA [Jordan]—Cocamidopropylamine oxide
Jordamox CDA-40 [Jordan]—Cetyl dimethyl amine oxide
Jordamox MDA [Jordan]—Myristyl dimethyl amine oxide
Jortaine COSB, CSB, CSB-50 [Jordan]—Cocamidopropyl hydroxysultaine
Jortaine LMAB [Jordan]—Lauramidopropyl betaine
Jortaine TM [Jordan]—Dihydroxyethyl tallow glycinate
Kadox-15, -25, -72, -215, -272, -515 [N.J. Zinc]—Zinc oxide (French process)
Kalcohl 80 [Kao Corp.]—Stearyl alcohol
Kelacid [Kelco Div. of Merck]—Alginic acid
Kelco-Gel HV, LV [Kelco/Div. of Merck & Co.]—Algin (specially clarified, low-
    calcium)
Kelco-Pac [Kelco/Div. of Merck & Co.]—Algin (specially clarified, low-calcium)
Kelcoloid, D, DH, DO, DSF, HUF, LUF, O, S [Kelco]—Propylene glycol alginate
Kelcosol [Kelco/Div. of Merck & Co.]—Algin
Kelgin F, HV, LV, MV, RL, XL [Kelco/Div. of Merck & Co.]—Algin (refined)
Kelgin QH, QL, QM [Kelco/Div. of Merck & Co.]—Algin (treated for improved
    dispersion)
Kelmar, Kelmar Improved [Kelco]—Potassium alginate
Keltex, P, S [Kelco/Div. of Merck & Co.]—Algin (industrial)
Keltone [Kelco/Div. of Merck & Co.]—Algin
Keltrol [Kelco]—Xanthan gum (food grade)
Keltrol F [Kelco]—Xanthan gum (food grade)
Kelvis [Kelco/Div. of Merck & Co.]—Algin (refined)
Kelzan D 35, S [Kelco]—Xanthan gum
Kelzan, D, M, XC Polymer [Kelco]—Xanthan gum (industrial grade)
Kemester 104, 105, 115, 205 [Humko]—Methyl oleate
Kemester 4000 [Humko]—Butyl oleate
Kessco Butyl Oleate [Armak]—Butyl oleate
Kessco Ethylene Glycol Monostearate 70 [Armak]—Ethylene glycol monostearate
Kessco Ethylene Glycol Monostearate [Armak]—Ethylene glycol monostearate
Kessco PEG 1000 Dilaurate [Armak]—POE (20) dilaurate
Kessco PEG 1000 Monolaurate [Armak]—POE (20) monolaurate
Kessco PEG 1540 Dilaurate [Armak]—POE (32) dilaurate
Kessco PEG 4000 Dilaurate [Armak]—POE (75) dilaurate
Kessco PEG 6000 Dilaurate [Armak]—POE (150) dilaurate
Kessco PEG 6000 Distearate [Armak]—POE (150) distearate
Kessco Triacetin [Armak]—Glyceryl triacetate
Klucel, 6, EF, GF, HF, J, LF, MF [Hercules]—Hydroxypropylcellulose
Kodaflex DBP [Eastman]—Dibutyl phthalate
Kodaflex DEP [Eastman]—Diethyl phthalate
Kodaflex DMP [Eastman]—Dimethyl phthalate
Kodaflex DOA [Eastman]—Dioctyl adipate

Kodaflex DOP [Eastman]—Dioctyl phthalate
Kodaflex TOTM [Eastman]—Trioctyl trimellitate
Kodaflex Triacetin [Eastman]—Glyceryl triacetate
Kronitex TCP [FMC Corp.]—Tricresyl phosphate
Kuplur SML [BASF Wyandotte]—Sorbitan monolaurate
Kuplur STS [BASF Wyandotte]—Sorbitan tristearate
Lakeway 100-CA [Bofors Lakeway]—Coconut diethanolamide
Lanexol AWS [Croda]—POP (12) POE (50) lanolin
Lanol S [Seppic]—Stearyl alcohol
Lauridit KD, KDG [Akzo Chemie]—Coconut diethanolamide
Lauridit LM [Akzo Chemie]—Lauric monoethanolamide
Lauridit OD [Akzo Chemie]—Oleic diethanolamide
Lexaine CSB-35, CSB-50 [Inolex]—Cocamidopropyl hydroxysultaine
Lexemul EGMS [Inolex]—Ethylene glycol monostearate
Lexol DIA [Inolex]—Diisopropyl adipate
Lindol [Harwick]—Tricresyl phosphate
Lipamide S [Lipo]—Stearic diethanolamide (1:1)
Lipo EGMS [Lipo]—Ethylene glycol monostearate
Liponic SO-20 [Lipo]—POE (20) sorbitol ether
Lipopeg 6000-DS [Lipo]—POE (150) distearate
Liposorb L [Lipo]—Sorbitan monolaurate
Liposorb TO [Lipo]—Sorbitan trioleate
Liposorb TS [Lipo]—Sorbitan tristearate
Locust Bean Gum FL 50-40, FL 50-50, FL 70-50 [Hercules]—Locust bean gum
Lonzaine CS [Lonza]—Cocamidopropyl hydroxysultaine
Lonzest STO [Lonza]—Sorbitan trioleate
Lonzest STS [Lonza]—Sorbitan tristearate
Loramine C212 [Dutton & Reinisch]—Coconut monoethanolamide
Loramine DC212/S, DC212/SE, DC220/SE [Dutton&Reinisch]—Coconut dietha-
    nolamide
Loramine DL 203/S [Dutton & Reinisch]—Lauric diethanolamide
Loramine DO280/SE [Dutton & Reinisch]—Oleic diethanolamide
Loramine IPE 280 [Dutton & Reinisch]—Oleic monoisopropanolamide
Loramine IPL203 [Dutton & Reinisch]—Lauric monoisopropanolamide
Loramine IPP 240 [Dutton & Reinisch]—Coconut monoisopropanolamide
Loramine L203 [Dutton & Reinisch]—Lauric monoethanolamide
Loramine S 280 [Dutton & Reinisch]—Stearic monoethanolamide
Loropan LD [Thomas Triantaphyllou]—Lauric diethanolamide
Loropan OD [Thomas Triantaphyllou]—Oleic diethanolamide
Lubral 710 [Alcoa]—Aluminum hydroxide
Macaloid [NL Chem.]—Magnesium aluminum silicate
Mackam HV [McIntyre]—Oleamidopropyl betaine

371

Mackam LMB [McIntyre]—Lauramidopropyl betaine
Mackam LMB-LS [McIntyre]—Lauramidopropyl betaine (low salt)
Mackamide 100-A, C, MC [McIntyre]—Coconut diethanolamide
Mackamide CMA [McIntyre]—Coconut monoethanolamide
Mackamide L-95, LLM, LMD [McIntyre]—Lauric diethanolamide
Mackamide LMM [McIntyre]—Lauric monoethanolamide
Mackamide O [McIntyre]—Oleic diethanolamide
Mackamide PK [McIntyre]—Palm kernel oil acid diethanolamide
Mackamide S [McIntyre]—Soya diethanolamide
Mackazoline O [McIntyre]—Oleyl hydroxyethyl imidazoline
Macol NP-30, NP-30(70) [Mazer]—POE (30) nonyl phenyl ether
Macol NP-100 [Mazer]—POE (100) nonyl phenyl ether
Makon 30 [Stepan]—POE (30) nonyl phenyl ether
Manro CD, CDS [Manro]—Coconut diethanolamide
Manro CMEA [Manro]—Coconut monoethanolamide
Mapeg 6000DS [Mazer]—POE (150) distearate
Mapeg EGMS [Mazer]—Ethylene glycol monostearate
Mark 99, 180, 1314, 1330C, 1413, TT, WS [Argus]—Barium-cadmium compound
Mark 503, 507, 755, 1014, 2109, 2114, 2115 Series, BB [Argus]—Barium-cadmium-zinc
    compound
Mark 1178, 1178B, TNPP [Argus]—Tris (nonylphenyl) phosphite
Mark 1413 [Argus]—2-Hydroxy-4-(octyloxy) benzophenone
Mark TPP [Argus]—Triphenyl phosphite
Marlamid D1218 [Chem.Werke Huls]—Coconut diethanolamide
Marlamid D1885 [Chemische Werke Huls AG]—Oleic diethanolamide
Marlamid M1218 [Chem.Werke Huls]—Coconut monoethanolamide
Marsamid 10, 40, 50, [Mars]—Coconut diethanolamide
Marvanol SCO 75% [Marlowe-Van Loan]—Castor oil, sulfated
Mazamide 70, 80, CA-20, CS-148 [Mazer]—Coconut diethanolamide
Mazamide C-5 [Mazer]—POE (6) coconut amide
Mazamide L-5 [Mazer]—POE (6) lauryl amide
Mazamide LM-20 [Mazer]—Lauric diethanolamide
Mazamide LS-173, LS-196 [Mazer]—Lauric diethanolamide
Mazamide O-20 [Mazer]—Oleic diethanolamide (2:1)
Mazamide SS-10 [Mazer]—Soya diethanolamide
Mazol PGO-104 [Mazer]—Decaglycerol tetraoleate
Merpinal EH 40, Q 147 [Kempen]—Sodium octyl sulfate
Merpinamid KD11, LD/E, LSD/E [Kempen]—Lauric diethanolamide
Merpinamid LMIPA [Kempen]—Lauric monoisopropanolamide
Merpinamid OD [Elektrochemische Fabrik Kempen]—Oleic diethanolamide
Merpoxen NO300 [Kempen]—POE (30) nonyl phenyl ether
Methocel A, A4C, A4M, A15-LV [Dow Chem.]—Methylcellulose

Methocel E, E4M, E5, E15-LV, E50-LV, F, F4M, F50-LV, J, K, K3, K4M, K15M, K35, K100-LV, K100M [Dow Chem.]—Hydroxypropyl methylcellulose
Micral 916, 932 [Solem]—Aluminum hydroxide
Minugel 400, LF [Floridin]—Attapulgite (colloidal)
Miramine OC [Miranol]—Oleyl hydroxyethyl imidazoline
Miramine TOC [Miranol]—Tall oil hydroxyethyl imidazoline
Miranol ISM [Miranol]—Isostearoamphopropionate
Mirataine BB [Miranol]—Lauramidopropyl betaine
Mirataine CBS [Miranol]—Cocamidopropyl hydroxysultaine
Mirataine TM [Miranol]—Dihydroxyethyl tallow glycinate
Mixxim LS-24 [Fairmount]—2,4-Dihydroxybenzophenone
Monamid 7-100, 7-153CS, 150-AD, 150-ADD, 150-DR, 759 [Mona]—Coconut diethanolamide
Monamid 15-70W, 150-ADY [Mona]—Linoleoyl diethanolamide (1:1)
Monamid 150-GLT, 150-LMW-C, 150-LW, 150-LWA, 716, 770 [Mona]—Lauric diethanolamide
Monamid 150-IS [Mona]—Isostearic diethanolamide (1:1)
Monamid 718 [Mona]—Stearic diethanolamide (1:1)
Monamid CMA [Mona]—Coconut monoethanolamide
Monamid LIPA [Mona]—Lauric monoisopropanolamide
Monamid LMA, LMMA [Mona]—Lauric monoethanolamide
Monamid LMIPA [Mona]—Lauric monoisopropanolamide
Monamid S [Mona]—Stearic monoethanolamide (1:1)
Monamide [Zohar Detergent]—Coconut monoethanolamide
Monamine AA-100, AD-100, ADD-100, ADD-100LE, ALX-80SS, ALX-100S, ARA-100, I-76 [Mona]—Coconut diethanolamide
Monamine ACO-100, LM-100 [Mona]—Lauric diethanolamide
Monamine ADY-100 [Mona]—Linoleoyl diethanolamide (1:2)
Monamine T-100 [Mona]—Tall oil acid diethanolamide (1:2)
Monateric ISA-35 [Mona]—Isostearoamphopropionate
Monateric LMAB [Mona]—Lauramidopropyl betaine
Monazoline CY [Mona]—Capryl hydroxyethyl imidazoline
Monazoline O [Mona]—Oleyl hydroxyethyl imidazoline
Monazoline T [Mona]—Tall oil hydroxyethyl imidazoline
Monomuls 60-30 [Chemische Fabrik Grunau GmbH]—Palm oil glycerides
Monoplex DDA [Rohm & Haas]—Diisodecyl adipate
Monoplex DIOA [Rohm & Haas]—Diisooctyl adipate
Monoplex DOA [C.P. Hall]—Dioctyl adipate
Monoplex DOS [C.P. Hall]—Dioctyl sebacate
Monoplex NODA [C.P. Hall; Rohm & Haas]—n-Octyl-n-decyl adipate
Monoplex S-73 [Rohm & Haas]—Epoxidized octyl tallate
Monopol Oil 75 [GAF]—Castor oil, sulfated

Montane 20 [Seppic]—Sorbitan monolaurate
Montane 65 [Seppic]—Sorbitan tristearate
Montane 85 [Seppic]—Sorbitan trioleate
Morflex 330 [Morflex]—Diisodecyl adipate
Naphthenic Oil 100 SUS, 150 SUS, 200 SUS, 1300 SUS, 2400 SUS, 6000 SUS [R.E.
    Carroll]—Naphthenic oil
Natrosol, Natrosol 250 [Hercules]—Hydroxyethylcellulose
Naturechem GMHS [CasChem]—Glyceryl hydroxystearate
Naturechem PGHS [CasChem]—Propylene glycol hydroxystearate
Naturechem PGR [CasChem]—Propylene glycol monoricinoleate
Naugard BHT-Food Grade [Uniroyal]—Butylated hydroxytoluene
Naugard BHT-Tech. [Uniroyal]—Butylated hydroxytoluene
Newcol 20 [Nippon Nyukazai]—Sorbitan monolaurate
Niaproof 08 [Niacet]—Sodium octyl sulfate
Nikkol SO-30 [Nikko]—Sorbitan trioleate
Nikkol SS-30 [Nikko]—Sorbitan tristearate
Nimco 1781 Lanolin Acids [Emery]—Lanolin acid (cosmetic grade)
Ninol 128 Extra, 2012 Extra [Stepan]—Coconut diethanolamide
Ninol 201 [Stepan]—Oleic diethanolamide
Ninol 4821, AA-62, AA-62 Extra, P-616, P-621 [Stepan]—Lauric diethanolamide
Ninol CNR [Stepan Europe]—Coconut monoethanolamide
Ninol SNR [Stepan]—Stearic monoethanolamide
Ninox CA [Stepan]—Cocamidopropylamine oxide
Ninox FCA [Stepan Europe]—Cocamidopropylamine oxide
Ninox M [Stepan]—Myristyl dimethyl amine oxide
Nissan Nonion LP-20R, LP-20RS [Nippon Oil & Fats]—Sorbitan monolaurate
Nissan Nonion OP-85·R [Nippon Oil & Fats]—Sorbitan trioleate
Nissan Stafoam MF [Nippon Oil & Fat]—Coconut monoethanolamide
Nitrene 11120, 11230, 13026, A-309, A-567, C, C-Extra [Henkel]—Coconut dietha-
    nolamide
Nitrene L-90 [Henkel]—Lauric diethanolamide
Nopcocastor, Nopcocastor L [Diamond Shamrock]—Castor oil, sulfated
Nopcogen 14-S [Diamond Shamrock]—Stearic diethanolamide
Nopcosulf CA-60, CA-70 [Diamond Shamrock]—Castor oil, sulfated
Nuostabe V-133, V-134, V-1099, V-1399, V-1728, V-1760, V-1764, V-1767, V-1786,
    V-1785 [Tenneco]—Barium-cadmium compound
Nuostabe V-1207, V-1541 [Tenneco]—Barium-cadmium-zinc compound
Onyxol 42 [Onyx]—Stearic diethanolamide (1:1)
Onyxol 336, 345 [Onyx]—Lauric diethanolamide
Onyxol 345 [Onyx]—Coconut diethanolamide
Ottalume 2100 [Ferro]—Zinc oxide (fluorescent)
Palatinol DBP [Badische]—Dibutyl phthalate

Palatinol DIDP [Badische]—Diisodecyl phthalate
Palatinol DIDP-E [Badische]—Diisodecyl phthalate (with 0.25% Bisphenol-A)
Palatinol DOA [Badische]—Dioctyl adipate
Palatinol DOP [Badische]—Dioctyl phthalate
Palatinol DOP-SG [Badische]—Dioctyl phthalate (special grade)
Palatinol N [Badische]—Diisononyl phthalate
Palatinol TOTM [Badische]—Trioctyl trimellitate
Palatinol TOTM-E [Badische]—Trioctyl trimellitate (with 0.5% Bisphenol A)
Paraplex G-60, G-62 [C.P. Hall; Rohm & Haas]—Epoxidized soybean oil
Paricin 9 [CasChem]—Propylene glycol hydroxystearate
Paricin 13 [CasChem]—Glyceryl hydroxystearate
Paroil 1160 [Dover]—Chlorinated paraffin
Peganol NP40, NP40 (70%) [Borg-Warner]—POE (40) nonyl phenyl ether
Peganol NP50, NP50 (70%) [Borg-Warner]—POE (50) nonyl phenyl ether
Peganol NP100, NP100 (70%) [Borg-Warner]—POE (100) nonyl phenyl ether
Pegosperse 50MS [Glyco]—Ethylene glycol monostearate
Pegosperse 6000-DS [Glyco]—POE (150) distearate
Pharmasorb Colloidal [Engelhard Min. & Chem.]—Attapulgite (colloidal)
Pharmasorb Reg. [Engelhard Min. & Chem.]—Attapulgite
Phosflex 179C [Harwick]—Tricresyl phosphate
Photox-7, -8, -77, -80, -85, -90, -801 [N.J. Zinc]—Zinc oxide (French process)
Plas-Chek 775 [Ferro]—Epoxidized soybean oil
Plasthall 503 [C.P. Hall]—Butyl oleate
Plasthall DIBA [C.P. Hall]—Diisobutyl adipate
Plasthall DIDA [C.P. Hall]—Diisodecyl adipate
Plasthall DOA [C.P. Hall]—Dioctyl adipate
Plasthall DOS [C.P. Hall]—Dioctyl sebacate
Plasthall DOZ [C.P. Hall]—Dioctyl azelate
Plasthall NODA [C.P. Hall]—n-Octyl-n-decyl adipate
Plasthall TIOTM [C.P. Hall]—Triisooctyl trimellitate
Plasthall TOTM [C.P. Hall]—Trioctyl trimellitate
Plastoflex 2307 [Interstab]—Epoxidized soybean oil
Plurest DS150 [BASF Wyandotte]—POE (150) distearate
Polycizer 162, DOP [Harwick]—Dioctyl phthalate
Polycizer Butyl Oleate [Harwick]—Butyl oleate
Polycizer DBP [Harwick]—Dibutyl phthalate
Polycizer DBS [Harwick]—Dibutyl sebacate
Polycizer DIDP [Harwick]—Diisodecyl phthalate
Polycizer DOA [Harwick]—Dioctyl adipate
Polycizer DOS [Harwick]—Dioctyl sebacate
Polycizer TOTM [Harwick]—Trioctyl trimellitate
Polygard [Uniroyal]—Tris (nonylphenyl) phosphite

375

Polystep A-18 [Stepan]—Sodium olefin sulfonate (C_{14-16})
Polystep F-9 [Stepan]—POE (30) nonyl phenyl ether
Polystep F-10 [Stepan]—POE (40) nonyl phenyl ether
Product LT 10-8-1 [Clintwood]—Oleic diethanolamide
Product LT 15-73-1 [Clintwood]—Linoleoyl diethanolamide
Product LT 18-48 [Clintwood]—Lauric diethanolamide
Profan AA62 Extra [Sanyo]—Lauric diethanolamide
Progacyl COS-1 [Lyndal]—Guar gum
Progacyl COS-20, COS-70 [Lyndal]—Hydroxypropyl guar
Protox-169 [N.J. Zinc]—Zinc oxide (surface treated, with propionic acid)
PX-104 [USS Chem.]—Dibutyl phthalate
PX-120 [USS Chem.]—Diisodecyl phthalate
PX-126 [USS Chem.]—Ditridecyl phthalate
PX-138 [USS Chem.]—Dioctyl phthalate
PX-209 [USS Chem.]—Diisononyl adipate
PX-238 [USS Chemical]—Dioctyl adipate
PX-338 [USS Chem.]—Trioctyl trimellitate
PX-504 [USS Chem.]—Dibutyl maleate
PX-538 [USS Chem.]—Dioctyl maleate
PX-914 [USS Chem.]—Butyl octyl phthalate
Radia 7040 [Oleofina S.A.]—Butyl oleate
Radia 7060 [Oleofina S.A.]—Methyl oleate
Radia 7194 [Oleofina S.A.]—Diisopropyl adipate
Radia 7197 [Oleofina]—Diisobutyl adipate
Radiamuls 125, SORB 2125 [Oleofina]—Sorbitan monolaurate
Radiamuls 345 [Oleofina S.A.]—Sorbitan tristearate
Radiamuls SORB 2345 [Oleofina S.A.]—Sorbitan tristearate
Radiamuls SORB 2355 [Oleofina]—Sorbitan trioleate
Radiasurf 7125 [Oleofina]—Sorbitan monolaurate
Radiasurf 7144 [Oleofina SA]—Glyceryl monococoate
Radiasurf 7153 [Oleofina]—Glyceryl monoricinoleate
Radiasurf 7270 [Oleofina]—Ethylene glycol monostearate
Refinex [Floridin]—Attapulgite
Renex 650 [ICI United States]—POE (30) nonyl phenyl ether
Rewomid 203/S, DL203 [Dutton & Reinisch]—Lauric diethanolamide
Rewomid C212 [Dutton & Reinisch]—Coconut monoethanolamide
Rewomid DC212/S, DC220/SE [Dutton & Reinisch]—Coconut diethanolamide
Rewomid DC212/SE, DC220/LS [Rewo Chem. Werke]—Coconut diethanolamide
Rewomid DO280SE [Dutton & Reinisch]—Oleic diethanolamide
Rewomid F [Rewo Chemische Werke GmbH]—Linoleoyl diethanolamide
Rewomid IPE 280 [Dutton & Reinisch]—Oleic monoisopropanolamide
Rewomid IPL203 [Dutton & Reinisch]—Lauric monoisopropanolamide

Rewomid IPP 240 [Dutton & Reinisch]—Coconut monoisopropanolamide
Rewomid L203 [Dutton & Reinisch]—Lauric monoethanolamide
Rewominoxid B 204 [Dutton & Reinisch]—Cocamidopropylamine oxide
Rewopal C-6 [Dutton & Reinisch]—POE (6) coconut amide
Rewopal PEG 6000 DS [Rewo Chemische Werke GmbH]—POE (150) distearate
Rewopol NEHS 40 [Rewo Chemische]—Sodium octyl sulfate
Rexol 25/30, 25/307 [Hart Chem. Ltd.]—POE (30) nonyl phenyl ether
Rexol 25/50, 25/507 [Hart Chem. Ltd.]—POE (50) nonyl phenyl ether
Rexol 25/100-70% [Hart Chem. Ltd.]—POE (100) nonyl phenyl ether
Rhodigel 23 [R.T. Vanderbilt]—Xanthan gum
Rhodopol [R.T. Vanderbilt]—Xanthan gum
Richamide 5085 [Richardson]—Oleic diethanolamide (2:1)
Richamide 6310 [Richardson]—Lauric diethanolamide
Richamide 6404, M-3 [Richardson]—Coconut diethanolamide
Richamide MX [Richardson]—Coconut monoethanolamide
Rilanit DBS [Henkel KGaA]—Dibutyl sebacate
Rilanit DEHS [Henkel KGaA]—Dioctyl sebacate
Rilanit DNOP [Henkel KGaA]—Dioctyl phthalate
Ritalafa [R.I.T.A.]—Lanolin acid
Ritawax ALA [Rita]—Acetylated lanolin alcohol
Rycel 100, 105 [Ryco]—Sodium carboxymethyl cellulose
Santicizer 8 [Monsanto]—Ethyl toluenesulfonamide
Santicizer 160 [Monsanto]—Butyl benzyl phthalate
Sarkosyl L, NL-30 [Ciba-Geigy]—Lauroyl sarcosine
Sarkosyl LC [Ciba-Geigy]—Cocoyl sarcosine
Sarkosyl O [Ciba Geigy]—Oleyl sarcosine
SB 100, 30-0, 400, 500, 600, 700 Series [Solem]—Aluminum hydroxide
Schercamox C-AA [Scher]—Cocamidopropylamine oxide
Schercamox DMA, DMM [Scher]—Myristyl dimethyl amine oxide
Schercemol DIA [Scher]—Diisopropyl adipate
Schercemol EGMS [Scher]—Ethylene glycol monostearate
Schercomid 1214, LD, SL-Extra, SLM, SLM-C, SL-ML, SLM-LC, SLM-S [Scher]—
    Lauric diethanolamide
Schercomid CCD, CDA, CDA-H, SCE, SCO-Extra [Scher]—Coconut diethanolamide
Schercomid CME [Scher]—Coconut monoethanolamide
Schercomid CMI [Scher]—Coconut monoisopropanolamide
Schercomid ID, SI-M [Scher]—Isostearic diethanolamide
Schercomid MME [Scher]—Myristic monoethanolamide
Schercomid ODA [Scher]—Oleic diethanolamide
Schercomid OMI [Scher]—Oleic monoisopropanolamide
Schercomid SI [Scher]—Isostearic diethanolamide (1:1)
Schercomid SLE, SLS [Scher]—Linoleoyl diethanolamide

Schercomid SME [Scher]—Stearic monoethanolamide
Schercomid SME-A, SME-M [Scher]—Stearic monoethanolamide (1:1)
Schercomid SME-S [Scher]—Stearic monoethanolamide stearate
Schercomid SO-A [Scher]—Oleic diethanolamide (1:1)
Schercomid SO-T, TO-2 [Scher]—Tall oil acid diethanolamide
Schercopearl EA-100 [Scher]—Stearic monoethanolamide
Schercotaine OAB [Scher]—Oleamidopropyl betaine
Schercotaine PAB [Scher]—Palmitamidopropyl betaine
Schercotaine SCAB [Scher]—Cocamidopropyl hydroxysultaine
Schercozoline O [Scher]—Oleyl hydroxyethyl imidazoline
Seachem 55 [Seaboard]—Potassium ricinoleate
Serdet DSK 40 [Servo BV]—Sodium octyl sulfate
Serdox NNP30/70 [Servo B.V.]—POE (30) nonyl phenyl ether
Sermul EN30/70, EN145 [Servo B.V.]—POE (30) nonyl phenyl ether
Sipex BOS, OLS [Alcolac]—Sodium octyl sulfate
Sipomide 843 [Alcolac]—Lauric diethanolamide
Sipomide 1500 [Alcolac]—Coconut diethanolamide
Siponate 301-10F [Alcolac]—Sodium olefin sulfonate (C_{14-16})
Siponate A-167, A-168 [Alcolac]—Sodium olefin sulfonate (C_{16-18})
Siponate A-246, A-246L, A246LX [Alcolac]—Sodium olefin sulfonate (C_{14-16})
Siponic NP40 [Alcolac]—POE (40) nonyl phenyl ether
Skliro [Croda]—Lanolin acid (cosmetic grade)
S-Maz 20, 20R [Mazer]—Sorbitan monolaurate
S-Maz 65 [Mazer]—Sorbitan tristearate
S-Maz 85 [Mazer]—Sorbitan trioleate
Softigen 701 [Dynamit Nobel]—Glyceryl monoricinoleate
Softigen 701 [Dynamit-Nobel]—Glyceryl hydroxystearate
Sole-Terge TS-2-S [Hodag]—Sodium octyl sulfate
Solricin 135 [NL Industries]—Potassium ricinoleate
Sol-Speedi-Dri, Auto-Dri [Engelhard Min. & Chem.]—Attapulgite
Soprofor S/20 [Mario Geronazzo]—Sorbitan monolaurate
Soprofor S-65 [Geronazzo SpA]—Sorbitan tristearate
Soprofor S/85 [Geronazzo SpA]—Sorbitan trioleate
Sorbax SML [Chemax]—Sorbitan monolaurate
Sorbax STO [Chemax]—Sorbitan trioleate
Sorbax STS [Chemax]—Sorbitan tristearate
Sorbon S-20 [Toho]—Sorbitan monolaurate
Sorgen 90 [Dai-ichi Kogyo Seiyaku]—Sorbitan monolaurate
Span 20 [ICI]—Sorbitan monolaurate
Span 65 [ICI United States]—Sorbitan tristearate
Span 85 [ICI United States]—Sorbitan trioleate
Staflex BOP [Reichhold]—Butyl octyl phthalate

Staflex DBM [Reichhold]—Dibutyl maleate
Staflex DBP [Reichhold]—Dibutyl phthalate
Staflex DIDA [Reichhold]—Diisodecyl adipate
Staflex DIDP [Reichhold]—Diisodecyl phthalate
Staflex DINA [Reichhold]—Diisononyl adipate
Staflex DIOA [Reichhold]—Diisooctyl adipate
Staflex DIOP [Reichhold]—Diisooctyl phthalate
Staflex DMP [Reichhold]—Dimethyl phthalate
Staflex DOA [Reichhold]—Dioctyl adipate
Staflex DOM [Reichhold]—Dioctyl maleate
Staflex DOP [Reichhold]—Dioctyl phthalate
Staflex DOZ [Reichhold]—Dioctyl azelate
Staflex DTDP [Reichhold]—Ditridecyl phthalate
Staflex NODA [Reichhold]—*n*-Octyl-*n*-decyl adipate
Staflex TIOTM [Reichhold]—Triisooctyl trimellitate
Staflex TOTM [Reichhold]—Trioctyl trimellitate
Stafoam DF-1, DF-4, F [Nippon Oil & Fats]—Coconut diethanolamide
Stafoam DL [Nippon Oil & Fats]—Lauric diethanolamide
Stanclere T-94 C, T-126, T-801 [Interstab]—Butyltin mercaptide
Standamid 100 [Henkel Argentina]—Coconut monoethanolamide
Standamid CM, CMG, KM, SM [Henkel]—Coconut monoethanolamide
Standamid ID [Henkel]—Isostearic diethanolamide (1:1)
Standamid KD, PD, SD [Henkel]—Coconut diethanolamide
Standamid LD, LD 80/20, LDM, LDO, LDS [Henkel]—Lauric diethanolamide
Standamid LM [Henkel]—Lauric monoethanolamide
Standamid SOD [Henkel]—Linoleoyl diethanolamide (1:1)
Standamid SOMD [Henkel]—Linoleoyl diethanolamide
Standamox CAW [Henkel]—Cocamidopropylamine oxide
Standapol SCO [Henkel]—Castor oil, sulfated
Stave TNPP [Stave]—Tris (nonylphenyl) phosphite
Stearal [Amerchol]—Stearyl alcohol (USP grade)
Steinamid 203/S [Dutton & Reinisch]—Lauric diethanolamide
Steinamid C212 [Dutton & Reinisch]—Coconut monoethanolamide
Steinamid DC212/S, DC220/SE [Dutton&Reinisch]—Coconut diethanolamide
Steinamid DO280/SE [Dutton & Reinisch]—Oleic diethanolamide
Steinamid IPE 280 [Dutton & Reinisch]—Oleic monoisopropanolamide
Steinamid IPL203 [Dutton & Reinisch]—Lauric monoisopropanolamide
Steinamid IPP 240 [Dutton & Reinisch]—Coconut monoisopropanolamide
Steinamid L203 [Dutton & Reinisch]—Lauric monoethanolamide
Stepan C68 [Stepan]—Methyl oleate
Stepantan 29N, 39N [Stepan]—Sodium olefin sulfonate (linear)
Sterling AOS [Canada Packers]—Sodium olefin sulfonate (C_{14-16})

Sterling DEA [Canada Packers]—Coconut diethanolamide
Sterling Granulated Wax [Canada Packers]—Coconut monoethanolamide
Sterling LDEA-90 [Canada Packers]—Lauric diethanolamide
Sulfotex OA [Henkel]—Sodium octyl sulfate
Sulframin AOS, AOS90 [Witco/Organics]—Sodium olefin sulfonate
Super Amide GR [Onyx]—Coconut diethanolamide
Super Amide L9, L9C, LL, LM [Onyx]—Lauric diethanolamide
Supercol U [Henkel]—Guar gum
Supratol VF [Hart]—Castor oil, sulfated
Surco 128-T, SR-200, WC Conc. [Onyx]—Coconut diethanolamide
Surco AOS [Onyx]—Sodium olefin sulfonate
Surco CMEA, CMEA Flake [Onyx]—Coconut monoethanolamide
Surco Coco Condensate [Onyx]—Coconut diethanolamide
Surfactol-13 [NL Industries]—Castor oil (modified)
Surfonic N-300, NB-5 [Jefferson]—POE (30) nonyl phenyl ether
Surfonic N-400, NB-14 [Jefferson]—POE (40) nonyl phenyl ether
Sustane BHA, BHA 1-F [UOP Process Div.]—Butylated hydroxyanisole
Sustane BHT [UOP Process]—Butylated hydroxytoluene (food grade)
Sustane PG [UOP]—Propyl gallate
Sustane TBHQ [UOP]—*t*-Butyl hydroquinone
Swanic 51 [Swastik]—Coconut monoethanolamide
Synotol 119N, CN60, CN80, CN90 [PVO]—Coconut diethanolamide
Synotol L60, LM60 [PVO]—Lauric diethanolamide
Synotol L90 [Drew Produtos]—Lauric diethanolamide
Synperonic NP30 [ICI Petrochem. Div.]—POE (30) nonyl phenyl ether
Synperonic NP40 [ICI Petrochem. Div.]—POE (40) nonyl phenyl ether
Synperonic NP50 [ICI Petrochem. Div.]—POE (50) nonyl phenyl ether
Synpron 1343, 1434 [Synthetic Prod.]—Barium-cadmium-zinc compound
Syntase 62 [Neville-Synthese]—2-Hydroxy-4-methoxybenzophenone
Syntase 100 [Neville]—2,4-Dihydroxybenzophenone
Syntase 230 [Neville]—2-Hydroxy-4-methoxybenzophenone-5-sulfonic acid
Syntase 800 [Neville]—2-Hydroxy-4-(octyloxy) benzophenone
TBC [Croda]—Tributyl citrate
T-Det N-30, N-307 [Thompson-Hayward]—POE (30) nonyl phenyl ether
T-Det N-40, N-407 [Thompson-Hayward]—POE (40) nonyl phenyl ether
T-Det N-50, N-507 [Thompson-Hayward]—POE (50) nonyl phenyl ether
T-Det N-100, N-1007 [Thompson-Hayward]—POE (100) nonyl phenyl ether
Techfill A-100 Series, A-200 Series [Great Lakes Minerals]—Aluminum hydroxide
Techfill AS-101, AS-1005 [Great Lakes Minerals]—Aluminum hydroxide (precipitated
    grades)
Tecquinol [Eastman]—Hydroquinone
Tegin G6100 [Th. Goldschmidt A.G.]—Ethylene glycol monostearate

380

Tenox BHA [Eastman Chem. Prod.]—Butylated hydroxyanisole
Tenox PG [Eastman]—Propyl gallate
Tenox TBHQ [Eastman]—*t*-Butyl hydroquinone
Tergitol NP-40, NP-40 (70% Aqueous), NP-44 [Union Carbide]—POE (40) nonyl phenyl
    ether
Teric 100 [ICI Australia Ltd.]—POE (100) nonyl phenyl ether
Teric CDE [ICI Australia]—Coconut diethanolamide
Teric N30 [ICI Australia Ltd.]—POE (30) nonyl phenyl ether
Teric N40 [ICI Australia Ltd.]—POE (40) nonyl phenyl ether
Therm-Chek 6-V-6A, 1292, 5918 [Ferro]—Barium-cadmium-zinc compound
Thixcin E, R [NL Industries]—Glyceryl tri(12-hydroxystearate)
T-Maz 28 [Mazer]—POE (80) sorbitan monolaurate
Tribase [Associated Lead]—Tribasic lead sulfate
Tribase E Special, EXL Special [Associated Lead]—Tribasic lead sulfate (modified)
Tribase XL [Associated Lead]—Tribasic lead sulfate (coated)
Triton N-302 [Rohm & Haas]—POE (30) nonyl phenyl ether
Triton N-401 [Rohm & Haas]—POE (40) nonyl phenyl ether
Triton N-998, N-998-70% [Rohm & Haas]—POE (100) nonyl phenyl ether
Trycol NP-30, NP-307 [Emery]—POE (30) nonyl phenyl ether
Trycol NP-40, NP-407 [Emery]—POE (40) nonyl phenyl ether
Trycol NP-50, NP-507 [Emery]—POE (50) nonyl phenyl ether
Trycol NP-1007 [Emery]—POE (100) nonyl phenyl ether
Trylox SS-20 [Emery]—POE (20) sorbitol ether
Tylose H Series [Hoechst AG]—Hydroxyethylcellulose
Ultrawet AOK [Arco]—Sodium olefin sulfonate (C_{14-16})
Unamide C-5 [Lonza]—POE (6) coconut amide
Unamide CDX, JJ-35, LDL, N-72-3 [Lonza]—Coconut diethanolamide
Unamide CMX [Lonza]—Coconut monoethanolamide
Unamide J-56, LDX [Lonza]—Lauric diethanolamide
Unamide L-5 [Lonza]—POE (6) lauryl amide
Unamide S [Lonza]—Stearic diethanolamide (1:1)
Unamine O [Lonza]—Oleyl hydroxyethyl imidazoline
Uniflex BYO [Union Camp]—Butyl oleate
Uniflex DBS [Union Camp]—Dibutyl sebacate
Uniflex DCA [Union Camp]—Dicapryl adipate
Uniflex DOA [Union Camp]—Dioctyl adipate
Uniflex DOS [Union Camp]—Dioctyl sebacate
Uniflex TOTM [Union Camp]—Trioctyl trimellitate
Unimoll DB [Bayer AG]—Dibutyl phthalate
UV-Absorber 325 [Bayer AG; Mobay]—2-Hydroxy-4-methoxybenzophenone
Uvasorb 20H [3-V Chem.]—2,4-Dihydroxybenzophenone
Uvasorb MET-3 [3-V Chem.]—2-Hydroxy-4-methoxybenzophenone

UV-Chek AM-300, AM-301 [Ferro]—2-Hydroxy-4-(octyloxy) benzophenone

Uvinul 400 [BASF Wyandotte]—2,4-Dihydroxybenzophenone

Uvinul 408 [BASF Wyandotte]—2-Hydroxy-4-(octyloxy) benzophenone

Uvinul M-40 [BASF-Wyandotte]—2-Hydroxy-4-methoxybenzophenone

Uvinul MS-40 [BASF Wyandotte]—2-Hydroxy-4-methoxybenzophenone-5-sulfonic acid

Vancide 89, 89RE [R.T. Vanderbilt]—Captan

Van Gel [R.T. Vanderbilt]—Magnesium aluminum silicate

Vanlube PC [R.T. Vanderbilt]—Butylated hydroxytoluene

Vanstay 162-B, 246, 3027, 6032, 6040, 6053, 6055, 6074, 6078, 6133, 6172, 6191, 6201, HA, HTF, RRE, RRZ [R.T. Vanderbilt]—Barium-cadmium-zinc compound

Vanstay 4017, 4030, 4039, 7024, 7025, 7032, HT, HTA, RR [R.T. Vanderbilt]—Barium-cadmium compound

Vanstay RZ-25 [R.T. Vanderbilt]—Barium-cadmium-zinc compound (with chelating agents)

Varamide A-2, A-10, A-12, A-83 [Sherex]—Coconut diethanolamide

Varamide A-7 [Sherex]—Oleic diethanolamide

Varamide L-1, ML-4, SL-9 [Sherex]—Lauric diethanolamide

Varox 1770 [Sherex]—Cocamidopropylamine oxide

Veegum, F, HS, HV, K, Neutral, S-728, WG [R.T. Vanderbilt]—Magnesium aluminum silicate

Veegum T [R.T. Vanderbilt]—Magnesium aluminum silicate (tech.)

Vulkanox KB [Mobay]—Butylated hydroxytoluene

Weston 399, 399B, TNPP [Borg Warner]—Tris (nonylphenyl) phosphite

Weston TPP [Borg Warner]—Triphenyl phosphite

Wickenol 158 [Wickhen Products]—Dioctyl adipate

Witcamide 61 [Witco]—Oleic monoisopropanolamide

Witcamide 70 [Witco/Organics]—Stearic monoethanolamide

Witcamide 82, 1017, 5130, 5133 [Witco/Organics]—Coconut diethanolamide

Witcamide 5195, 6310, STD-HP [Witco/Organics]—Lauric diethanolamide

Witcamide MAS [Witco/Organics]—Stearic monoethanolamide stearate

Witcamide MM [Witco]—Myristic monoethanolamide

Witco Aluminum Monostearate USP [Witco/Organics]—Aluminum monostearate

Witco Aluminum Stearate 18 [Witco/Organics]—Aluminum di/tristearate

Witco Aluminum Stearate 22, 30 [Witco/Organics]—Aluminum distearate

Witco Aluminum Stearate EA [Witco/Organics]—Aluminum distearate (food grade)

Witcolate D-510 [Witco]—Sodium octyl sulfate

Witconate AOS [Witco/Org.]—Sodium olefin sulfonate (C_{14-16})

Wytox 312 [Olin]—Tris (nonylphenyl) phosphite

Xanflood [Kelco]—Xanthan gum (industrial grade)

Zinc Oxide 35 [Akron]—Zinc oxide (precipitated)

Zinc Oxide 500, 911 [C.P. Hall]—Zinc oxide (French process)

Zinc Oxide 500-36, 911-36 [C.P. Hall]—Zinc oxide (French process, surface treated)
Zinox-Grade 350 [Amer. Chemet]—Zinc oxide (precipitated)
Zoco 100, 103, 104 [Zochey]—Zinc oxide (French process)
Zoco 172 [Zochey]—Zinc oxide (French process, surface treated)

GENERIC CHEMICAL SYNONYMS
AND CROSS REFERENCES

Acetic, 1,2,3-propanetriyl ester. See Glyceryl triacetate

Acetyl monoethanolamide. See Book IV

2-(Acetyloxy)-1,2,3-propanetricarboxylic acid, tributyl ester. See Acetyl tributyl citrate

2-(Acetyloxy)-1,2,3-propanetricarboxylic acid, triethyl ester. See Acetyl triethyl citrate

Acrylamides copolymer. See Book VI

Acrylates/steareth-20 methacrylate copolymer. See Book VI

Acrylonitrile butadiene copolymer. See Book VI

Algaroba. See Locust bean gum

Alginic acid, ester with 1,2-propanediol. See Propylene glycol alginate

Alginic acid, potassium salt. See Potassium alginate

Alginic acid, sodium salt. See Algin

Alizarin assistant. See Castor oil, sulfated

Alizarin oil. See Castor oil, sulfated

D-alpha tocopherol. See Tocopherol (CTFA)

DL-alpha tocopherol. See Tocopherol (CTFA)

Alumina, hydrated. See Aluminum hyroxide

Alumina trihydrate. See Aluminum hyroxide

Aluminosilicic acid, magnesium salt. See Magnesium aluminum silicate

Aluminum, dihyroxy (octadecanoate-O)-. See Aluminum monostearate

Aluminum hydrate. See Aluminum hyroxide

Aluminum, hydroxybis(octadecanoato-O)-. See Aluminum distearate

Aluminum magnesium silicate. See Magnesium aluminum silicate

Aluminum stearate (CTFA). See Aluminum monostearate

Amides, coco, N,N-bis (2-hydroxyethyl)-. See Coconut diethanolamide

Amides, coco, N-[3-(dimethylamino) propyl], N-oxide. See Cocamidopropylamine oxide

Amides, coco, N-(2-hydroxyethyl)-. See Coconut monoethanolamide

Amides, coco, N-(2-hydroxypropyl)-. See Coconut monoisopropanolamide

Amides, coconut oil. See Coconut oil amide

Amides, soya, N,N-bis (hydroxyethyl)-. See Soya diethanolamide

Amides, tall oil fatty, N,N-bis (2-hydroxyethyl)-. See Tall oil acid diethanolamide

Amides, tallow, hydrogenated. See Hydrogenated tallow amide

Aminomethyl propanol. See Book II

Ammonium cumenesulfonate. See Book II

Ammonium lauryl ether sulfate. See Book I

Ammonium xylenesulfonate. See Book II

1,4-Anhydro-D-glucitol, 6 dodecanoate. See Sorbitan monolaurate

Anhydrosorbitol trioleate. See Sorbitan trioleate

Anhydrosorbitol tristearate. See Sorbitan tristearate

Anhydrosorbol monolaurate. See Sorbitan monolaurate
Ba-Cd stabilizer. See Barium-cadmium compound
Ba-Cd-Zn stabilizer. See Barium-cadmium-zinc compound
1,2-Benzenedicarboxylic acid, butyl phenylmethyl ester. See Butyl benzyl phthalate
1,2-Benzenedicarboxylic acid, dibutyl ester. See Dibutyl phthalate
1,2-Benzenedicarboxylic acid, diethyl ester. See Diethyl phthalate
1,2-Benzenedicarboxylic acid, dimethyl ester. See Dimethyl phthalate
1,4-Benzenediol. See Hydroquinone
1,4-Benzenediol, 2-(1,1-dimethylethyl)-. See t-Butyl hydroquinone
Benzenesulfonic acid, 5-benzoyl-4-hydroxy-2-methoxy-. See 2-Hydroxy-4-methoxybenzophenone-5-sulfonic acid
Benzoic acid, 4-[bis (2-hydroxypropyl) amino]-, ethyl ester. See Ethyl dihydroxypropyl PABA
Benzoic acid, 3,4,5-trihydroxy-, propyl ester. See Propyl gallate
Benzophenone-1 (CTFA). See 2,4-Dihydroxybenzophenone
Benzophenone-3 (CTFA). See 2-Hydroxy-4-methoxybenzophenone
Benzophenone-4 (CTFA). See 2-Hydroxy-4-methoxybenzophenone-5-sulfonic acid
Benzophenone-8 (CTFA). See 2,2´-Dihydroxy-4-methoxybenzophenone
Benzophenone-12 (CTFA). See 2-Hydroxy-4-(octyloxy) benzophenone
2H-1-Benzopyran-6-ol, 3,4-dihydro-2,5,7,8-tetramethyl-2-(4,8,12-trimethyltridecyl)-. See Tocopherol (CTFA)
Benzoresorcinol. See 2,4-Dihydroxybenzophenone
5-Benzoyl-4-hydroxy-2-methoxybenzene sulfonic acid. See 2-Hydroxy-4-methoxybenzophenone-5-sulfonic acid
4-Benzoyl resorcinol. See 2,4-Dihydroxybenzophenone
BHA (CTFA). See Butylated hydroxyanisole
BHT (CTFA). See Butylated hydroxytoluene
2,6-Bis (1,1-dimethylethyl)-4-methylphenol. See Butylated hydroxytoluene
Bis (2-ethylhexyl) adipate. See Dioctyl adipate
Bis (2-ethylhexyl) 2-butenedioate. See Dioctyl maleate
Bis (2-ethylhexyl) decanedioate. See Dioctyl sebacate
Bis (2-ethylhexyl) hexanedioate. See Dioctyl adipate
N,N-Bis (2-hydroxyethyl) coco amides. See Coconut diethanolamide
N,N-Bis (2-hydroxyethyl) cocoamine oxide. See Book IV
N,N-Bis (2-hydroxyethyl) coco fatty acid amide. See Coconut diethanolamide
N,N-Bis (2-hydroxyethyl) dodecanamide. See Lauric diethanolamide
N,N-Bis (2-hydroxyethyl) isooctadecanamide. See Isostearic diethanolamide
N,N-Bis (2-hydroxyethyl) lauramide. See Lauric diethanolamide
N,N-Bis (2-hydroxyethyl) linoleamide. See Linoleoyl diethanolamide
N,N-Bis (2-hydroxyethyl)-9,12-octadecadienamide. See Linoleoyl diethanolamide
N,N-Bis (2-hydroxyethyl) octadecanamide. See Stearic diethanolamide
N,N-Bis (2-hydroxyethyl)-9-octadecenamide. See Oleic diethanolamide

N,N-Bis (2-hydroxyethyl) oleamide. See Oleic diethanolamide

N,N-Bis (2-hydroxyethyl) palm kernel oil acid amide. See Palm kernel oil acid diethanolamide

N,N-Bis (hydroxyethyl) soya amides. See Soya diethanolamide

N,N-Bis (2-hydroxyethyl) stearamide. See Stearic diethanolamide

N,N-Bis (2-hydroxyethyl) tall oil acid amide. See Tall oil acid diethanolamide

N,N-Bis (2-hydroxyethyl) tall oil fatty amides. See Tall oil acid diethanolamide

Bis (2-hydroxyethyl) tallow amine oxide. See Book IV

4-[Bis (2-hydroxypropyl) amino] benzoic acid, ethyl ester. See Ethyl dihydroxypropyl PABA

Bis (1-methylethyl) hexanedioate. See Diisopropyl adipate

Bis (2-methylpropyl) hexanedioate. See Diisobutyl adipate

2-Butenedioic acid, bis (2-ethylhexyl) ester. See Dioctyl maleate

Butyl 9-octadecenoate. See Butyl oleate

Butyl stearate. See Book IV

C1 77947. See Zinc oxide

Calcium lignosulfonate. See Book II

Calcium stearate. See Book IV

Capric acid diethanolamide. See Book I

Caprylic/capric triglyceride. See Book II

Carbomer 910. See Book VI

Carbomer 934. See Book VI

Carbomer 934P. See Book VI

Carbomer 940. See Book VI

Carbomer 941. See Book VI

Carboxymethyl cellulose. See Sodium carboxymethyl cellulose

N-(Carboxymethyl) -N,N- dimethyl -3- [(1-oxododecyl) amino] -1- propanaminium hyroxide, inner salt. See Lauramidopropyl betaine

N-(Carboxymethyl)-N,N-dimethyl-3-[(1-oxohexadecyl) amino]-1-propanaminium hydroxide, inner salt. See Palmitamidopropyl betaine

N-(Carboxymethyl)-N,N-dimethyl-3-[(1-oxooctadecenyl) amino]-1-propanaminium hydroxide, inner salt

Carob bean gum. See Locust bean gum

Carob flour. See Locust bean gum

Castor oil, soluble. See Castor oil, sulfated

Castor oil, sulfonated. See Castor oil, sulfated

Cellulose, carboxymethyl ether. See Sodium carboxymethyl cellulose

Cellulose gum (CTFA). See Sodium carboxymethyl cellulose

Cellulose, 2-hydroxyethyl ether. See Hydroxyethylcellulose

Cellulose, 2-hydroxypropyl ether. See Hydroxypropylcellulose

Cellulose, 2-hydroxypropyl methyl ether. See Hydroxypropyl methylcellulose

Cellulose, methyl ether. See Methylcellulose

Ceratonia. See Locust bean gum
Cetamine oxide. See Cetyl dimethyl amine oxide
Cetyl alcohol. See Book IV
Cetyl lactate. See Book IV
Chinese bean oil, epoxidized. See Epoxidized soybean oil
Chlorinated alkane. See Chlorinated paraffin
Chondrus. See Carrageenan
Citrus pectin. See Pectin
CMC. See Sodium carboxymethyl cellulose
Cocamide (CTFA). See Coconut oil amide
Cocamide DEA (CTFA). See Coconut diethanolamide
Cocamide MEA (CTFA). See Coconut monoethanolamide
Cocamide MIPA (CTFA). See Coconut monoisopropanolamide
(3-Cocamidopropyl)(2-hydroxy-3-sulfopropyl) dimethyl quaternary ammonium com-
     pounds, hydroxide, inner salt. See Cocamidopropyl hydroxysultaine
Cocaminobutyric acid. See Book II
Cocoamide. See Coconut oil amide
Coco amides, N,N-bis (2-hydroxyethyl)-. See Coconut diethanolamide
Coco amides, N-[3-(dimethylamino) propyl], N-oxide. See Cocamidopropylamine oxide
Coco amidopropyl betaine. See Book IV
Coco amido propyl dimethyl amine oxide. See Cocamidopropylamine oxide
Coco amido sulfobetaine. See Cocamidopropyl hydroxysultaine
Coco amine oxide. See Book IV
Cocoamphopropionate. See Book I
Coco betaine. See Book IV
Coco diethanolamide. See Coconut diethanolamide
Coco fatty acid diethanolamide. See Coconut diethanolamide
Coco monoethanolamide. See Coconut monoethanolamide
Coconut acid amide. See Coconut oil amide
Coconut amide polyglycolether (6 moles EO). See POE (6) coconut amide
Coconut fatty acid diethanolamide. See Coconut diethanolamide
Coconut fatty acid isopropanolamide. See Coconut monoisopropanolamide
Coconut fatty acid, monoethanolamide. See Coconut monoethanolamide
Coconut isopropanolamide. See Coconut monoisopropanolamide
Coconut monoethanolamide ethoxylate (6 moles EO). See POE (6) coconut amide
Coconut oil diethanolamide. See Coconut diethanolamide
Coconut oil fatty acid diethanolamide. See Coconut diethanolamide
Coco oil diethanolamide. See Coconut diethanolamide
Cocoyl diethanolamide. See Coconut diethanolamide
Cocoyl imidazoline. See Book II
N-Cocoyl-N-methyl glycine. See Cocoyl sarcosine
Cocoyl monoethanolamine. See Coconut monoethanolamide

N-Cocoyl sarcosine. See Cocoyl sarcosine
Colophony. See Rosin
Corn sugar gum. See Xanthan gum
C-Weiss 1 (Germany). See Aluminum hyroxide
DBP. See Dibutyl phthalate
DBPC. See Butylated hydroxytoluene
DBS. See Dibutyl sebacate
Decaglycerol decaoleate. See Book II
Decaglyceryl tetraoleate. See Decaglycerol tetraoleate
Decanedioic acid, bis (2-ethylhexyl) ester. See Dioctyl sebacate
Decanedioic acid, dibutyl ester. See Dibutyl sebacate
DEP. See Diethyl phthalate
DIBA. See Diisobutyl adipate
Dibutyl 1,2-benzenedicarboxylate. See Dibutyl phthalate
2,6-Di-*t*-butyl-*p*-cresol. See Butylatehydroxytoluene
Dibutyl decanedioate. See Dibutyl sebacate
DIDA. See Diisodecyl adipate
Didecyl hexanedioate. See Dicapryl adipate
Didodecyl 3,3´-thiodipropionate. See Dilauryl thiodipropionate
DIDP. See Diisodecyl phthalate
Diester of isopropyl alcohol and adipic acid. See Diisopropyl adipate
Diethanolamine coconut fatty acid condensate. See Coconut diethanolamide
Diethanolamine lauric acid amide. See Lauric diethanolamide
Diethanolamine lauryl sulfate. See Book I
Diethanolamine linoleic acid amide. See Linoleoyl diethanolamide
Diethanolamine oleic acid amide. See Oleic diethanolamide
Diethanolamine palm kernel oil acid amide. See Palm kernel oil acid diethanolamide
Diethanolamine stearic acid amide. See Stearic diethanolamide
Diethanolamine tall oil acid amide. See Tall oil acid diethanolamide
2-(Diethylamino) ethyl octadecanoate. See Diethylaminoethyl stearate
Diethyl, 1,2-benzenedicarboxylate. See Diethyl phthalate
Diethylene glycol monobutyl ether. See Book II
Diethylene glycol monolaurate. See Book I
Diethylene glycol monooleate. See Book I
Diethylene glycol monoricinoleate. See Book II
Diethylene glycol monostearate. See Book I
Di (2-ethylhexyl) adipate. See Dioctyl adipate
Di (2-ethylhexyl) azelate. See Dioctyl azelate
Di (2-ethylhexyl) maleate. See Dioctyl maleate
Di (2-ethylhexyl) phthalate. See Dioctyl phthalate
Di (2-ethylhexyl) sebacate. See Dioctyl sebacate
4,5-Dihydro-7-nortall oil-1H-imidazole-1-ethanol. See Tall oil hydroxyethyl imidazo-
    line

3,4-Dihydro-2,5,7,8-tetramethyl-2-(4,8,12-trimethyltridecyl)-2H-1-benzopyran-6-ol.
    See Tocopherol (CTFA)
p-Dihydroxy benzene. See Hydroquinone
Dihydroxy (octanoateo-O) aluminum. See Aluminum monostearate
(2,4-Dihydroxyphenyl) phenylmethanone. See 2,4-Dihydroxybenzophenone
Diisobutyl hexanedioate. See Diisobutyl adipate
Dilaurin. See Glyceryl dilaurate
N-[3-(Dimethylamino) propyl] coco amides-N-oxide. See Cocamidopropylamine oxide
Dimethyl 1,2-benzenedicarboxylate. See Dimethyl phthalate
2-(1,1-Dimethylethyl)-1,4-benzenediol-. See t-Butyl hydroquinone
(1,1-Dimethylethyl)-4-methoxyphenol. See Butylatehydroxyanisole
N,N-Dimethyl-1-hexadecanamine-N-oxide. See Cetyl dimethyl amine oxide
Dimethyl myristyl amine oxide. See Myristyl dimethyl amine oxide
N,N-Dimethyl-1-tetradecanamine-N-oxide. See Myristyl dimethyl amine oxide
DINA. See Diisononyl adipate
DINP. See Diisononyl phthalate
DIOA. See Diisooctyl adipate
DIOP. See Diisooctyl phthalate
Dioxybenzone. See 2,2´-Dihydroxy-4-methoxybenzophenone
Dipropylene glycol monomethyl ether. See Book II
Disodium deceth-6 sulfosuccinate. See Book I
Disodium laurethsulfosuccinate. See Book I
Disodium lauryl sulfosuccinate. See Book I
Disodium N-oleyl sulfosuccinamate. See Book I
Ditetradecyl 3,3´-thiobispropanoate. See Dimyristyl thiodipropionate
Ditridecyl adipate. See Book IV
Di (tridecyl) thiodipropionate. See Ditridecyl thiodipropionate
DLTDP. See Dilauryl thiodipropionate
DOA. See Dioctyl adipate
Dodecahydro-1,4a-dimethyl-7-(1-methylethyl)-1-phenanthrenemethanol. See Dihy-
    droabietyl alcohol
Dodecanamide, N-(2-hydroxyethyl)-. See Lauric monoethanolamide
Dodecanamide, N-(2-hydroxyproyl)-. See Lauric monoisopropanolamide
n-Dodecanoic acid. See Lauric acid
Dodecanoic acid, diester with 1,2,3-propanetriol. See Glyceryl dilaurate
DOM. See Dioctyl maleate
DOP. See Dioctyl phthalate
DOZ. See Dioctyl azelate
DTDP. See Ditridecyl phthalate
DTDTDP. See Ditridecyl thiodipropionate
EGMS. See Ethylene glycol monostearate
Erucic acid amide. See Book IV

ESO. See Epoxidized soybean oil
Ethyl dihydroxypropyl p-aminobenzoate. See Ethyl dihydroxypropyl PABA
N,N´-ethylene bisstearamide. See Book IV
Ethylene glycol distearate. See Book I
Ethylene glycol monobutyl ether. See Book II
Ethylene glycol monoethyl ether. See Book II
Ethylene/maleic anhydride copolymer. See Book VI
Ethyl phthalate. See Diethyl phthalate
N-Ethyl o,p-toluene sulfonamide. See Ethyl toluenesulfonamide
Fatty acids, lanolin. See Lanolin acid
Germany C-Weiss 8. See Zinc oxide
D-Glucitol, 1,4-anhydro-, 6 dodecanoate. See Sorbitan monolaurate
Glycerides, coconut oil mono-. See Glyceryl monococoate
Glycerides, palm oil mono-, di-, and tri-. See Palm oil glycerides
Glycerine. See Glycerin
Glycerol. See Glycerin
Glycerol mono coconut oil. See Glyceryl monococoate
Glycerol monoricinoleate. See Glyceryl monoricinoleate
Glycerol triacetate. See Glyceryl triacetate
Glyceryl cocoate (CTFA). See Glyceryl monococoate
Glyceryl coconate. See Glyceryl monococoate
Glyceryl monococoate. See Glyceryl monococoate
Glyceryl monolaurate. See Book I
Glyceryl monooleate. See Book I
Glyceryl monostearate. See Book I
Glyceryl ricinoleate (CTFA). See Glyceryl monoricinoleate
Glyceryl trioleate. See Book IV
Glycine, N-methyl-, N-coco acyl derivs.. See Cocoyl sarcosine
Glycine, N-methyl-N-(1-oxococonut alkyl)-. See Cocoyl sarcosine
Glycine, N-methyl-N-(1-oxododecyl)-. See Lauroyl sarcosine
Glycine, N-methyl-N-(1-oxo-9-octadecenyl)-. See Oleyl sarcosine
Glycol monostearate. See Ethylene glycol monostearate
Glycol stearate (CTFA). See Ethylene glycol monostearate
Glycyl alcohol. See Glycerin
Guar flour. See Guar gum
Guar gum, 2-hydroxypropyl ether. See Hydroxypropyl guar
Guar, gum, 2-hydroxy-3-(trimethylammonio) propyl ether, chloride. See Guar hydroxypropyltrimonium chloride
Guar hydroxypropyl trimethyl ammonium chloride. See Guar hydroxypropyltrimonium chloride
Gum rosin. See Rosin
H. E. Cellulose. See Hydroxyethylcellulose

2-(8-Heptadecenyl)-4,5-dihydro-1H-imidazole-1-ethanol. See Oleyl hydroxyethyl imidazoline
1-Hexadecanamine, N,N-dimethyl-, N-oxide. See Cetyl dimethyl amine oxide
Hexanedioic acid, bis (2-ethylhexyl) ester. See Dioctyl adipate
Hexanedioic acid, bis (1-methylethyl) ester. See Diisopropyl adipate
Hexanedioic acid, bis (2-methylpropyl) ester. See Diisobutyl adipate
Hexanedioic acid, didecyl ester. See Dicapryl adipate
Hexanedioic acid, diisobutyl ester. See Diisobutyl adipate
Hydrated alumina. See Aluminum hyroxide
Hydrated aluminum oxide. See Aluminum hyroxide
Hydroabietyl alcohol. See Dihydroabietyl alcohol
Hydrochinone. See Hydroquinone
Hydroquinol. See Hydroquinone
Hydroxybis(octadecanoato-O) aluminum. See Aluminum distearate
N-(2-Hydroxyethyl) coco fatty acid amide. See Coconut monoethanolamide
N-(2-Hydroxyethyl) dodecanamide. See Lauric monoethanolamide
N-(2-Hydroxyethyl) lauramide. See Lauric monoethanolamide
N-(2-Hydroxyethyl) octadecanamide. See Stearic monoethanolamide
2-Hydroxy-ethyl octadecanoate. See Ethylene glycol monostearate
N-(2-Hydroxyethyl) stearamide. See Stearic monoethanolamide
N-(2-Hydroxyethyl) tetradecanamide. See Myristic monoethanolamide
Hydroxylated lanolin. See Book IV
(2-Hydroxy-4-methoxyphenyl) (2-hydroxyphenyl) methanone. See 2,2´-Dihydroxy-4-methoxybenzophenone
(2-Hydroxy-4-methoxyphenyl) phenylmethanone. See 2-Hydroxy-4-methoxybenzophenone
12-Hydroxyoctadecanoic acid, 1,2,3-propanetriol ester. See Glyceryl tri(12-hydroxystearate)
12-Hydroxy-9-octadecenoic acid, monoester with 1,2-propanediol. See Propylene glycol monoricinoleate
12-Hydroxy-9-octadecenoic acid, monoester with 1,2,3-propanetriol. See Glyceryl monoricinoleate
12-Hydroxy-9-octadecenoic acid, monopotassium salt. See Potassium monoricinoleate
2-Hydroxy-4-n-octoxybenzophenone. See 2-Hydroxy-4-(octyloxy) benzophenone
[2-Hydroxy-4-(octyloxy) phenyl] phenylmethanone. See 2-Hydroxy-4-(octyloxy) benzophenone
p-Hydroxyphenol. See Hydroquinone
2-Hydroxy-1,2,3-propanetricarboxylic acid, tributyl ester. See Tributyl citrate
2-Hydroxy-1,2,3-propanetricarboxylic acid, triethyl ester. See Triethyl citrate
N-(2-Hydroxypropyl) dodecanamide. See Lauric monoisopropanolamide
N-(2-Hydroxypropyl)-9-octadecenamide. See Oleic monoisopropanolamide
12-Hydroxystearic acid, beta-hydroxyoctacosanyl ester. See Hydroxyoctacosanyl hy-

droxystearate

Hydroxystearic acid, monoester with glycerol. See Glyceryl hydroxystearate

Hypromellose. See Hydroxypropyl methylcellulose

3H-Imidazole-1-ethanol, 4,5-dihydro-2-nonyl-. See Capryl hydroxyethyl imidazoline

1H-Imidazole-1-ethanol, 4,5-dihydro-, 2-nortall oil. See Tall oil hydroxyethyl imidazoline

1H-Imidazole-1-ethanol, 2-(8-heptadecenyl)-4,5-dihydro-. See Oleyl hydroxyethyl imidazoline

1H-Imidazolium, 1-(2-carboxyethyl)-4,5-dihydro-3-(2-hydroxyethyl)-2-isoheptadecyl-, hydroxide, inner salt. See Isostearoamphopropionate

Irish moss extract. See Carrageenan

1H-Isoindole-1,3(2H)-dione, 3a,4,7,7a-tetrahydro-2-[(trichloromethyl)thio]-. See Captan

Isooctadecanamide, N,N-bis (2-hydroxyethyl)-. See Isostearic diethanolamide

Isopropyl myristate. See Book IV

Isopropyl palmitate. See Book IV

Isostearamide DEA (CTFA). See Isostearic diethanolamide

Isostearamidopropyl betaine. See Book IV

Isostearyl alcohol. See Book IV

Isostearyl neopentanoate. See Book IV

Lanolic acids. See Lanolin acid

Lanolin. See Book IV

Lanolin alcohol. See Book IV

Lanolin, alcohols, acetates. See Acetylated lanolin alcohol

Lanolin fatty acids. See Lanolin acid

Lanolin oil. See Book IV

Lauramide DEA (CTFA). See Lauric diethanolamide

Lauramide MEA (CTFA). See Lauric monoethanolamide

Lauramide MIPA (CTFA). See Lauric monoisopropanolamide

Lauric acid diethanolamide. See Lauric diethanolamide

Lauric acid monoethanolamide. See Lauric monoethanolamide

Lauric DEA. See Lauric diethanolamide

Lauric fatty acid diethanolamide. See Lauric diethanolamide

Lauric fatty acid monoisopropanolamide. See Lauric monoisopropanolamide

Lauric isopropanolamide. See Lauric monoisopropanolamide

Lauric MIPA. See Lauric monoisopropanolamide

Lauroyl diethanolamide. See Lauric diethanolamide

Lauroyl isopropanolamide. See Lauric monoisopropanolamide

Lauroyl monoethanolamide. See Lauric monoethanolamide

Lauryl betaine. See Book I

Lauryl dimethyl amine oxide. See Book IV

Lecithin. See Book II

Linoleamide DEA (CTFA). See Linoleoyl diethanolamide
Meroxapol 105. See Book VI
Meroxapol 108. See Book VI
Meroxapol 171. See Book VI
Meroxapol 172. See Book VI
Meroxapol 174. See Book VI
Meroxapol 178. See Book VI
Meroxapol 251. See Book VI
Meroxapol 252. See Book VI
Meroxapol 254. See Book VI
Meroxapol 255. See Book VI
Meroxapol 258. See Book VI
Meroxapol 311. See Book VI
Meroxapol 312. See Book VI
Meroxapol 314. See Book VI
Methanone, (2,4-dihydroxyphenyl) phenyl-. See 2,4-Dihydroxybenzophenone
Methanone, (2-hydroxy-4-methoxyphenyl) (2-hydroxyphenyl)-. See 2,2′-Dihydroxy-4-methoxybenzophenone
Methanone, (2-hydroxy-4-methoxyphenyl) phenyl-. See 2-Hydroxy-4-methoxybenzophenone
Methanone, [2-hydroxy-4-(octyloxy) phenyl] phenyl-. See 2-Hydroxy-4-(octyloxy)benzophenone
N-Methylglycine, N-coco acyl derivs.. See Cocoyl sarcosine
Methyl hydroxypropyl cellulose. See Hydroxypropyl methylcellulose
Methyl 9-octadecenoate. See Methyl oleate
N-Methyl-N-(1-oxococonut alkyl) glycine. See Cocoyl sarcosine
N-Methyl-N-(1-oxododecyl) glycine. See Lauroyl sarcosine
N-Methyl-N-(1-oxo-9-octadecenyl) glycine. See Oleyl sarcosine
Mineral oil. See Book IV
Mixed tocopherols. See Tocopherol (CTFA)
Monoethanolamine coconut acid amide. See Coconut monoethanolamide
Monoethanolamine lauric acid amide. See Lauric monoethanolamide
Monoethanolamine myristic acid condensate. See Myristic monoethanolamide
Monoethanolamine stearic acid amide. See Stearic monoethanolamide
Monoisopropanolamine coconut acid amide. See Coconut monoisopropanolamide
Monoisopropanolamine lauric acid amide. See Lauric monoisopropanolamide
Monoisopropanolamine oleic acid amide. See Oleic monoisopropanolamide
Monoricinolein. See Glyceryl monoricinoleate
Myristamide MEA (CTFA). See Myristic monoethanolamide
Myristamine oxide (CTFA). See Myristyl dimethyl amine oxide
Myristic diethanolamide. See Book I
Myristoyl monoethanolamide. See Myristic monoethanolamide

Myristyl lactate. See Book IV
Myristyl myristate. See Book IV
Naphthenic petroleum oil ASTM D2226 Type 103. See Naphthenic oil
NODA. See n-Octyl-n decyl adipate
Nonoxynol-30 (CTFA). See POE (30) nonyl phenyl ether
Nonoxynol-40 (CTFA). See POE (40) nonyl phenyl ether
Nonoxynol-50 (CTFA). See POE (50) nonyl phenyl ether
Nonoxynol-100 (CTFA). See POE (100) nonyl phenyl ether
2-Nonyl-4,5-dihydro-1H-imidazole-1-ethanol. See Capryl hydroxyethyl imidazoline
Nonyl phenol 30 polyglycol ether. See POE (30) nonyl phenyl ether
Nonyl phenol 40 polyglycol ether. See POE (40) nonyl phenyl ether
Nonyl phenol 50 polyglycol ether. See POE (50) nonyl phenyl ether
Nonyl phenol 100 polyglycol ether. See POE (100) nonyl phenyl ether
Nonyl phenol ethoxylate (30 moles EO). See POE (30) nonyl phenyl ether
Nonyl phenol ethoxylate (40 moles EO). See POE (40) nonyl phenyl ether
Nonyl phenol ethoxylate (50 moles EO). See POE (50) nonyl phenyl ether
Nonyl phenol ethoxylate (100 moles EO). See POE (100) nonyl phenyl ether
Nonylphenoxy polyethoxy ethanol (30 moles EO). See POE (30) nonyl phenyl ether
Nonylphenoxy polyethoxy ethanol (40 moles EO). See POE (40) nonyl phenyl ether
Nonylphenoxy polyethoxy ethanol (50 moles EO). See POE (50) nonyl phenyl ether
Nonylphenoxy polyethoxy ethanol (100 moles EO). See POE (100) nonyl phenyl ether
Nonylphenoxy poly(ethyleneoxy) ethanol (30 moles EO). See POE (30) nonyl phenyl
    ether
Nonylphenoxy poly(ethyleneoxy) ethanol (40 moles EO). See POE (40) nonyl phenyl
    ether
Nonylphenoxy poly(ethyleneoxy) ethanol (50 moles EO). See POE (50) nonyl phenyl
    ether
Nonylphenoxy poly(ethyleneoxy) ethanol (100 moles EO). See POE (100) nonyl phenyl
    ether
Nonylphenyl phosphite (3:1). See Tris (nonylphenyl) phosphite
Norgine. See Alginic acid
2-Nortall oil-1H-imidazole-1-ethanol, 4,5-dihydro-. See Tall oil hydroxyethyl imidazo-
    line
Octabenzone. See 2-Hydroxy-4-(octyloxy) benzophenone
9,12-Octadecadienamide, N,N-bis (2-hydroxyethyl)-. See Linoleoyl diethanolamide
Octadecanamide, N,N-bis (2-hydroxyethyl)-. See Stearic diethanolamide
Octadecanamide, N-(2-hydroxyethyl)-. See Stearic monoethanolamide
Octadecanoic acid, aluminum salt. See Aluminum monostearate
Octadecanoic acid, 2-(diethylamino) ethyl ester. See Diethylaminoethyl stearate
Octadecanoic acid, 2-hydroxyethyl ester. See Ethylene glycol monostearate
Octadecanoic acid, 12-hydroxy-, monoester with 1,2-propanediol. See Propylene glycol
    hydroxystearate

Octadecanoic acid, 12-hydroxy-1,2,3,-propanetriol ester. See Glyceryl tri(12-hydroxys-tearate)

Octadecanoic acid, 2-[(1-oxooctadecyl) amino] ethyl ester. See Stearic monoetha-nolamide stearate

1-Octadecanol. See Stearyl alcohol

9-Octadecenamide, N,N-bis (2-hydroxyethyl)-. See Oleic diethanolamide

9-Octadecenamide, N-2-(hydroxypropyl)-. See Oleic monoisopropanolamide

9-Octadecenoic acid. See Oleic acid

9-Octadecenoic acid, 12-(acetyloxy)-, 1,2,3-propanetriol ester. See Glyceryl triacetyl ricinoleate

9-Octadecenoic acid, butyl ester. See Butyl oleate

9-Octadecenoic acid, 12-hydroxy-, monoester with 1,2-propanediol. See Propylene glycol monoricinoleate

9-Octadecenoic acid, 12-hydroxy-, monoester with 1,2,3-propanetriol. See Glyceryl monoricinoleate

9-Octadecenoic acid, 12-hydroxy-, monopotassium salt. See Potassium monoricinoleate

9-Octadecenoic acid, methyl ester. See Methyl oleate

9-Octadecenoic acid, tetraester with decaglycerol. See Decaglycerol tetraoleate

Octadecyl alcohol. See Stearyl alcohol

Octyl epoxy tallate. See Epoxidized octyl tallate

Oleamide DEA (CTFA). See Oleic diethanolamide

Oleamide MIPA (CTFA). See Oleic monoisopropanolamide

Oleamido betaine. See Oleamidopropyl betaine

Oleamidopropyl dimethyl glycine. See Oleamidopropyl betaine

Oleic acid diethanolamide. See Oleic diethanolamide

Oleic fatty acid diethanolamide. See Oleic diethanolamide

Oleic isopropanolamide. See Oleic monoisopropanolamide

Oleoyl sarcosine (CTFA). See Oleyl sarcosine

Oleyl alcohol. See Book IV

Oleyl amide. See Book IV

Oleyl amido betaine. See Oleamidopropyl betaine

Oleyl amidopropyl betaine. See Oleamidopropyl betaine

Oleyl amine acetate. See Book I

Oleyl imidazoline. See Oleyl hydroxyethyl imidazoline

Oleyl N-methylaminoacetic acid. See Oleyl sarcosine

Oleyl methylaminoethanoic acid. See Oleyl sarcosine

Oleyl N-methylglycine. See Oleyl sarcosine

2-[(1-Oxooctadecyl) amino] octadecanoic acid, ethyl ester. See Stearic monoetha-nolamide stearate

Oxybenzone. See 2-Hydroxy-4-methoxybenzophenone

3,3′-Oxydyl-1-propanol dibenzoate. See Dipropylene glycol dibenzoate

Palm kernelamide DEA (CTFA). See Palm kernel oil acid diethanolamide

Palm kernel oil acid amide, N,N-bis (2-hydroxyethyl)-. See Palm kernel oil acid
    diethanolamide
Palmitamine oxide (CTFA). See Cetyl dimethyl amine oxide
Palmityl dimethylamine oxide. See Cetyl dimethyl amine oxide
PEG-9. See Book VI
PEG-14. See Book VI
PEG-32. See Book VI
PEG-6 cocamide (CTFA). See POE (6) coconut amide
PEG 300 coconut amide. See POE (6) coconut amide
PEG 100 dibenzoate. See Diethylene glycol dibenzoate
PEG-20 dilaurate (CTFA). See POE (20) dilaurate
PEG-32 dilaurate (CTFA). See POE (32) dilaurate
PEG-75 dilaurate (CTFA). See POE (75) dilaurate
PEG-150 dilaurate (CTFA). See POE (150) dilaurate
PEG 1000 dilaurate. See POE (20) dilaurate
PEG 1540 dilaurate. See POE (32) dilaurate
PEG 4000 dilaurate. See POE (75) dilaurate
PEG 6000 dilaurate. See POE (150) dilaurate
PEG-150 distearate (CTFA). See POE (150) distearate
PEG 6000 distearate. See POE (150) distearate
PEG-45/dodecyl glycol copolymer. See Book VI
PEG-6 lauramide (CTFA). See POE (6) lauryl amide
PEG-20 laurate (CTFA). See POE (20) monolaurate
PEG 300 lauryl amide. See POE (6) lauryl amide
PEG-1 lauryl ether. See Book I
PEG-120 methyl glucose dioleate (CTFA). See POE (120) methyl glucose dioleate
PEG (120) methyl glucose dioleate. See POE (120) methyl glucose dioleate
PEG 1000 monolaurate. See POE (20) monolaurate
PEG-30 nonyl phenyl ether. See POE (30) nonyl phenyl ether
PEG (30) nonyl phenyl ether. See POE (30) nonyl phenyl ether
PEG-40 nonyl phenyl ether. See POE (40) nonyl phenyl ether
PEG-50 nonyl phenyl ether. See POE (50) nonyl phenyl ether
PEG (50) nonyl phenyl ether. See POE (50) nonyl phenyl ether
PEG-100 nonyl phenyl ether. See POE (100) nonyl phenyl ether
PEG (100) nonyl phenyl ether. See POE (100) nonyl phenyl ether
PEG 2000 nonyl phenyl ether. See POE (40) nonyl phenyl ether
PEG-80 sorbitan laurate (CTFA). See POE (80) sorbitan monolaurate
PEG (80) sorbitan monolaurate. See POE (80) sorbitan monolaurate
PEG-20 sorbitol ether. See POE (20) sorbitol ether
PEG 1000 sorbitol ether. See POE (20) sorbitol ether
Pentaerythritol tetraoleate. See Book IV
Pentaerythritol tetrastearate. See Book IV

Petrolatum. See Book IV

1-Phenanthrenemethanol, dodecahydro-1,4a-dimethyl-7-(1-methylethyl)-. See Dihydroabietyl alcohol

Phenol, 2,6-bis (1,1-dimethylethyl)-4-methyl-. See Butylated hydroxytoluene

Phenol, (1,1-dimethylethyl)-4-methoxy-. See Butylated hydroxyanisole

Phenol, nonyl-, phosphite (3:1). See Tris (nonylphenyl) phosphite

Phosphoric acid, tris (methylphenyl) ester. See Tricresyl phosphate

POE (5) castor oil. See Book II

POE (10) castor oil. See Book II

POE (30) castor oil. See Book II

POE (40) castor oil. See Book II

POE (2) cetyl ether. See Book I

POE (10) cetyl/stearyl ether. See Book II

POE (15) cetyl/stearyl ether. See Book II

POE (20) cetyl/stearyl ether. See Book II

POE 6 coco monoethanolamide. See POE (6) coconut amide

POE (2) coconut amine. See Book I

POE (5) coconut amine. See Book I

POE (15) coconut amine. See Book I

POE (2) dibenzoate. See Diethylene glycol dibenzoate

POE (4) dilaurate. See Book II

POE (6) dilaurate. See Book II

POE (12) dilaurate. See Book II

POE (150) dinonyl phenyl ether. See Book II

POE (4) dioleate. See Book I

POE (8) dioleate. See Book I

POE (12) dioleate. See Book I

POE (4) distearate. See Book I

POE (8) distearate. See Book I

POE (12) distearate. See Book I

POE (32) distearate. See Book II

POE (12) ditallate. See Book II

POE (40) hydrogenated castor oil. See Book II

POE (60) hydrogenated castor oil. See Book II

POE (50) hydrogenated tallow amide. See Book II

POE (2) isostearyl ether. See Book I

POE (10) isostearyl ether. See Book I

POE (20) isostearyl ether. See Book I

POE (75) lanolin. See Book IV

POE (20) lanolin ether. See Book IV

POE (4) lauryl ether. See Book I

POE (7) lauryl ether. See Book II

POE (12) lauryl ether. See Book I
POE (23) lauryl ether. See Book I
POE (120) methyl glucoside dioleate. See POE (120) methyl glucose dioleate
POE (8) monococoate. See Book I
POE (4) monolaurate. See Book I
POE (6) monolaurate. See Book I
POE (8) monolaurate. See Book I
POE (12) monolaurate. See Book I
POE (4) monooleate. See Book I
POE (8) monooleate. See Book I
POE (12) monooleate. See Book I
POE (20) monooleate. See Book I
POE (4) monostearate. See Book I
POE (8) monostearate. See Book I
POE (12) monostearate. See Book I
POE (20) monostearate. See Book I
POE (40) monostearate. See Book I
POE (75) monostearate. See Book I
POE (150) monostearate. See Book II
POE (1500) monostearate. See Book I
POE (1) nonyl phenyl ether. See Book II
POE (2) nonyl phenyl ether. See Book II
POE (4) nonyl phenyl ether. See Book II
POE (6) nonyl phenyl ether. See Book II
POE (8) nonyl phenyl ether. See Book II
POE (9) nonyl phenyl ether. See Book II
POE (10) nonyl phenyl ether. See Book II
POE (11) nonyl phenyl ether. See Book II
POE (12) nonyl phenyl ether. See Book II
POE (15) nonyl phenyl ether. See Book II
POE (20) nonyl phenyl ether. See Book II
POE (30) nonyl phenol. See POE (30) nonyl phenyl ether
POE (40) nonyl phenol. See POE (40) nonyl phenyl ether
POE (50) nonyl phenol. See POE (50) nonyl phenyl ether
POE (100) nonyl phenol. See POE (100) nonyl phenyl ether
POE (7) octyl phenyl ether. See Book II
POE (40) octyl phenyl ether. See Book II
POE (70) octyl phenyl ether. See Book II
POE (2) oleyl ether. See Book I
POE (4) oleyl ether. See Book I
POE (5) oleyl ether. See Book II
POE (10) oleyl ether. See Book I

POE (20) oleyl ether. See Book I

POE (50) POP (12) lanolin. See POP (12) POE (50) lanolin

POE (20) sorbitan monoisostearate. See Book II

POE (20) sorbitan monolaurate. See Book I

POE (20) sorbitan monooleate. See Book I

POE (20) sorbitan monopalmitate. See Book I

POE (4) sorbitan monostearate. See Book I

POE (20) sorbitan monostearate. See Book I

POE (20) sorbitan trioleate. See Book I

POE (20) sorbitan tristearate. See Book I

POE (5) soya sterol. See Book IV

POE (10) soya sterol. See Book IV

POE (2) stearyl ether. See Book II

POE (20) stearyl ether. See Book II

POE (50) tallow amine. See Book II

POE (3) tridecyl ether. See Book I

POE (6) tridecyl ether. See Book I

POE (9) tridecyl ether. See Book II

POE (10) tridecyl ether. See Book II

POE (15) tridecyl ether. See Book II

POE (12) tridecyl ether. See Book I

POE (4) dilaurate. See Book II

Poloxamine 707. See Book VI

Poloxamine 908. See Book VI

Polyacrylamide. See Book VI

Polyacrylic acid. See Book VI

Polybutene. See Book VI

Polyethoxylated nonylphenol (40 EO). See POE (40) nonyl phenyl ether

Polyglyceryl-10 tetraoleate (CTFA). See Decaglycerol tetraoleate

Poly (vinyl isobutyl ether). See Book VI

Polyvinylpyrrolidone. See Book VI

POP (2) dibenzoate. See Dipropylene glycol dibenzoate

Potassium ricinoleate (CTFA). See Potassium monoricinoleate

PPG-9. See Book VI

PPG-17. See Book VI

PPG-26. See Book VI

PPG-30. See Book VI

PPG-2 dibenzoate (CTFA). See Dipropylene glycol dibenzoate

PPG (2) dibenzoate. See Dipropylene glycol dibenzoate

PPG-12-PEG-50 lanolin (CTFA). See POP (12) POE (50) lanolin

1-Propanaminium, N-(carboxymethyl)-N-N-dimethyl-3-[(1-oxododecyl)amino]-, hy-
    droxide, inner salt. See Lauramidopropyl betaine

1-Propanaminium, N-(carboxymethyl)-N,N-dimethyl-3-[(1-oxohexadecyl)amino]-, hydroxide, inner salt. See Palmitamidopropyl betaine

1-Propanaminium, N-(carboxymethyl)-N,N-dimethyl-3-[(1-oxooctadecenyl) amino]-hydroxide, inner salt. See Oleamidopropyl betaine

1,2,3-Propanetricarboxylic acid, 2-(acetyloxy)-, tributyl ester. See Acetyl tributyl citrate

1,2,3-Propanetricarboxylic acid, 2-(acetyloxy)-, triethyl ester. See Acetyl triethyl citrate

1,2,3-Propanetricarboxylic acid, 2-hydroxy-, tributyl ester. See Tributyl citrate

1,2,3-Propanetricarboxylic acid, 2-hydroxy-, triethyl ester. See Triethyl citrate

1,2,3-Propanetriol. See Glycerin

1,2,3-Propanetriol, triacetate. See Glyceryl triacetate

1,2,3-Propanetriyl 12-(acetoxy)-9-octadecenoate. See Glyceryl triacetyl ricinoleate

Propanoic acid, 3,3'-thiobis-, didodecyl ester. See Dilauryl thiodipropionate

Propanoic acid, 3,3'-thiobis-, dihexadecyl ester. See Dicetyl thiodipropionate

Propanoic acid, 3,3'-thiobis-, dioctadecyl ester. See Distearyl thiodipropionate

Propanoic acid, 3,3'-thiobis-, ditetradecyl ester. See Dimyristyl thiodipropionate

Propanoic acid, 3,3'-thiobis-, ditridecyl ester. See Ditridecyl thiodipropionate

Propoxylated (2 moles) ethyl para-aminobenzoate. See Ethyl dihydroxypropyl PABA

Propylene glycol dicaprylate/dicaprate. See Book II

Propylene glycol monolaurate. See Book I

Propylene glycol monomethyl ether. See Book II

Propylene glycol monooleate. See Book I

Propylene glycol monostearate. See Book I

Propylene glycol myristyl ether acetate. See Book IV

Propylene glycol ricinoleate (CTFA). See Propylene glycol monoricinoleate

n-Propyl 3,4,5-trihydroxybenzoate. See Propyl gallate

PVM/MA Copolymer. See Book VI

Quaternary ammonium compounds, bis (hydrogenated tallow alkyl) dimethyl, chlorides, reaction product with bentonite. See Quaternium-18 bentonite

Quaternary ammonium compounds, bis (hydrogenated tallow alkyl) dimethyl, chlorides, reaction product with hectorite). See Quaternium-18 hectorite

Quaternary ammonium compounds, (3-cocamidopropyl) (2-hydroxy-3-sulfopropyl) dimethyl, hydroxide, inner salt. See Cocamidopropyl hydroxysultaine

Quinol. See Hydroquinone

Ricinolein, 1-mono-. See Glyceryl monoricinoleate

Ricinus oil. See Castor oil

Rosin gum. See Rosin

Sodium alginate. See Algin

Sodium alpha-olefin sulfonate. See Sodium olefin sulfonate

Sodium CMC. See Sodium carboxymethyl cellulose

Sodium C$_{14-16}$ olefin sulfonate (CTFA). See Sodium olefin sulfonate

Sodium C$_{16-18}$ olefin sulfonate (CTFA). See Sodium olefin sulfonate

Sodium cumenesulfonate. See Book II

Sodium dicyclohexyl sulfosuccinate. See Book II
Sodium dihexyl sulfosuccinate. See Book II
Sodium dioctyl sulfosuccinate. See Book II
Sodium dodecylbenzenesulfonate. See Book I
Sodium 2-ethylhexyl sulfate. See Sodium octyl sulfate
Sodium hexametaphosphate. See Book II
Sodium lauryl sulfate. See Book I
Sodium lignosulfonate. See Book II
Sodium methyl oleoyl taurate. See Book II
Sodium α-olefin sulfonate. See Sodium olefin sulfonate
Sodium polyacrylate. See Book VI
Sodium polymethacrylate. See Book VI
Sodium polynaphthalene sulfonate. See Book VI
Sodium stearate. See Book IV
Sodium stearoyl lactylate. See Book I
Sodium toluenesulfonate. See Book II
Sodium xylene sulfonate. See Book II
Sorbeth-20 (CTFA). See POE (20) sorbitol ether
Sorbitan laurate (CTFA). See Sorbitan monolaurate
Sorbitan, monododecanoate. See Sorbitan monolaurate
Sorbitan monooleate. See Book IV
Sorbitan monopalmitate. See Book II
Sorbitan sesquioleate. See Book IV
Sorbitan, trioctadecanoate. See Sorbitan tristearate
Sorbitan, tri-9-octadecenoate. See Sorbitan trioleate
Soya amides, N,N-bis (hydroxyethyl)-. See Soya diethanolamide
Soyabean oil, epoxidized. See Epoxidized soybean oil
Soya DEA. See Soya diethanolamide
Soyamide DEA (CTFA). See Soya diethanolamide
Soya oil, epoxidized. See Epoxidized soybean oil
Soybean oil, epoxidized. See Epoxidized soybean oil
Soy oil, epoxidized. See Epoxidized soybean oil
Soy sterol. See Book IV
St. John's bread. See Locust bean gum
Stearamide MEA (CTFA). See Stearic monoethanolamide
Stearamide MEA-stearate (CTFA). See Stearic monoethanolamide stearate
Stearic acid. See Book IV
Stearic acid amide. See Book IV
Stearic acid diethanolamide. See Stearic diethanolamide
Stearic acid, hydroxy-, monoester with glycerol. See Glyceryl hydroxystearate
Stearic acid monoethanolamide. See Stearic monoethanolamide
Stearic DEA. See Stearic diethanolamide

Stearic MEA. See Stearic monoethanolamide
Stearoyl diethanolamide. See Stearic diethanolamide
Stearoyl monoethanolamide. See Stearic monoethanolamide
Stearyl dimethyl amine oxide. See Book IV
Styrene/PVP copolymer. See Book VI
Sucrose monostearate. See Book II
Sulfated castor oil (CTFA). See Castor oil, sulfated
Sulfuric acid, mono (2-ethylhexyl) ester, sodium salt. See Sodium octyl sulfate
Sulisobenzone. See 2-Hydroxy-4-methoxybenzophenone-5-sulfonic acid
Synthetic spermaceti. See Book IV
Tallamide DEA (CTFA). See Tall oil acid diethanolamide
Tallow amides, hydrogenated. See Hydrogenated tallow amide
Tallow dihydroxyethyl betaine. See Dihydroxyethyl tallow glycinate
Tallow dimethyl amine oxide. See Book IV
TBHQ. See *t*-Butyl hydroquinone
TCP. See Tricresyl phosphate
Tetradecanamide, N-(2-hydroxyethyl)-. See Myristic monoethanolamide
Tetradecanamine, N,N-dimethyl-, N-oxide. See Myristyl dimethyl amine oxide
Tetradecyl dimethyl amine oxide. See Myristyl dimethyl amine oxide
3a,4,7,7a-Tetrahydro-2-[(trichloromethyl) thio]-1H-isoindole-1,3(2H)-dione. See Captan
3,3′-Tetramethylnonyl thiodipropionate. See Ditridecyl thiodipropionate
Tetrapotassium pyrophosphate. See Book II
Tetrasodium dicarboxyethyl stearyl sulfosuccinamate. See Book II
Tetrasodium ethylene diamine tetraacetate. See Book I
3,3′-Thiobispropanoic acid, dihexadecyl ester. See Dicetyl thiodipropionate
3,3′-Thiobispropanoic acid, dioctadecyl ester. See Distearyl thiodipropionate
3,3′-Thiobispropanoic acid, ditridecyl ester. See Ditridecyl thiodipropionate
TIOTM. See Triisooctyl trimellitate
TOTM. See Trioctyl trimellitate
Triacetin (CTFA). See Glyceryl triacetate
N-Trichloromethylthio-4-cyclohexene-1,2-dicarboximide. See Captan
N-Trichloromethylthiotetrahydro-phthalimide. See Captan
Tri-2-ethylhexyl trimellitate. See Trioctyl trimellitate
3,4,5-Trihydroxybenzoic acid, propyl ester. See Propyl gallate
Trihydroxystearin (CTFA). See Glyceryl tri(12-hydroxystearate)
Tripropylene glycol monomethyl ether. See Book II
Turkey red oil. See Castor oil, sulfated
Vitamin E. See Tocopherol (CTFA)
WW Wood rosin. See Rosin
Xanthan. See Xanthan gum

TRADENAME PRODUCT MANUFACTURERS

Aceto Chemical Co., Inc.
126-02 Northern Blvd.
Flushing, NY 11368

Acme Hardesty, Inc.
Benjamin Fox Pavilion
Jenkintown, PA 19046

Akron Chemical Co.
255 Fountain St.
Akron, OH 44304

Akzo Chemie America/Armak Chemical
300 S. Riverside
Chicago, IL 60606

Akzo Chemie BV
Stationsstraat 48, POB 247
3800 AE-Amersfoort, Netherlands

Albright & Wilson (Australia) Ltd.
610 St. Kilda Rd., PO Box 4544
Melbourne 3001, Australia

Albright & Wilson/
Detergents Div., Marchon Works
PO Box 15, Whitehaven
Cumbria CA28 9QQ, UK

Alcoa
1501 Alcoa Bldg.
Pittsburgh, PA 15219

Alcolac Inc.
3440 Fairfield Rd.
Baltimore, MD 21226

Alkaril Chemicals Inc.
Industrial Pkwy., PO Box 1010
Winder, GA 30680

Amerchol Corp.
PO Box 4051, 136 Talmadge Rd.
Edison, NJ 08818

Amerchol Europe
Havenstraat 84, B-1800
Vilvoorde, Belgium

American Chemet Corp.
400 County Line Rd., PO Box 437
Deerfield, IL 60015

American Cyanamid Co.
Berdan Ave.
Wayne, NJ 07470

American Smelting & Refining Co.
120 Braodway
New York, NY 10271

Arco Chemical Co.
1500 Market St.
Philadelphia, PA 19101

Ardmore Chemical Co.
29 Riverside Ave., Bldg. #14
Newark, NJ 07104

Argus Chemical Corp./
Div. of Witco Chemical Corp.
633 Court St.
Brooklyn, NY 11231

Armak Industrial Chem. Div./
Akzo Chemie America
300 S. Riverside Dr.
Chicago, IL 60606

Armstrong Chemical Co.
1530 South Jackson St.
Janesville, WI 53545

Ashland Chemical Co.
Box 2219
Columbus, OH 43216

Associated Lead Inc.
PO Box 3728
Philadelphia, PA 19125

Atlas Refinery Inc.
142 Lockwood St.
Newark, NJ 07105

Badische Corp.
PO Box 405
Bridgeport, NJ 08014

BASF Canada Ltd.
PO Box 430
Montral, Quebec H4L 4V8, Canada

BASF Wyandotte Corp.
100 Cherry Hill Rd.
Parsippany, NJ 07054

Bayer AG
Sitz der Gesellschaft, Leverkusen,
Eintragung, Amstgericht
Leverkusen HRB 1122, Germany

Bofors Lakeway
5025 Evanston Ave., PO Box 328
Muskegon, MI 49443

Borg-Warner Chemicals Inc.
International Center
Parkersburg, WV 26102

Canada Packers Ltd.
5100 Timberlea Blvd.
Mississauga, Ontario L4W 2S5, Canada

Capital City Products Co.
PO Box 569
Columbus, OH 43216

R.E. Carroll Inc.
1570 N. Olden Ave., PO Box 139
Trenton, NJ 08601

Carson Chemicals. See Lonza

CasChem Inc.
40 Avenue A
Bayonne, NJ 07002

Chemax, Inc.
POB 6067, Highway 25 South
Greenville, SC 29606

Chemform Corp.
141 S.W. 18th St.
Pompano Beach, FL 33061

Ciba-Geigy Corp.
PO Box 18300
Greensboro, NC 27419

Cincinnati Milacron Chem. Inc.
West St.
Reading, OH 45215

Clintwood Chemical Co.
4342 S. Wolcott Ave.
Chicago, IL 60609

Consos Inc.
PO Box 34186
Charlotte, NC 28234

Continental Oil Co.
See Vista Chemical Co.

Croda Chemicals Ltd.
Cowick Hall, Snaith Goole
North Humberside DN14 9AA, UK

Croda Inc.
51 Madison Ave.
New York, NY 10010

Croda Surfactants Inc.
183 Madison Ave.
New York, NY 10016

Cyclo Chemical Corp.
7500 N.W. 66th St.
Miami, FL 33166

Dai-ichi Kogyo Seiyaku Co., Ltd.
Miki Building, 3-12-1, Nihombashi,
Chuo-ku
Tokyo 103, Japan

Diamond Shamrock/
Process Chem. Div.
350 Mt. Kemble Ave.
Morristown, NJ 07960

Domtar Inc./CDC Div.
1136 Matheson Blvd.
Mississauga, Ontario L4W 2V4
Canada

Dover Chemical Corp.
W. 15th & Davis Sts., PO Box 40
Dover, OH 44622

Dow Chem. Co.
1703 S. Saginaw Rd.
Midland, MI 48640

Drew Produtos Quimicos Ltds.
Rua Sampaio Viana, 425
04004, Sao Paulo, Brazil-CP4885

E.I. DuPont de Nemours & Co.
Nemours Bldg.
Wilmington, DE 19898

Durkee Industrial Foods/SCM Corp.
900 Union Commerce Bldg.
Cleveland, OH 44115

Dutton & Reinisch Ltd.
Crown House, London Rd., Morden
Surrey SM45DU, UK

Dynamit Nobel
10 Link Dr.
Rockleigh, NJ 07647

Eagle Picher Industries Inc.
580 Walnut St.
Cincinnati, OH 45202

Eastman Chemical Products, Inc.
PO Box 431
Kingsport, TN 37662

Emery Industries Inc.
1501 W. Elizabeth Ave.
Linden, NJ 07036

Emulsion Systems Inc.
215 Kent Ave.
Brooklyn, NY 11211

Engelhard Minerals & Chem. Corp.
Menlo Park, PO Box 2900
Edison, NJ 08818

Essential Chemicals Corp.
28391 Essential Rd.
Merton, WI 53056

Evans Chemetics/W.R. Grace & Co.
90 Tokeneke Rd.
Darien, CT 06820

Exxon Chemical Americas
13501 Katy Fwy., PO Box 3272
Houston, TX 77001

Fairmount Chemical Co., Inc.
117 Blanchard St.
Newark, NJ 07105

The Fanning Corp.
3117 North Clybourn Ave.
Chicago, IL 60618

Ferro Corp.
7050 Krick Rd.
Bedford, OH 44146

Finetex Inc.
418 Falmouth Ave.
Elmwood Park, NJ 07407

Floridin Co.
701 McKnight Park Dr.
Pittsburgh, PA 15237

FMC Corp.
2000 Market St.
Philadelphia, PA 19103

GAF Corp.
1361 Alps Rd.
Wayne, NJ 07470

Mario Geronazzo Ind. Chim. SpA
78, Ospiate Di Bollate
Milano, Italy 20021

Glyco Chemicals Inc.
PO Box 700, 51 Weaver St.
Greenwich, CT 06830

Th. Goldschmidt AG
Goldschmidtstr. 100
4300 Essen 1, Postfach 101461
West Germany

W.R. Grace & Co.
55 Hayden Ave.
Lexington, MA 02173

Graden Chem. Co., Inc.
426 Bryan St.
Havertown, PA 19083

Great Lakes Minerals Co.
2855 Coolidge Hwy., Suite 112
Troy, MI 48084

Grinsted Products Inc.
201 Industrial Pkwy., PO Box 26
Industrial Airport, KS 66031

A. Gross & Co./
Div. of Millmaster Onyx Corp.
652 Doremus Ave., PO Box 818
Newark, NJ 07101

Chemische Fabrik Grunau GmbH
Robert-Hansen Str. 1, Postfach 1063
D-7918 Jllertissen
Bavaria, West Germany

C.P. Hall Co.
7300 South Central Ave.
Chicago, IL 60638

Halstab Div./
Hammond Lead Products Inc.
PO Box 6408
Hammond, IN 46325-6408

Hart Chem. Ltd.
256 Victoria Rd. South
Guelph, Ontario N1H 6K8, Canada

Hart Products Corp.
173 Sussex St.
Jersey City, NJ 07302

406

Harwick Chemical Corp.
60 South Seiberling St.
Akron, OH 44305

Hatco Chemical Corp.
King George Post Rd.
Fords, NJ 08863

Hefti Ltd.
PO Box 1623, CH-8048
Zurich, Switzerland

Henkel Argentina S.A.
Avda. E. Madero Piso 14
1106 Capital Federal
Argentina

Henkel Chem. (Canada) Ltd.
9550 Ray Lawson Blvd.
Ville d'Anjou, Quebec H1J 1L3, Canada

Henkel Inc.
480 Alfred Ave.
Teaneck, NJ 07666

Henkel KGaA
Postfach 1100, D-4000
Dusseldorf 1, West Germany

Hercules BV
8 Veraartlaan, PO Box 5822
2280 HV Rijswijk, The Netherlands

Hercules Inc.
Hercules Plaza
Wilmington, DE 19894

Heterene Chemical Co., Inc.
POB 247, 792 21 Ave.
Paterson, NJ 07513

Hexagon Enterprises Inc.
60 Midvale Rd.
Mountain Lake, NJ 07046

High Point Chem. Corp.
601 Taylor St., PO Box 2316
High Point, NC 27261

Hodag Chem. Corp.
7247 N. Central Park Ave.
Skokie, IL 60076

Hoechst AG
Verhaufkanststoffe, D-6230
Frankfurt (M) 80, West Germany

Housmex Inc.
2603 W. Market St., Suite 100
Akron, OH 44313

Hüls AG, Chemische Werke
Postfach 1320 D-4370
Marl 1, West Germany

Humko Chem./Div. Witco Chem.
PO Box 125, 755 Crossover Lane
Memphis, TN 38101

ICI Americas Inc.
New Murphy Rd. & Concord Pike
Wilmington, DE 19897

ICI Australia Ltd.
ICI House, 1 Nicholson St.
Melbourne 3000, Australia

ICI Specialty Chemicals
Everslann 45 B-3078
Kortenberg, Belgium

Inolex Chem.Co.
4221 S. Western Blvd.
Chicago, IL 60609

Interstab Chemicals, Inc.
500 Jersey Ave.
New Brunswick, NJ 08903

Jefferson Chemical Co. Inc.
PO Box 4128
Austin, TX 78765

Jordan Chem. Co.
1830 Columbia Ave.
Folcroft, PA 19032

Kaiser Chemicals
300 Lakeside Dr.
Oakland, CA 94643

Kao Corp.
14-10 Nihonbashi, Kayabacho 1-chome,
Chuo-ku
Tokyo 103, Japan

Kelco Co./Div. of Merck & Co.
8355 Aero Dr.
San Diego, CA 92123

Elektrochemische Fabrik Kempen GmbH
Postfach 100 260
D-4152 Kempen 1, West Germany

Kenrich Petrochemicals
140 E. 22nd St., PO Box 64
Bayonne, NJ 07002

Lankro Chem. Ltd.
PO Box 1, Eccles
Manchester M30 0BH, UK

La Tassilchimica
Bergamo, Italy

Lipo Chemicals Inc.
207 19th Ave.
Paterson, NJ 07504

Lonza Inc.
22-10 Route 208
Fair Lawn, NJ 07410

Lyndal Chem. Co.
PO Box 1740
Dalton, GA 30720

Manro Products Ltd.
Bridge St., Stalybridge
Cheshire SK15 1PH, UK

Marlowe-Van Loan Corp.
PO Box 1851
High Point, NC 27261

Mars Chem. Corp.
762 Marietta Blvd. N.W.
Atlanta, GA 30318

Mazer Chem. Inc.
3938 Porett Dr.
Gurnee, IL 60031

McIntyre Chem. Co., Ltd.
4851 S. St. Louis Ave.
Chicago, IL 60632

Miranol Chem. Corp.
68 Culver Rd., PO Box 411
Dayton, NJ 08810

Mobay Chemical Corp.
Penn Lincoln Parkway West
Pittsburgh, PA 15205

Mona Industries, Inc.
PO Box 425, 76 E. 24th St
Paterson, NJ 07544

Monsanto Co.
800 N. Lindbergh Blvd.
St. Louis, MO 63167

Morflex Chemical Co. Inc.
2110 High Point Rd.
Greensboro, NC 27403

Murphy-Phoenix Co.
PO Box 22930
Beachwood, OH 44122

Neville Chemical Co.
Neville Island
Pittsburgh, PA 15225

Neville-Synthèse Organics Inc.
Neville Island
Pittsburgh, PA 15225

New Jersey Zinc Co.
Fourth & Delaware
Palmerton, PA 18071

Niacet Corp.
47th St. & Niagara Falls Blvd.
Niagara Falls, NY 14304

Nikko Chem. Co., Ltd.
1-4-8 Nihonbashi-Bakurocho, Chuo-ku
Tokyo 103, Japan

Nippon Nyukazai Co., Ltd.
19-9, 3-chome, Ginza, Chuo-ku
Tokyo 104, Japan

Nippon Oil & Fats Co., Ltd.
5-1 chome, Yurakucho, Chiyoda-ku
Tokyo, Japan

NL Industries
PO Box 700
Hightstown, NJ 08520

Nuodex Inc.
Turner Pl.
Piscataway, NJ 08854

Oleofina S.A.
Rue de Science 37 Wetenschapsstraat
1040 Brussels, Belgium

Olin Chemicals
120 Long Ridge Rd.
Stamford, CT 06904

Onyx Chem. Co./
Millmaster Onyx Group
190 Warren St.
Jersey City, NJ 07302

Pearsall Chem. Div./
Witco Chem. Corp.
PO Box 437, 2519 Fairway Park Dr.
Houston, TX 77001

Pilot Chem. Co.
11756 Burke St.
Santa Fe Springs, CA 90670

PMC Specialties Group, Inc.
101 Prospect Ave. N.W.
Cleveland, OH 44115

Polymerics Inc.
2828 Second St.
Cuyahoga Falls, OH 44221

PVO International Inc.
416 Division St.
Boonton, NJ 07005

Reichhold Chemicals Inc.
RCI Bldg.
White Plains, NY 10603

Rewo Chemicals Inc.
107B Allen Blvd.
E. Farmingdale, NY 11735

Rewo Chemische Werke GmbH
Postfach 1160, Industriegebiet West
D-6497 Steinau an der Strasse
West Germany

Rhodia Inc.
PO Box 125
Monmouth Junction, NJ 08852

Richardson Co.
2400 Devon Ave.
Des Plaines, IL 60018

RITA Corp.
PO Box 556
Crystal Lake, IL 60014

Rohm & Haas Co.
Independence Mall West
Philadelphia, PA 19105

Ryco Inc./
Div. Reilly-Whiteman Inc.
801 Washington St.
Conshohocken, PA 19428

Santech Inc.
150 Norseman St.
Toronto, Ontario M8Z 5M4 Canada

Sanyo Chem. Industries Ltd.
11-1 Ikkyo Nomoto-cho Higashiyama-ku
Kyoto 605, Japan

Scher Chem. Inc.
Industrial West & Styertowne Rd.
Clifton, NJ 07012

Seaboard Chemicals Inc.
30 Foster St., PO Box 707
Salem, MA 01970

Seppic
70 Champs Elysees
75008 Paris, France

Servo Chemische Fabriek B.V.
PO Box 1, 7490 AA
Delden, Holland

Shell Chem. Co.
One Shell Plaza
Houston, TX 77001

Sherex Chem. Co.
5777 Frantz Rd., PO Box 646
Dublin, OH 43017

Solem Industries Inc.
5824-D Peachtree Corners East
Norcross, GA 30092

Southland Corp.
7666 W. 63rd St.
Summit, IL 60501

Stave Chemical Co., Inc.
20 Marilyn St.
Basking Ridge, NJ 07920

Stepan Co.
Edens & Winnetka Rds.
Northfield, IL 60093

Stepan Europe
BP127
38340 Voreppe, France

Summit Chemical Co.
2108 Braewick Circle
Akron, OH 44313

Swastik Household & Industrial Products
  Ltd.
Shahibag House, PB 362
13 Walchand Hirachand Marg
Ballard Estate
Bombay 400 038, India

Swift Chem. Co.
1211 West 22nd St.
Oak Brook, IL 60521

Synthetic Products Co.
16601 St. Clair Ave.
Cleveland, OH 44110

Taiwan Surfactant Corp.
8-1 Floor, No. 106, Sec. 2
Changan East Road
Taipei, Taiwan, R.O.C.

Tenneco Chemicals Inc.
Turner Place, PO Box 365
Piscataway, NJ 08854

Thompson-Hayward Chemical Co.
PO Box 2383
5200 Speaker Rd.
Kansas City, KS 66110

Toho Chem. Industry Co., Ltd.
14-9, 1-chome, Kakigara-cho
Nihonbashi, Chuo-ku
Tokyo 103, Japan

Thomas Triantaphyllou S.A.
405 Tatoiou Av., TK 136 71, Acharnes
Athens, Greece

Unichema International
Postfach 1280
D-4240 Emmerich, West Germany

Union Camp Corp.
1600 Valley Rd.
Wayne, NJ 07470

Union Carbide Corp.
39 Old Ridgebury Rd.
Danbury, CT 06817

Uniroyal Inc.
World Headquarters
Middlebury, CT 06749

Unitex Chemical Corp.
PO Box 16344, 520 Broome Rd.
Greensboro, NC 27406

Universal Chemicals Corp.
1224 Mendon Rd.
Ashton, RI 02864

UOP Process Div./
Universal Oil Products Inc.
20 U.O.P. Plaza
Des Plaines, IL 60016

USS Chemicals
600 Grant St., Rm. 2858
Pittsburgh, PA 15230

3-V Chemical Corp.
4 Woodlawn Green, Ste. 229
Charlotte, NC 28210

Van Dyk & Co., Inc.
11 Williams St.
Belleville, NJ 07109

R.T. Vanderbilt & Co., Inc.
30 Winfield St.
Norwalk, CT 06855

Velsicol Chemical Corp.
341 E. Ohio St.
Chicago, IL 60068

Vista Chemical Co.
15990 N. Barkers Landing Rd.
PO Box 19029
Houston, TX 77079

Wickhen Products Inc.
Big Pond Rd.
Huguenot, NY 12746

Witco Chemical Corp./Argus Chem. Div.
633 Court St.
Brooklyn, NY 11231

Witco Chemical Corp.
Organics Div. & Sonneborn Div.
277 Park Ave.
New York, NY 10017

Zohar Detergent Factory
Kibbutz Dalia, Israel 18920

www.ingramcontent.com/pod-product-compliance
Lightning Source LLC
Chambersburg PA
CBHW060747220326
41598CB00022B/2359